Why Indoor Chemistr

T0073814

Committee on Emerging Science on
Indoor Chemistry

Board on Chemical Sciences and
Technology

Division on Earth and Life Studies

Consensus Study Report

NATIONAL ACADEMIES PRESS 500 Fifth Street, NW Washington, DC 20001

This study was sponsored by the Alfred P. Sloan Foundation, the Centers for Disease Control and Prevention, the Environmental Protection Agency, and the National Institute of Environmental Health Sciences. This activity was supported by contracts between the National Academies of Sciences, Engineering, and Medicine and in whole or in part with Federal funds from the National Institutes of Health, Department of Health and Human Services, under Contract No. HHSN263201800029I Task Order No. 75N98020F00009. Any opinions, findings, conclusions, or recommendations expressed in this publication do not necessarily reflect the views of any organization or agency that provided support for the project.

International Standard Book Number-13: 978-0-309-08399-7
International Standard Book Number-10: 0-309-08399-0
Digital Object Identifier: https://doi.org/10.17226/26228
Library of Congress Control Number: 2022946407

This publication is available from the National Academies Press, 500 Fifth Street, NW, Keck 360, Washington, DC 20001; (800) 624-6242 or (202) 334-3313; http://www.nap.edu.

Suggested citation: National Academies of Sciences, Engineering, and Medicine. 2022. *Why Indoor Chemistry Matters*. Washington, DC: The National Academies Press. https://doi.org/10.17226/26228.

The **National Academy of Sciences** was established in 1863 by an Act of Congress, signed by President Lincoln, as a private, nongovernmental institution to advise the nation on issues related to science and technology. Members are elected by their peers for outstanding contributions to research. Dr. Marcia McNutt is president.

The **National Academy of Engineering** was established in 1964 under the charter of the National Academy of Sciences to bring the practices of engineering to advising the nation. Members are elected by their peers for extraordinary contributions to engineering. Dr. John L. Anderson is president.

The **National Academy of Medicine** (formerly the Institute of Medicine) was established in 1970 under the charter of the National Academy of Sciences to advise the nation on medical and health issues. Members are elected by their peers for distinguished contributions to medicine and health. Dr. Victor J. Dzau is president.

The three Academies work together as the **National Academies of Sciences, Engineering, and Medicine** to provide independent, objective analysis and advice to the nation and conduct other activities to solve complex problems and inform public policy decisions. The National Academies also encourage education and research, recognize outstanding contributions to knowledge, and increase public understanding in matters of science, engineering, and medicine.

Learn more about the National Academies of Sciences, Engineering, and Medicine at **www.national academies.org**.

COMMITTEE ON EMERGING SCIENCE ON INDOOR CHEMISTRY

Members

DAVID C. DORMAN (*Chair*), North Carolina State University
JONATHAN ABBATT, University of Toronto
WILLIAM P. BAHNFLETH, Pennsylvania State University
ELLISON CARTER, Colorado State University
DELPHINE FARMER, Colorado State University
GILLIAN GAWNE-MITTELSTAEDT, Partnership for Air Matters/Tribal Healthy
    Homes Network
ALLEN H. GOLDSTEIN, University of California, Berkeley
VICKI H. GRASSIAN, University of California, San Diego
RIMA HABRE, University of Southern California
GLENN MORRISON, University of North Carolina, Chapel Hill
JORDAN PECCIA, Yale University
DUSTIN POPPENDIECK, National Institute of Standards and Technology
KIMBERLY A. PRATHER (NAS/NAE), University of California, San Diego
MANABU SHIRAIWA, University of California, Irvine
HEATHER M. STAPLETON, Duke University (since February 2021)
MEREDITH WILLIAMS, California Department of Toxic Substances Control

Staff

MEGAN E. HARRIES, Study Director
MICHELLE BAILEY, Program Assistant (until July 2021)
KESIAH CLEMENT, Research Associate (until July 2021)
MEGHAN HARRISON, Senior Program Officer (until July 2021)
ELLEN K. MANTUS, Scholar (until April 2021)
EMMA SCHULMAN, Program Assistant (until March 2022)
MARILEE SHELTON-DAVENPORT, Senior Program Officer (until January 2021)
ABIGAIL ULMAN, Research Assistant (until November 2021)
BENJAMIN ULRICH, Senior Program Assistant (until August 2021)

Reviewers

This Consensus Study Report was reviewed in draft form by individuals chosen for their diverse perspectives and technical expertise. The purpose of this independent review is to provide candid and critical comments that will assist the National Academies of Sciences, Engineering, and Medicine in making each published report as sound as possible and to ensure that it meets the institutional standards for quality, objectivity, evidence, and responsiveness to the study charge. The review comments and draft manuscript remain confidential to protect the integrity of the deliberative process.

We thank the following individuals for their review of this report:

Linda Birnbaum (NAM), National Institute of Environmental Health Sciences (retired)
Robin Dodson, Silent Spring Institute
Sharon Haynie, DuPont (retired)
William Nazaroff, University of California Berkeley (retired)
Paula Olsiewski, Johns Hopkins University
Veronica Vaida (NAS), University of Colorado Boulder
Charles Weschler, Rutgers University
Jonathan Williams, Max Planck Institute

Although the reviewers listed above provided many constructive comments and suggestions, they were not asked to endorse the conclusions or recommendations of this report nor did they see the final draft before its release. The review of this report was overseen by **Cynthia Beall** (NAS), Case Western Reserve University, and **Teresa Fryberger**, Consultant, Chemical Sciences and Policy. They were responsible for making certain that an independent examination of this report was carried out in accordance with the standards of the National Academies and that all review comments were carefully considered. Responsibility for the final content rests entirely with the authoring committee and the National Academies.

Acronyms

AMS	aerosol mass spectrometers
ANSI	American National Standards Institute
CADR	clean air delivery rate
CARB	California Air Resources Board
CFD	computational fluid dynamics
CIMS	chemical ionization mass spectrometry
EC	elemental carbon
EDC	endocrine-disrupting chemical
EJ	environmental justice
EPA	U.S. Environmental Protection Agency
ESP	electrostatic precipitator
HEPA	high-efficiency particulate air
HOM	highly oxygenated organic molecule
HOMEChem	House Observations of Microbial and Environmental Chemistry
HVAC	heating, ventilation, and air conditioning
IOM	Institute of Medicine
MERV	Minimum Efficiency Reporting Value
mVOC	microbial volatile organic compound
NHANES	National Health and Nutrition Examination Survey
NHAPS	National Human Activity Patterns Survey
NRC	National Research Council

OA	organic aerosol
PAH	polycyclic aromatic hydrocarbon
PBDE	polybrominated diphenyl ether
PCB	polychlorinated biphenyl
PCO	photocatalytic oxidation
PFAS	per- and polyfluoroalkyl substances
PM	particulate matter
$PM_{2.5}$	fine particulate matter
PMF	positive matrix factorization
PPE	personal protective equipment
PTR-MS	proton-transfer-reaction mass spectrometer
RIOPA	Relationship of Indoor, Outdoor, and Personal Air study
ROS	reactive oxygen species
SDO	standards developing organization
SDS	Safety Data Sheet
SOA	secondary organic aerosol
SVOC	semivolatile organic compound
SV-TAG	semivolatile thermal desorption aerosol gas chromatography
TAG	thermal desorption aerosol gas chromatography
TCE	trichloroethylene
TEG	triethylene glycol
TSCA	Toxic Substances Control Act
UFP	ultrafine particle
UHI	urban heat islands
UV	ultraviolet
UVGI	ultraviolet germicidal irradiation
VCP	volatile chemical product
VOC	volatile organic compound

CHEMICAL FORMULAS

Cl_2	chlorine
$ClNO_2$	nitryl chloride
CO	carbon monoxide
CO_2	carbon dioxide
HO_2	hydroperoxy
H_2O_2	hydrogen peroxide
HOCl	hypochlorous acid
HONO	nitrous acid
H_2SO_3	sulfurous acid

NH_3	ammonia
NO	nitric oxide
NO_2	nitrogen dioxide
NOx	nitrogen oxides
O_3	ozone
OH	hydroxyl
SiO_2	silica or silicon dioxide
TiO_2	titanium dioxide

Contents

Summary

hemicals found indoors are a significant risk factor that can modify or degrade the indoor environment. Direct emission of chemicals into the indoor environment can occur from various sources, including building materials, paints, stoves, cleaning products, pesticides, furnishings, electronics, and personal care products. Biologic sources including microorganisms, plants, pets, and other animals can also contribute to indoor chemistry.

Why does indoor chemistry matter? It matters because people spend most of their time at home or in other indoor locations. Complex mixtures of chemicals in indoor environments may adversely impact indoor air quality and human health. Whether exposures to indoor chemicals result in an adverse effect is dependent on exposure duration and additional factors, including the inherent toxicity of the chemical mixture, chemical concentrations in the environment, the route of exposure, and the susceptibility of the person.

This report explores indoor chemistry from different perspectives including sources and reservoirs of indoor chemicals and the ability of these chemicals to undergo transformations and partitioning in the indoor environment. In some cases, indoor chemistry can result in the creation of potentially more or less toxic products, reactive intermediates, or products with different physiochemical properties.

This study provides a status report for indoor chemistry research. The overarching goal of this project was to identify new findings about previously under-reported chemical species, chemical reactions, and sources of chemicals, as well as the distribution of chemicals; and improve our understanding of how indoor chemistry is linked with chemical exposure, air quality, and human health. The report also identifies future research needs.

PRIMARY SOURCES AND RESERVOIRS OF CHEMICALS INDOORS

Thousands of chemical compounds have been detected in the indoor environment, including in air, particles, and settled dust, or on surfaces. These chemicals are intermittently or chronically emitted into the indoor environment from primary sources that originate either indoors or outdoors. Materials used to construct buildings, the furnishings that are brought into those buildings, and building occupants and their activities are important indoor sources of chemicals. Cooking is one of the most notable sources of indoor chemicals, with emissions varying depending on cooking style and ingredients. Cooking generates gases and particles, both from the heat source and the food. The use of personal care and consumer products contributes chemicals to the indoor environment, as does the use of disinfectants and cleaning products in health care facilities and schools.

Over the past few decades, new materials and chemicals have been introduced in the indoor environment—for example, new types of building insulation and greater use of electronics and smart devices in indoor environments, particularly in homes. These devices and materials can be sources of chemicals to the indoor environment, such as plastics, plasticizers, antioxidants, ultraviolet stabilizers, and flame retardants. Increased use of cleaning and disinfection agents, particularly during the COVID-19 pandemic, has led to increased levels of some chemicals, such as quaternary ammonium compounds, in the indoor environment. Over time, some chemicals that were emitted by primary sources in the indoor environment have been phased out of use owing to concerns about their elevated exposure and/or toxicity. However, recycling of certain materials may lead to ongoing exposure to phased-out chemicals. For example, polyurethane foam from discarded furniture is sometimes recycled into bonded carpet padding. Unfortunately, phased-out chemicals are sometimes replaced with chemicals that have less data available on their emissions, exposure, and potential hazards. Some legacy contaminants that were phased out, such as polychlorinated biphenyls, chlordane, and chlorpyrifos, are persistent in indoor environments and contribute to prolonged and chronic exposure in older buildings and apartments, which in some cases can lead to social justice issues.

A wide range of analytical techniques are currently being used to identify new chemicals, in both bottom-up and top-down approaches, that may be released into the indoor environment, but these approaches are costly and time-consuming. Challenges in identifying chemical sources in the indoor environment, including a lack of transparency in chemical use in consumer products, continues to be a major obstacle to chemical inventory and risk evaluation. Another risk management challenge relates to the lack of information on the health effects or distribution of many of the chemicals found in the indoor environment.

To address these challenges, the committee identified a need to prioritize research linking sources with exposures and to characterize the impacts of mixtures on health. To support discovery, improved analytical methods and non-targeted approaches are needed, as well as harmonized databases for chemical information. Although the study of indoor environments has recently increased, data are lacking for nonresidential settings and underrepresented countries and contexts.

PARTITIONING OF CHEMICALS IN INDOOR ENVIRONMENTS

Partitioning of chemicals plays an important role in indoor chemistry and indoor air quality. Partitioning refers to both the thermodynamic state of chemicals distributed among phases in a system and the processes that transfer chemicals among phases, generally with a net tendency to approach equilibrium. At the thermodynamic state of equilibrium, partitioning determines the concentration of a chemical in air, on surfaces, or elsewhere.

Chemicals in indoor environments are often not at equilibrium, and net transfer can occur between phases. Partitioning distributes chemicals from their initial sources throughout indoor spaces to air, building materials, furnishings, dust, and so forth. For example, phthalates emitted from plastics can partition to surfaces, porous materials, settled dust, and other compartments. These compartments buffer the air concentrations of chemicals, reducing the short-term effectiveness of controls by ventilation or filtration. Partitioning also influences occupant exposure to chemicals. For example, partitioning of indoor chemicals to aerosols increases inhalation exposure, while partitioning to dust and surfaces increases ingestion exposure, especially by toddlers.

Chemicals can have a greater affinity for one compartment than another, and this affinity is characterized by a "partition coefficient." Because of the very high surface-area-to-volume ratio of indoor environments, partitioning influences the manner, extent, and duration over which occupants are exposed to contaminants. For example, nicotine has a high affinity for indoor material surfaces; therefore, more nicotine is found on surfaces than in the air. Many molecules that are entirely volatile in the outdoor environment behave like semivolatile organic compounds indoors. Compartment size is also important. By combining partition coefficients with the compartment size, one can determine where most of the mass of a chemical contaminant will be at equilibrium. Equilibrium conditions may change with time as environmental parameters such as temperature and relative humidity change. Partitioning does not occur instantaneously, and it can take time, sometimes years, for compartments to approach equilibrium due to slow rates of molecular transport. Thus, partitioning of semivolatile and low-volatility molecules to indoor surfaces can increase indoor residence times. Partitioning can also result in chemicals moving to locations and conditions favorable for chemical transformations.

Challenges remain in understanding partitioning from the molecular to whole building scales and how partitioning influences exposure and chemistry. These challenges include limited information on the detailed composition of building materials across the building stock; complexity of spatial and temporal variations in environmental conditions; and poor understanding of surfaces and indoor materials at molecular and nanometer length scales, as well as of the molecular interactions that determine the partitioning of a chemical between phases.

Despite a rapidly growing base of knowledge about indoor partitioning, important data gaps remain. The materials that are present in buildings, or comprise buildings, are not physically or chemically well characterized. Partition coefficients have been measured for very few chemical contaminants and materials. Models to predict thermodynamic parameters exist, but their application to real indoor materials has not been widely demonstrated. Furthermore, models have not been successfully applied to some chemical classes important in indoor environments, such as surfactants. The extent to which environmental and other building factors, occupant activities, and control systems influence partitioning and exposure remains to be explored. There is also a need to improve our understanding of partitioning at the molecular level. Addressing these needs should lead to improved predictive models of partitioning and, by extension, exposure.

CHEMICAL TRANSFORMATIONS

Chemical transformations are chemical processes that lead to the loss or removal of certain substances (e.g., reactants) and the generation or formation of new substances (e.g., products). The products that arise from these reactions frequently are very different from the reactants in terms of their partitioning, toxicity, and other properties. For example, chlorinated chemicals found in cleaning products can react with unsaturated organic compounds to produce higher molecular weight products that can contribute to film growth or secondary organic aerosol (SOA) formation.

Different types of chemical reactions are relevant indoors, including photolysis, hydrolysis, acid-base reactions, and redox reactions. Some of these processes are irreversible, leading to a permanent loss of species, while others are reversible, resulting in the temporary loss and eventual regeneration of reactants. These chemical processes are complex and extensive, with numerous species involved as precursors, intermediates, or products.

Chemical transformations occur at different locations indoors, including the gas phase, airborne particles, and indoor surfaces, as well as hidden places such as ducts and the heating, ventilation, and air-conditioning (HVAC) system. Surface-adsorbed molecules may diffuse into the bulk of indoor surfaces and materials, where they may undergo chemical transformations. The relative rates of ventilation, gas-phase loss, and loss to surfaces are important to compare when evaluating the fate of an indoor air molecule. Reactions on surfaces can be very important, even if relatively slow, if the species preferentially partitions to the surface.

Major findings from the past several years illustrate the complexity of chemical reactions that occur in indoor environments. In particular, gas-phase oxidation reactions, some occurring via auto-oxidation mechanisms, lead to the formation of a suite of highly oxygenated gas-phase species which may form SOA. In addition, much of reactive indoor chemistry occurs on surfaces, via multiphase chemistry. Although long acknowledged to be important, ozonolysis reactions of unsaturated organics have recently been demonstrated to form highly oxygenated species, such as secondary ozonides and volatile oxygenates, on surfaces. This chemistry is known to occur on humans, their clothing, and other surfaces contaminated by cooking or smoking emissions.

The complexity of such reactions presently precludes a quantitative understanding of these processes under actual indoor conditions, where substrate composition and environmental parameters (e.g., relative humidity) have been shown to affect the mechanisms and kinetics. The identities and amounts of many indoor chemicals, especially in surface reservoirs, remain incompletely understood. This data gap can lead to incomplete toxicological and epidemiological evaluation of chemical dose and health outcomes in indoor environments. Furthermore, such uncertainties in reactive chemistry, when coupled with uncertainties in partitioning, make it challenging to determine the relative importance of the major exposure pathways for many indoor chemicals.

New chemistry has been identified recently when chemical cleaning agents, such as chlorine bleach, are used on indoor surfaces. The suite of chemical products that arises from such use is only just starting to be studied. The reactive chemistry that occurs with some other common cleaning agents, such as hydrogen peroxide, has yet to be investigated under indoor conditions.

Photochemistry largely occurs in genuine indoor settings where air or surfaces are directly illuminated with sunlight. While infrequent in many indoor settings, high levels of oxidants can be generated, and other reactive photochemistry can occur in such situations. It is possible that important, yet slow, photochemistry occurs elsewhere on indoor surfaces that are not exposed to direct sunlight, but this has yet to be confirmed.

Condensed-phase water is an important medium for facilitating indoor chemical transformations. These can include acid-base reactions, slow hydrolysis of organic compounds such as esters, reactions with Criegee intermediates that form during ozonolysis of unsaturated organics, and the nitrogen dioxide disproportionation reaction that forms nitric and nitrous acids.

Important progress has been made in the past few years to develop models that integrate the growing knowledge of chemical transformations, partitioning between different indoor reservoirs, mass transfer, and indoor-outdoor air exchange. However, these models remain limited in their predictive capabilities owing to uncertainties in the underlying fundamental chemistry, especially on surfaces. Elucidation of the role of occupants on indoor chemistry remains a research need. Addressing these data gaps may require the application of advanced instrumentation and analytical techniques to study chemistry taking place in buildings and on surfaces.

MANAGEMENT OF CHEMICALS IN INDOOR ENVIRONMENTS

Effective management of chemicals in the indoor environment is critical to human health. The management of chemical contaminants in indoor environments includes removal (through ventilation, filtration, sorption, physical cleaning, and passive surface removal) and chemical transformations (including photolysis, ionizers, chemical additions, photocatalysis). No single management approach can remove all contaminants that are present indoors; therefore, source elimination is always the preferred method of control. However, combinations of management approaches can also be effective at reducing exposure, as can situation-specific choices, such as increasing ventilation to reduce air contaminant exposure. Every management approach has different chemical consequences; for example, approaches that include oxidation are particularly prone to generating products of concern.

Several knowledge gaps remain in the scientific community's understanding of underlying physical and chemical principles of air cleaning, including the fundamental chemistry of many air-cleaning technologies. Except for ventilation, particle filtration, and sorption, few air-cleaning approaches have been tested in real-world environments, which contain a far more complicated mixture of compounds than most laboratories. Chemical reactions in indoor environments can follow complex mechanisms and result in numerous products. This makes predicting chemical reactions and the efficacy of air-cleaning devices challenging and highlights the need for better testing standards for air-cleaning efficacy and chemistry that account for this complexity. There is insufficient research to truly understand and prioritize chemical byproducts in terms of toxicology and health effects, and to identify safe and effective levels of chemical additives for air-cleaning technologies. The committee also identified real-world testing of management approaches that have potential to induce chemistry as a future research need.

Given the recent increased public interest in indoor air quality, driven in part by COVID-19, device manufacturers, researchers, and public health professionals need to communicate clearly to consumers about the efficacy and chemical consequences of different air-cleaning approaches. The lack of testing and regulation has led to rampant unsubstantiated claims about efficacy and health benefits of devices. The potential health risks and benefits resulting from their use warrant further investigation and potential certification or regulatory oversight. Based on the current state of knowledge, the committee cautions against approaches that induce secondary chemistry in occupied settings, unless the benefits demonstrably outweigh the risks of exposure to chemical reactants and byproducts.

INDOOR CHEMISTRY AND EXPOSURE

To date, the foremost goal of exposure science has been to identify and characterize the inhalation, ingestion, and dermal uptake by people of harmful chemicals that can cause acute or chronic health effects. The application of exposure science to the study of indoor environments and exposures that occur therein is relatively nascent but rapidly evolving. Cost-effective policies and guidance suitable for diverse indoor environments and indoor-dwelling populations demand a thorough understanding of indoor exposure profiles. Understanding large differences in indoor exposures requires deeper insight on the societal and systemic context in which exposures occur in residential and nonresidential environments. Environmental health disparities that are persistently observed in the United States and around the world too often remain understudied.

The evidence base and toolkits for developing a robust and comprehensive understanding of indoor exposure profiles are growing rapidly. This evidence base has grown through multiple research channels, including field-based, laboratory-based, and modeling studies. Among field-based studies, emergent tools are addressing long-standing challenges of assessing spatial and temporal resolution on concentrations of airborne hazards, as well as diversity of chemical species in indoor air. Consumer-grade measurement tools and research-grade, high-resolution instrumentation are achieving wider use in indoor environments.

Researchers are working to understand exposure to chemical mixtures. These efforts complement strategic priorities of federal agencies, such as the National Institutes of Health. For example, the National Institute of Environmental Health Sciences has identified strengthening understanding of combined exposures as a strategic priority. Measurement science advances applied to indoor environments and personal sampling are helping to better understand discrepancies—for example, between personal exposures and stationary monitors or indoor and outdoor area concentrations. Yet, inconsistency in chemical identifiers remains a challenge. Exposure data are collected across diverse sampling platforms, ranging from transient, short-duration exposures to chronic and longitudinal exposures, among populations that vary greatly in size and participant composition. This leads to diverse data that are not standardized and therefore not readily available to support modeling efforts.

One of the most important and fundamental needs for improving the utility of exposure models is to form linkages between physical process and exposure and uptake models. Integrating frameworks can lead to an improved understanding of the relationship among indoor air chemistry, exposure, and internal dose. Exposure model improvements will rest upon an enhanced understanding of indoor exposures to chemicals and a deeper understanding of human behavior as it relates to indoor chemistry.

A PATH FORWARD FOR INDOOR CHEMISTRY

This report focuses on different aspects of indoor chemistry, including new findings related to under-reported chemical species, chemical reactions, and sources of chemicals and their distribution in indoor spaces. An understanding of how indoor chemistry fits into the context of what is known about the links among chemical exposure, air quality, and human health continues to evolve. The committee provides its recommendations for critical needs to advance research, enhance coordination and collaboration, and overcome barriers for implementation of new research findings into practice in indoor environments. A critical cornerstone of the committee's vision for the future of indoor chemistry research is increased awareness on the part of the scientific community of the challenges and opportunities for innovation in indoor chemistry research, as well as the need to fund research in indoor chemistry. It is critical to translate the emerging science on indoor chemistry into practice that benefits public health and the environment.

Chemical Complexity in the Indoor Environment

An emerging theme in indoor chemistry is the high degree of chemical complexity in indoor environments. People are often in close proximity to sources and processes that, respectively, emit and transform chemicals. Environmental conditions and indoor chemistry vary between buildings depending in part on the building use, design, construction materials, operation, maintenance, occupant density, chemical use, contents, and air handling and purification systems. For example, chemical emissions from cooking are common in restaurants and residences but are uncommon in other public buildings.

Despite the importance of indoor exposure, researchers know very little about how humans are exposed to multiple indoor chemicals across phases and pathways, how these joint exposures interact across timescales, and the cumulative and long-term impacts of the indoor chemical environment on human health. Humans are in contact with mixtures of chemicals with potentially synergistic or antagonistic modes of action and effects on health. Many of these mixtures are not chemically characterized or quantified, nor have their chemical transformations or partitioning between different indoor reservoirs been studied. Studies of exposure to mixtures in the indoor environment and their health effects are lacking, in part due to the complexity and dynamics of indoor chemistry.

Recommendation 1: Researchers should further investigate the chemical composition of complex mixtures present indoors in a wide range of residential and nonresidential settings and how these mixtures impact chemical exposure and health.

Recommendation 4: All stakeholders should proactively engage across disciplines to further the development of knowledge on the fundamental aspects of complex indoor chemistry and its impact on indoor environmental quality, exposure assessment, and human health.

Indoor Chemistry in a Changing World

Unprecedented changes are occurring to the outdoor environment owing to climate change, wildfires, and urbanization, standing in contrast to improvements derived from environmental regulations and advancements in technology. These changes have impacts on indoor environments, many of which have yet to be fully characterized.

Recommendation 5: Researchers who study toxicology and epidemiology and their funders should prioritize resources toward understanding indoor exposures to contaminants, including those of outdoor origin that undergo subsequent transformations indoors.

Recommendation 6: Researchers and their funders should devote resources to creating emissions inventories specific to building types and to identifying indoor transformations that impact outdoor air quality.

Recommendation 7: Researchers and engineers should integrate indoor chemistry considerations into their building system design and mitigation approaches. This can be accomplished in different ways, including by consulting with indoor air scientists.

Future Investments in Research

The emerging picture of indoor environments indicates chemical complexity in gas, particle, and surface phases. Although new analytical tools have been instrumental in improving understanding of indoor chemistry, several key challenges remain that will require strategic investments.

Recommendation 8: Given the challenges, complexity, knowledge gaps, and importance of indoor chemistry, federal agencies and others that fund research should make the study of indoor chemistry and its impact on indoor air quality and public health a national priority.

Recommendation 11: Federal agencies should design and regularly implement an updated National Human Activity Pattern Survey. Federal and state agencies should add survey questions in existing surveys that capture people's activities in indoor environments as they relate to indoor chemistry and indoor chemical exposures.

Communicating Science and Risks

To many stakeholders, science concerns itself mainly with discovery ("what?" and "how?") and leaves questions of relevance ("why does it matter?") and application ("how can it be used?") to others. The process of creating scientific knowledge and transferring it into the spheres of practice and policy can be inefficient and slow. The monumental effort during the COVID-19 pandemic to bring scientific tools to bear on means of mitigating its effects exemplifies what can happen when this connection is made. Making the same connection between science and application is essential in indoor chemistry.

Recommendation 12: Researchers should proactively engage in links that connect research to application throughout the indoor chemistry research process—for example, at the dissemination stage, by engaging with technical and standard-writing committees, presenting at conferences attended by practitioners, and disseminating the significance of research findings in social and mass media.

Recommendation 13: Researchers and practitioners should include environmental justice communities in the wide range of indoor environments they study and engage these communities in formulating research priorities and recommendations for future indoor air quality standards.

Recommendation 14: Funding agencies should support interdisciplinary research to investigate the impact of products and services on indoor chemistry, especially under realistic conditions. There is also a need to determine how occupant access to air quality data leads to behavior that influences indoor chemistry.

Conclusion 1: Standardized consensus test methods could enable potential certification programs for air-cleaning products and services. Such test methods could help regulators determine whether action on these products and services is warranted.

Recommendation 15: Researchers and their funders should prioritize understanding the health impacts from exposure to specific classes and mixtures of chemicals in a wide range of indoor settings. Such understanding is needed to inform any future standards, guidelines, or regulatory efforts.

There is a growing awareness that exposure to environmental contaminants contributes to the burden of human disease. For decades, much of the attention of the scientific and regulatory communities has focused on outdoor air quality and drinking water. These efforts have contributed to improvements in outdoor air and water quality that have measurably protected human health and the environment.

As more attention is deservedly focused on indoor chemistry, more chemicals and their reactions will be identified, adding to an already complex problem. Many of these chemicals may have little to no information regarding their toxicity, either as individual agents or in combination with other chemicals present in the environment. Mitigating chemical hazards will depend on many factors and needs to be done in a manner that considers the impacts of any mitigation strategy itself on the indoor environment. This will require efforts in changing building design and operation, altering the use and contents of products and materials, and addressing the impact of human activity on indoor chemistry.

The immensity of this daunting task need not lead to inaction. Rather, investment at this time in a holistic approach that considers chemistry, biology, and social contributions to health will pay dividends in the future.

1

Introduction

T his chapter provides information about the motivation for the report and the committee's conduct of the study. It begins with an overview of why indoor chemistry is a subject that deserves examination. This report builds on prior efforts in the area, which are summarized in this chapter to help frame the issue. It provides the statement of task for the National Academies of Sciences, Engineering, and Medicine (the National Academies) committee responsible for this report and the committee's approach to completing its task. The chapter concludes with a guide to the organization of the report.

WHY INDOOR CHEMISTRY MATTERS

Indoor chemistry describes the complex collection of reactions involving chemicals indoors (Weschler and Carslaw, 2018). This includes chemical reactions occurring in the gas phase, on particles, or on surfaces (Weschler et al., 2006). Understanding this complex chemistry matters for three prime reasons: (1) indoor chemistry is complex and may lead to the creation of new chemicals of concern, (2) indoor chemicals can adversely impact indoor air quality and the indoor environment, and (3) indoor chemicals are an important source of human exposure that may result in adverse health effects.

Indoor Chemistry Is Complex

Indoors, chemicals can undergo oxidative and other types of reactions with other chemicals in the air or on surfaces, creating potentially more or less toxic products. These chemicals can then move within the indoor environment. For example, chemicals emitted into the gas phase can partition to particles suspended in the air or collect on surfaces. Under different environmental conditions (e.g., temperature and relative humidity), this process may be reversed, leading to re-emission of chemicals into the gas phase. Both environmental conditions and indoor chemistry can vary between buildings depending in part on the building use, design, construction materials, operation, maintenance, occupant density, chemical use, contents, and air handling and purification systems.

For example, ventilation can vary between different types of buildings. Mechanical ventilation is commonly used in office buildings in the United States, while many homes are naturally ventilated through windows, doors, and infiltration (IOM, 2011). Buildings also differ in the types of chemicals that may be emitted owing to the types of materials present and the various activities that occur in different types of spaces. For example, chemical emissions from cooking and candles are common in restaurants and residences but are uncommon in other public buildings. The use of disinfectants and cleaning products occurs more often in health care facilities and schools than in many other types of buildings. Each type of indoor space could therefore have very different indoor chemistry.

Indoor Chemistry Influences the Indoor Environment

Chemicals found indoors are a significant risk factor that can modify or degrade the indoor environment. Direct emission of chemicals into the indoor environment can occur from various sources, including building materials, paints, stoves, cleaning products, pesticides, furnishings, electronics, and personal care products (see Chapter 2 and Figure 2-1). Biologic sources including microorganisms, plants, pets, and other animals can also contribute to indoor chemistry. Some chemicals present indoors can be from outdoor sources, often because of ventilation. Chemicals that contaminate occupants' skin, clothing, or belongings can be brought indoors. People are also an important source of indoor aerosolized particles, carbon dioxide, ammonia, methane, and other organic chemicals (Ampollini et al., 2019; Li et al., 2020; Wang et al., 2021; Yang et al., 2021). Collectively these chemicals can affect indoor air quality and degrade the indoor environment in other ways.

Indoor Chemistry Can Impact Human Health

The indoor environment influences the behavior, comfort, productivity, and health of its occupants. The opportunity for human exposure to indoor chemicals is significant: exposure is the product of chemical concentration and time, and people spend most of their time indoors. Data from the National Human Activity Pattern Survey showed that Americans spent an average of nearly 69 percent of a 24-hour day in a residence and more than 18 percent in other indoor locations (Klepeis et al., 2001). A Canadian counterpart, the Canadian Human Activity Pattern Survey, revealed similar activity and location patterns except for seasonal differences. Canadians spend less time outdoors in winter and less time indoors in summer than their U.S. counterparts (Leech et al., 2002). On daily average, German children spend 65 percent of a 24-hour day in a residence and nearly 20 percent in other indoor locations (Conrad et al., 2013). Whether exposure to an indoor chemical results in an adverse effect is dependent on exposure duration and additional factors, including the inherent toxicity of the chemical (or mixture), its concentration in the environment, the route of exposure, and the susceptibility of the person. Box 1-1 provides several examples in which indoor chemistry is associated with adverse human health effects. Exposure to indoor-sourced chemicals can generally be reduced by increasing ventilation, using appropriate air-cleaning processes, selecting low-emission materials for building surfaces and furnishings, and avoiding certain cleaning and personal care products. Exposure can also be impacted by human factors, including the timing of when a building is occupied, occupant density, and occupant behaviors such as cooking and cleaning that may influence both chemical sources and exposure.

PRIOR EFFORTS ON WHICH THIS REPORT BUILDS

Research into indoor chemistry extends back decades and has often leveraged advances in atmospheric chemistry. Many initial investigations focused on the measurement of individual chemicals in indoor air and the release of chemicals from building materials and contents into the

---

BOX 1-1
Why Indoor Chemistry Matters: Indoor Air Quality and Human Health

Numerous examples demonstrate well-established associations between adverse effects on human health and indoor air quality. For example, indoor levels of radon contribute to the incidence of lung cancer (Cheng et al., 2021). Likewise, environmental tobacco smoke, particulate matter, and biologic materials (e.g., dust mites, cockroach allergens, pet dander) can precipitate asthmatic episodes in susceptible people (Kanchongkittiphon et al., 2015). Legionnaires' disease is closely associated with water to air transfer of *Legionella* in buildings with poorly maintained cooling towers (Prussin et al., 2017). And exposure to polybrominated diphenyl ethers, flame retardants that were applied to residential furniture, electronics, and insulation, has been linked to neurodevelopmental deficits in children and thyroid disease in adults (Allen et al., 2016; Chen et al., 2014).

Indoor carbon monoxide exposure remains a significant public health concern. Carbon monoxide is produced indoors by cooking, heating, and other combustion sources. Infiltration of carbon monoxide from outdoor air into the indoor environment also occurs and can be exacerbated by wildfires (Shrestha et al., 2019). Human fatalities from high levels of carbon monoxide indoors have prompted the widespread use of consumer-grade sensors that can alert the occupant when high levels of carbon monoxide are present indoors. Chronic exposure to lower levels of carbon monoxide may also contribute to the burden of disease in people (Townsend and Maynard, 2002).

Indoor sources of acrolein arise from thermal processes including those associated with tobacco smoke, gas stoves, and wood-burning fireplaces. Acrolein is also formed during cooking and deep frying, resulting in the decomposition of food-based fatty acids, glycerides, and carbohydrates (Schieweck et al., 2021). This indoor chemistry can result in indoor levels of acrolein exceeding those found outdoors. Exposure to acrolein has been linked with eye and respiratory irritation, respiratory tract injury, and cardiovascular effects. As is often the case for indoor chemicals, few epidemiological studies have examined human health and indoor acrolein exposure. Health Canada has released a Residential Indoor Air Quality Guideline for acrolein.

---

indoor environment. Studies performed in the mid-20th century examined lead, radon, asbestos, and cigarette smoke levels in homes (Samet and Spengler, 2003). One of the first studies showed that sulfur dioxide concentrations found in Dutch homes with gas-fired combustion devices differed from those found in outdoor air (Biersteker, 1965). Since that time, interest has grown not only in the presence or absence of individual chemicals but also in the various chemical processes that occur indoors. Researchers recognized that hydrolysis chemistry taking place in building materials increased emissions of formaldehyde as far back as 1962 (Wittmann, 1962). Studies examining the reaction of ozone with surfaces in the indoor environment began in the early 1970s (Mueller et al., 1973; Sabersky et al., 1973; Shair and Heitner, 1974). Comprehensive chemistry models of indoor air (Nazaroff and Cass, 1986), evidence of gas-phase chemistry (Weschler et al., 1994; Weschler and Shields, 1996, 1999), and surface reaction products (Pitts et al., 1985; Weschler et al., 1992), as well as indoor surface reservoirs (Tichenor et al., 1991), followed. The U.S. Environmental Protection Agency (EPA) launched the Total Exposure Assessment Methodology (TEAM) Project in 1979. This project focused on volatile organic compounds (VOCs), carbon monoxide, and pesticides (Wallace, 1989; Wallace et al., 1987). This study used a probabilistic sampling approach and personal air monitors and breath analysis to measure daily personal exposures in air and drinking water. This study showed that most exposures occurred indoors, and pollutant levels in the indoor environment were higher than levels measured outdoors. In 1990, EPA partnered with the California Air Resources Board to conduct a follow-up study measuring personal exposure to particles (PM_{10}). This Particle Total Exposure Assessment Methodology (PTEAM) study found that vacuuming, dusting, cooking, and living with a smoker were associated with higher PM_{10} exposures (Clayton et al., 1993). The Relationship of Indoor, Outdoor, and Personal Air (RIOPA) study measured

VOCs, semivolatile organic compounds, and fine particulate matter ($PM_{2.5}$) outdoors and in homes (Weisel et al., 2005). Additionally, it was shown that indoor concentrations of formaldehyde and acetaldehyde were higher indoors than outdoors, while acrolein and crotonaldehyde found in homes without tobacco use came primarily from outdoor air (Liu et al., 2006). The goal of the RIOPA study was to assess exposure, not to understand chemistry; it relied on collecting time-integrated samples that provided detailed snapshots of chemical composition of personal and indoor air but did not capture real-time changes in indoor chemistry.

This report builds on the results of a number of prior studies supported by foundations, agencies, and societies, including EPA, the National Science Foundation, the National Institute of Standards and Technology, the National Institute for Occupational Safety and Health, the National Institutes of Health, and ASHRAE, among others in the United States and internationally (Andersen and Gyntelberg, 2011; WHO, 2010). The Alfred P. Sloan Foundation (Sloan Foundation) launched the Microbiology of the Built Environment and Chemistry of Indoor Environments programs in 2004 and 2016, respectively. Unlike federal funders of research, who typically fund research within a single discipline or related group of disciplines, the Sloan Foundation had the flexibility to fund highly interdisciplinary research, aiming to break down siloes that commonly limit research progress in many fields. The foundation's ability to support multi-year programs led to the creation of a community of investigators with an interest in indoor chemistry who have formed consortia and partnered to conduct large field campaigns. The SURFace Consortium for Chemistry of Indoor Environments provides a forum for discussing research on the mechanisms and kinetics of chemical reactions that occur on indoor surfaces (Ault et al., 2020). The House Observations of Microbial and Environmental Chemistry (HOMEChem) experiment explored how human activities affected indoor sources of hydroxyl radicals, ozone, and other chemical oxidants and the formation of organic compounds (Farmer et al., 2019). The Modelling Consortium for Chemistry of Indoor Environments (MOCCIE) develops quantitative physical-chemical models that describe gas-phase and surface chemistry in the indoor environment and consider how occupants, indoor activities, and building conditions influence various molecular processes occurring indoors (Shiraiwa et al., 2019). In March 2022, the White House announced the Clean Air in Buildings Challenge, an EPA-led initiative to implement strategies for improving indoor air quality.

Associations among indoor chemistry, exposures, and adverse health effects have also been considered for several decades (Spengler and Sexton, 1983). A National Research Council (NRC) report entitled *Indoor Pollutants* (NRC, 1981) largely focused on indoor chemicals with a suspected association with negative health effects. This NRC report evaluated asbestos and other fibers, radon, formaldehyde, tobacco smoke, combustion products, and microorganisms, with a focus on allergens. This report also noted that headaches, mucous-membrane irritation, and difficulty concentrating and other signs of "sick building syndrome" were associated with occupancy of certain buildings (NRC, 1981). A later report by the Institute of Medicine (IOM) concluded that indoor allergens in the United States primarily come from house dust mites and cockroaches, microorganisms, and companion animals; exposure to these indoor allergens could contribute to the incidence of asthma and allergies (IOM, 1993). A 2016 National Academies workshop focused on particulate matter (PM) and a growing body of research that shows that human exposure to PM indoors can exceed outdoor levels and may represent an important risk factor (NASEM, 2016). The burning of solid fuel for cooking and heating in the United States (Rogalsky et al., 2014) and in developing countries is an important source of PM and other indoor chemicals and may account for 4 percent of the global burden of disease (Bennitt et al., 2021). The use of portable high-efficiency particulate air (HEPA) filters in homes located in a community with widespread residential wood combustion reduced indoor PM levels and improved microvascular endothelial function (Allen et al., 2011). The indoor environment has been associated with lung cancer, allergies, sick building syndrome and other hypersensitivity reactions, and respiratory infections (Sundell, 2004; Tran et al., 2020).

Finally, the worldwide pandemic of coronavirus disease 2019 (COVID-19) highlighted the importance of ventilation and filtration in reducing airborne disease transmission indoors. While much of the research to date about the relationship between chemical exposures and human health has focused on understanding acute health effects, associations with chronic and delayed illnesses are also important.

Against that backdrop, EPA, the Sloan Foundation, the National Institute of Environmental Health Sciences, and the Centers for Disease Control and Prevention approached the National Academies requesting an assessment of the state of the science regarding chemicals in indoor air in the nonindustrial indoor environment. The Committee on Emerging Science on Indoor Chemistry was formed to respond to that request. The overarching goal of this project is to identify new findings about previously under-reported chemical species, chemical reactions, and sources of chemicals, as well as the distribution of chemicals; and improve our understanding of how indoor chemistry is linked with chemical exposure, indoor air quality, and human health.

STATEMENT OF TASK

Faced with the challenges described above, the National Academies convened an ad hoc committee to summarize the state of the science regarding chemicals in indoor environments. Biographical information on the committee members is presented in Appendix B. The statement of task is provided in Box 1-2.

COMMITTEE'S APPROACH

The statement of task had several components that could be interpreted in a variety of ways. This section provides the committee's interpretation of the task, including how it approached and addressed each of the components. The committee interpreted "chemicals in indoor air" to include gas- and particle-phase species, elemental species, and PM. The committee discussed the extent to which it should also include toxins of biological origin. Because this topic was the subject of a recent National Academies report (NASEM, 2017), the committee chose to focus its primary attention on other chemical sources. The committee debated whether nonindustrial exposures

---

BOX 1-2
Statement of Task

The National Academies of Sciences, Engineering, and Medicine will convene an ad hoc committee of scientific experts and leaders to consider the state-of-the-science regarding chemicals in indoor air. Specifically, the committee will focus on: (1) new findings about previously under-reported chemical species, chemical reactions, and sources of chemicals, as well as the distribution of chemicals; and (2) how indoor chemistry findings fit into context of what is already known about the link between chemical exposure, air quality, and human health.

The committee's consideration of this information will lead to a report with findings and recommendations regarding: (1) key implications of the scientific research, including potential near-term opportunities for incorporating what is known into practice; and (2) where additional chemistry research will be most critical for understanding the chemical composition of indoor air and adverse exposures. As appropriate, opportunities for advancing such research by addressing methodological or technological barriers or enhancing coordination or collaboration will be noted. The committee will also provide recommendations for communicating its findings to affected stakeholders. The indoor environments focused on in this study will be limited to non-industrial exposure within buildings.

applied to all buildings associated with occupational exposures. The committee decided to include certain indoor environments, including schools, office buildings, and medical facilities, where occupational exposures occur. The committee also considered what "new findings about previously under-reported chemical species" would constitute. Although no strict time frame was enforced, the committee generally considered work published in the past decade to be "new." Likewise, the committee also considered the relevancy of certain well-studied chemicals (e.g., ozone, radon). The committee included well-studied chemicals when there was emerging evidence of their role in indoor chemistry.

The committee stayed informed about several ongoing National Academies activities, including the work of the Committee on Indoor Exposure to Fine Particulate Matter and Practical Mitigation Approaches. This National Academies effort resulted in several workshops given in 2021 that discussed the state-of-the-science on exposure to $PM_{2.5}$ indoors, its health impacts, and engineering approaches and interventions to reduce exposure risks, including practical mitigation solutions in residential settings (NASEM, 2021).

To address the statement of task, the committee consulted several sources of information on buildings, indoor chemistry, and occupant health. A comprehensive discussion of indoor chemicals and human health was considered beyond the scope of this report. Instead, the committee focused on providing emerging data showing associations between under-reported chemical species and human health. The primary source used for health outcomes was human epidemiologic studies that examined exposures to chemicals in homes and other nonindustrial settings. A systematic review of the research literature was deemed an undertaking beyond the scope of this report. Committee members independently compiled lists of potential citations based on their scientific expertise. The committee considered the most influential scientific work available at the time it completed its task in early 2022, encompassing observations in real indoor environments, laboratory experiments, and theoretical modeling studies. This is often referred to as the "three-legged stool" model and has been a very successful approach in outdoor atmospheric chemistry (Box 1-3). The committee also referred to the research and conclusions of prior National Academies committees that addressed indoor environments and health issues. The 2007 NRC report *Green Schools: Attributes for Health and Learning* (NRC, 2007); the 2011 IOM report *Climate Change, the Indoor Environment, and Health* (IOM, 2011); and the 2017 National Academies report *Microbiomes of the Built Environment: A Research Agenda for Indoor Microbiology, Human Health, and Buildings* (NASEM, 2017) were influential. Finally, the committee organized a public virtual information-gathering workshop that was held on April 5, 2021. The agenda for this workshop is available in Appendix C.

ORGANIZATION OF THIS CONSENSUS REPORT

The subsequent chapters of this report explore aspects of the dynamic, interacting systems connecting humans, nonindustrial built environments, and indoor chemistry. Chapter 2 provides an overview of major primary sources and reservoirs of chemicals in the indoor environment, their emission rates, and emerging science on detecting and quantifying them. Chemical emissions are linked to both chronic and episodic sources and are modified by behavioral, environmental, and physical factors. Building materials, consumer products, infiltration of outdoor air, and human behavior (particularly cooking) can strongly influence the chemicals present in the indoor environment. Dust and indoor surfaces also serve as reservoirs of chemicals that are slowly released from primary sources, and they play a role in understanding human exposure in the indoor environment.

Chapter 3 focuses on chemical partitioning between surfaces and the air. Partitioning refers to the transfer of molecules from one phase to another, such as from air to an indoor surface. This process is often reversible, and a material can act as a reservoir that can absorb or release chemicals depending on conditions. In addition to building materials, furnishings, and personal

---

BOX 1-3
Outdoor Air and Indoor Chemistry

Although outdoor air represents an important source of indoor chemicals, the reader of this report should not consider indoor air and outdoor air to be equivalent. The indoor environment provides additional sources and reservoirs of chemicals that can undergo partitioning and chemical reactions forming new products. This chemistry modifies indoor air. In many cases, chemicals in indoor air can be found at much higher concentrations than those observed in outdoor air. In other cases, chemicals present indoors can also contribute to outdoor air pollution. Methods used to detect chemicals in the outdoor air and predict their behavior with computational models have provided scientists with a useful starting point for research efforts conducted indoors. However, unique methods, approaches, and models are needed to study indoor chemistry.

Air pollution has been recognized as an important risk factor for human disease. Emission of certain hazardous chemicals is regulated by federal and state agencies in the United States. Similar regulatory efforts occur worldwide. These regulations have resulted in improved outdoor air quality and decreased the human burden of disease (Samet et al., 2017). Comparable regulations for the indoor environment are generally lacking and are only recently emerging internationally. This lack of regulations does not suggest that indoor air quality is not a concern. Indeed, many people spend most of their time indoors, and this can have significant implications for chemical exposure and health. There is also a paucity of human epidemiology studies that have considered indoor air exposures, further slowing recognition of the contribution that indoor chemistry and indoor air quality may make to the human burden of disease. This is especially important for children and other potentially susceptible subpopulations, including individuals experiencing environmental health disparities.

The reader is directed to later chapters in this report where the committee considers these and other topics. Ultimately, indoor chemistry matters for public health, and this report highlights ways indoor chemistry science could advance, thereby protecting human health and the environment.

---

items, chemicals can partition into dust, condensed water, and the organic films that coat indoor surfaces. Partitioning strongly influences the timing and intensity of occupant exposure to chemicals and can limit efforts to reduce exposure by increasing building air exchange or the use of active air cleaning.

Chapter 4 addresses recent advances in our understanding of the nature of chemical reactions that occur in indoor environments, both in the air and on surfaces. These chemical transformations lead not only to the loss of reactants but also to the formation of secondary products that are potentially more reactive and/or harmful than their precursors. Although early studies in the field had highlighted specific classes of reactions as being important in indoor spaces, recent work has delved into determining their nature in sufficient detail and new assessments of the impacts on human exposure can now be made.

Chapter 5 addresses the active management of chemicals in indoor environments using interventions to clean air and surfaces. The chapter first provides an overview of the hierarchy of controls as a framework for considering risk-reduction strategies. With this framework in mind, the chapter next considers management approaches that result in minimal changes in indoor chemistry followed by management approaches that induce chemical transformations. This chapter also describes environmental factors, human behavior, and other considerations to keep in mind when selecting and using management approaches.

Chapter 6 places the emerging science on the chemistry of indoor environments in context with exposure and the environmental health paradigm. It begins with an overview of fundamental exposure concepts and further describes features of indoor settings that influence exposure levels, timing, and duration. The chapter summarizes a broad range of exposure determinants, with a focus on factors that may drive or contribute to environmental exposure and health disparities. This chapter further considers the intersection of indoor chemistry with exposure modeling, advances in exposure assessment through measurement science, and emerging science on exposure to mixtures as it relates to indoor environments.

Chapters 2 through 6 each conclude with a short list of research recommendations that identify potential opportunities to advance the science. Finally, Chapter 7 organizes the committee's vision for the future of indoor chemistry as a field of study under four major themes: (1) chemical complexity in the indoor environment, (2) indoor chemistry in a changing world, (3) future investments in research, and (4) communicating science and risks. This final chapter of the report considers not only what research advances are highest priority but also how new findings can be translated into practice to reduce adverse indoor exposures and improve public health.

Appendix A contains a glossary of terms used in the report. Biographies of the committee members responsible for this study are provided in Appendix B. The agenda for the public workshop held by the committee can be found in Appendix C. A table summarizing a set of exposure models commonly used for predicting near-field exposures is presented in Appendix D.

REFERENCES

Allen, J. G., S. Gale, R. T. Zoeller, J. D. Spengler, L. Birnbaum, and E. McNeely. 2016. PBDE flame retardants, thyroid disease, and menopausal status in U.S. women. *Environmental Health* 15(1):60. https://doi.org/10.1186/s12940-016-0141-0.

Allen, R. W., C. Carlsten, B. Karlen, S. Leckie, S. van Eeden, S. Vedal, I. Wong, and M. Brauer. 2011. An air filter intervention study of endothelial function among healthy adults in a woodsmoke-impacted community. *American Journal of Respiratory and Critical Care Medicine* 183(9):1222–1230. https://doi.org/10.1164/rccm.201010-1572OC.

Ampollini, L., E. F. Katz, S. Bourne, Y. Tian, A. Novoselac, A. H. Goldstein, G. Lucic, M. S. Waring, and P. F. DeCarlo. 2019. Observations and contributions of real-time indoor ammonia concentrations during HOMEChem. *Environmental Science & Technology* 53(15):8591–8598.

Andersen, I., and F. Gyntelberg. 2011. Modern indoor climate research in Denmark from 1962 to the early 1990s: An eyewitness report. *Indoor Air* 21:182–190.

Ault, A. P., V. H. Grassian, N. Carslaw, D. B. Collins, H. Destaillats, D. J. Donaldson, D. K. Farmer, J. L. Jimenez, V. F. McNeill, G. C. Morrison, R. E. O'Brien, M. Shiraiwa, M. E. Vance, J. R. Wells, and X. Wiong. 2020. Indoor surface chemistry: Developing a molecular picture of reactions on indoor interfaces. *Chem* 6(12):3203–3218. https://doi.org/10.1016/j.chempr.2020.08.023.

Bennitt, F., S. Wozniak, K. Causey, K. Burkart, and M. Brauer. 2021. Estimating disease burden attributable to household air pollution: New methods within the Global Burden of Disease Study. *The Lancet Global Health* 9:S18. https://doi.org/10.1016/S2214-109X(21)00126-1.

Biersteker, K., H. De Graaf, and C. A. Nass. 1965. Indoor air pollution in Rotterdam homes. *Air and Water Pollution* 9:343–350.

Chen, A., K. Yolton, S. A. Rauch, G. M. Webster, R. Hornung, A. Sjödin, K. N. Dietrich, and B. P. Lanphear. 2014. Prenatal polybrominated diphenyl ether exposures and neurodevelopment in U.S. children through 5 years of age: The HOME study. *Environmental Health Perspectives* 122(8):856–862. https://doi.org/10.1289/ehp.1307562.

Cheng, E. S., S. Egger, S. Hughes, M. Weber, J. Steinberg, B. Rahman, H. Worth, A. Ruano-Ravina, P. Rawstorne, and X. Q. Yu. 2021. Systematic review and meta-analysis of residential radon and lung cancer in never-smokers. *European Respiratory Review* 30(159):200230. https://doi.org/10.1183/16000617.0230-2020.

Clayton, C. A., R. L. Perritt, E. D. Pellizzari, K. W. Thomas, R. W. Whitmore, L. A. Wallace, H. Ozkaynak, and J. D. Spengler. 1993. Particle Total Exposure Assessment Methodology (PTEAM) study: Distributions of aerosol and elemental concentrations in personal, indoor, and outdoor air samples in a southern California community. *Journal of Exposure Science and Environmental Epidemiology* 3(2):227–50. https://pubmed.ncbi.nlm.nih.gov/7694700/.

Conrad, A., M. Seiwert, A. Hunken, D. Quarcoo, M. Schlaud, and D. Groneberg. 2013. The German Environmental Survey for Children (GerES IV): Reference values and distributions for time-location patterns of German children. *International Journal of Hygiene and Environmental Health* 216(1):25–34. https://doi.org/10.1016/j.ijheh.2012.02.004.

Farmer, D. K., M. E. Vance, J. P. D. Abbatt, A. Abeleira, M. R. Alves, C. Arata, E. Boedicker, S. Bourne, F. Cardoso-Saldana, R. Corsi, P. F. DeCarlo, A. H. Goldstein, V. H. Grassian, L. Hildebrandt Ruiz, J. L. Jimenez, T. F. Kahan, E. F. Katz, J. M. Mattila, W. W. Nazaroff, A. Novoselac, R. E. O'Brien, V. W. Or, S. Patel, S. Sankhyan, P. S. Stevens, Y. Tian, M. Wade, C. Wang, S. Zhou, and Y. Zhou. 2019. Overview of HOMEChem: House Observations of Microbial and Environmental Chemistry. *Environmental Science: Processes & Impacts* 21(8):1280–1300. https://doi.org/10.1039/c9em00228f.

IOM (Institute of Medicine). 1993. *Indoor Allergens: Assessing and Controlling Adverse Health Effects.* Washington, DC: The National Academies Press.

IOM. 2011. *Climate Change, the Indoor Environment, and Health.* Washington, DC: The National Academies Press.

Kanchongkittiphon, W., M. J. Mendell, J. M. Gaffin, G. Wang, and W. Phipatanakul. 2015. Indoor environmental exposures and exacerbation of asthma: An update to the 2000 review by the Institute of Medicine. *Environmental Health Perspectives* 123(1):6–20. https://doi.org/10.1289/ehp.1307922.

Klepeis, N. E., W. C. Nelson, W. R. Ott, J. P. Robinson, A. M. Tsang, P. Switzer, J. V. Behar, S. C. Hern, and W. H. Engelmann. 2001. The National Human Activity Pattern Survey (NHAPS): A resource for assessing exposure to environmental pollutants. *Journal of Exposure Analysis and Environmental Epidemiology* 11(3):231–252. https://doi.org/10.1038/sj.jea.7500165.

Leech, J. A., W. C. Nelson, R. T. Burnett, S. Aaron, and M. E. Raizenne. 2002. It's about time: A comparison of Canadian and American time-activity patterns. *Journal of Exposure Analysis and Environmental Epidemiology* 12(6):427–432. https://doi.org/10.1038/sj.jea.7500244.

Li, M., C. J. Weschler, G. Beko, P. Wargocki, G. Lucic, and J. Williams. 2020. Human ammonia emission rates under various indoor environmental conditions. *Environmental Science & Technology* 54(9):5419–5428. doi.org/10.1021/acs.est.0c00094.

Liu, W., J. Zhang, L. Zhang, B. J. Turpin, C. P. Weisel, M. T. Morandi, T. H. Stock, S. Colome, and L. R. Korn. 2006. Estimating contributions of indoor and outdoor sources to indoor carbonyl concentrations in three urban areas of the United States. *Atmospheric Environment* 40(12):2202–2214. https://doi.org/10.1016/j.atmosenv.2005.12.005.

Mueller, F. X., L. Loeb, and W. H. Mapes. 1973. Decomposition rates of ozone in living areas. *Environmental Science & Technology* 7(4):342–346. https://doi.org/10.1021/es60076a003.

NASEM (National Academies of Sciences, Engineering, and Medicine). 2016. *Health Risks of Indoor Exposure to Particulate Matter: Workshop Summary.* Washington, DC: The National Academies Press.

NASEM. 2017. *Microbiomes of the Built Environment: A Research Agenda for Indoor Microbiology, Human Health, and Buildings.* Washington, DC: The National Academies Press.

NASEM. 2021. *Indoor Exposure to Fine Particulate Matter and Practical Mitigation Approaches: Proceedings of a Workshop.* Washington, DC: The National Academies Press.

Nazaroff, W. W., and G. R. Cass. 1986. Mathematical modeling of chemically reactive pollutants in indoor air. *Environmental Science & Technology* 20(9):924–934. https://doi.org/10.1021/es00151a012.

NRC (National Research Council). 1981. *Indoor Pollutants.* Washington, DC: The National Academies Press.

NRC. 2007. *Green Schools: Attributes for Health and Learning.* Washington, DC: The National Academies Press.

Pitts, J. N., T. J. Wallington, H. W. Biermann, and A. M. Winer. 1985. Identification and measurement of nitrous acid in an indoor environment. *Atmospheric Environment (1967)* 19(5):763–767. https://doi.org/10.1016/0004-6981(85)90064-2.

Prussin, A. J. 2nd, D. O. Schwake, and L. C. Marr. 2017. Ten questions concerning the aerosolization and transmission of Legionella in the built environment. *Building and Environment* 123:684–695. https://doi.org/10.1016/j.buildenv.2017.06.024.

Rogalsky, D. K., P. Mendola, T. A. Metts, and W. J. Martin, 2nd. 2014. Estimating the number of low-income Americans exposed to household air pollution from burning solid fuels. *Environmental Health Perspectives* 122(8):806–810. https://doi.org/10.1289/ehp.1306709.

Sabersky, R. H., D. A. Sinema, and F. H. Shair. 1973. Concentrations, decay rates, and removal of ozone and their relation to establishing clean indoor air. *Environmental Science & Technology* 7(4):347–353. https://doi.org/10.1021/es60076a001.

Samet, J. M., T. A. Burke, and B. D. Goldstein. 2017. The Trump Administration and the environment: Heed the science. *New England Journal of Medicine* 376(12):1182–1188. https://doi.org/10.1056/NEJMms1615242.

Samet, J. M., and J. D. Spengler. 2003. Indoor environments and health: Moving into the 21st century. *American Journal of Public Health* 93(9):1489–1493. https://doi.org/10.2105/ajph.93.9.1489.

Schieweck, A., E. Uhde, and T. Salthammer. 2021. Determination of acrolein in ambient air and in the atmosphere of environmental test chambers. *Environmental Science: Processes & Impacts* 23(11):1729–1746. https://doi.org/10.1039/d1em00221j.

Shair, F. H., and K. L. Heitner. 1974. Theoretical model for relating indoor pollutant concentrations to those outside. *Environmental Science & Technology* 8:444–451.

Shiraiwa, M., N. Carslaw, D. J. Tobias, M. S. Waring, D. Rim, G. Morrison, P. S. J. Lakey, M. Kruza, M. von Domaros, B. E. Cummings, and Y. Won. 2019. Modelling Consortium for Chemistry of Indoor Environments (MOCCIE): Integrating chemical processes from molecular to room scales. *Environmental Science: Processes & Impacts* 21(8):1240–1254. https://doi.org/10.1039/c9em00123a.

Shrestha, P. M., J. L. Humphrey, E. J. Carlton, J. L. Adgate, K. E. Barton, E. D. Root, and S. L. Miller. 2019. Impact of outdoor air pollution on indoor air quality in low-income homes during wildfire seasons. *International Journal of Environmental Research and Public Health* 16(19):3535. https://doi.org/10.3390/ijerph16193535.

Spengler, J. D., and K. Sexton. 1983. Indoor air pollution: A public health perspective. *Science* 221(4605): 9–17. https://doi.org/10.1126/science.6857273.

Sundell, J. 2004. On the history of indoor air quality and health. *Indoor Air* 14 (Suppl 7):51–58. https://doi.org/10.1111/j.1600-0668.2004.00273.x.

Tichenor, B. A., Z. Guo, J. E. Dunn, L. Sparks, and M. A. Mason. 1991. The interaction of vapor phase organic compounds with indoor sinks. *Indoor Air* 1(1):23–35. https://doi.org/10.1111/j.1600-0668.1991.03-11.x.

Townsend, C. L., and R. L. Maynard. 2002. Effects on health of prolonged exposure to low concentrations of carbon monoxide. *Occupational and Environmental Medicine* 59(10):708–711. https://doi.org/10.1136/oem.59.10.708.

Tran, V. V., D. Park, and Y. C. Lee. 2020. Indoor air pollution, related human diseases, and recent trends in the control and improvement of indoor air quality. *International Journal of Environmental Research and Public Health* 17(8). https://doi.org/10.3390/ijerph17082927.

Wallace, L. A. 1989. The Total Exposure Assessment Methodology (TEAM) Study: An analysis of exposures, sources, and risks associated with four volatile organic chemicals. *Journal of the American College of Toxicology* 8(5):883–895. https://doi.org/10.3109/10915818909018049.

Wallace, L. A., E. D. Pellizzari, T. D. Hartwell, C. Sparacino, R. Whitmore, L. Sheldon, H. Zelon, and R. Perritt. 1987. The TEAM (Total Exposure Assessment Methodology) Study: Personal exposures to toxic substances in air, drinking water, and breath of 400 residents of New Jersey, North Carolina, and North Dakota. *Environmental Research* 43(2):290–307. https://doi.org/10.1016/s0013-9351(87)80030-0.

Wang, N., N. Zannoni, L. Ernle, G. Beko, P. Wargocki, M. Li, C. J. Weschler, and J. Williams. 2021. Total OH reactivity of emissions from humans: In situ measurements and budget analysis. *Environmental Science & Technology* 55(1):149–159. https://doi.org/10.1021/acs.est.0c04206.

Weisel, C. P., J. Zhang, B. J. Turpin, M. T. Morandi, S. Colome, T. H. Stock, D. M. Spektor, L. Korn, A. Winer, S. Alimokhtari, J. Kwon, K. Mohan, R. Harrington, R. Giovanetti, W. Cui, M. Afshar, S. Maberti, and D. Shendell. 2005. Relationship of Indoor, Outdoor and Personal Air (RIOPA) study: Study design, methods and quality assurance/control results. *Journal of Exposure Analysis and Environmental Epidemiology* 15(2):123–137. https://doi.org/10.1038/sj.jea.7500379.

Weschler, C. J., and N. Carslaw. 2018. Indoor chemistry. *Environmental Science & Technology* 52(5):2419–2428. https://doi.org/10.1021/acs.est.7b06387.

Weschler, C. J., A. T. Hodgson, and J. D. Wooley. 1992. Indoor chemistry: Ozone, volatile organic compounds, and carpets. *Environmental Science & Technology* 26(12):2371–2377. https://doi.org/10.1021/es00036a006.

Weschler, C. J., and H. C. Shields. 1996. Production of the hydroxyl radical in indoor air. *Environmental Science & Technology* 30(11):3250–3258. https://doi.org/10.1021/es960032f.

Weschler, C. J., and H. C. Shields. 1999. Indoor ozone/terpene reactions as a source of indoor particles. *Atmospheric Environment* 33(15):2301–2312. https://doi.org/10.1016/S1352-2310(99)00083-7.

Weschler, C. J., H. C. Shields, and D. V. Naik. 1994. Indoor chemistry involving O_3, NO, and NO_2 as evidenced by 14 months of measurements at a site in Southern California. *Environmental Science & Technology* 28(12):2120–2132. https://doi.org/10.1021/es00061a021.

Weschler, C. J., J. R. Wells, D. Poppendieck, H. Hubbard, and T. A. Pearce. 2006. Workgroup report: Indoor chemistry and health. *Environmental Health Perspectives* 114(3):442–446. https://doi.org/10.1289/ehp.8271.

WHO (World Health Organization). 2010. *WHO Guidelines for Indoor Air Quality: Selected Pollutants*. Copenhagen: World Health Organization, Regional Office for Europe.

Wittmann, O. 1962. The subsequent dissociation of formaldehyde from particle board. *Holz Roh Werkst* 20(6):221–224.

Yang, S., G. Beko, P. Wargocki, J. Williams, and D. Licina. 2021. Human emissions of size-resolved fluorescent-aerosol particles: Influence of personal and environmental factors. *Environmental Science & Technology* 55(1):509–518. https://doi.org/10.1021/acs.est.0c06304.

2

Primary Sources and Reservoirs of Chemicals Indoors

Thousands of chemical compounds have been documented to exist at measurable levels in the indoor environment, including in air, particles, and settled dust, or on surfaces. These chemicals are emitted into the indoor environment from primary sources that can originate both indoors and outdoors. Primary sources are defined as those that directly emit or release chemicals into the indoor environment, as opposed to secondary sources, which create new chemicals based on reactions between primary precursors. Reservoirs are surfaces or materials that can act as both primary sources and sinks of chemicals in the indoor environment (see Chapter 3). For example, materials used to construct buildings (e.g., wood, insulation, wall coverings, flooring, and wiring) along with the products that are brought into those buildings (e.g., furniture, furnishings, electronics, personal care products, and food) are all indoor sources of chemicals. Cooking generates gases and particles, both from the heat source (e.g., combustion or electric coils) and the food preparation itself, which can differ based on the cooking style (e.g., frying or boiling) and ingredients (e.g., sauces, oils, spices, protein, and vegetables). Other occupant behaviors, such as the use of personal care (e.g., fragrances) and consumer products, as well as the occupants themselves (bioeffluents, which are chemicals of biological origin emitted from humans via breath, desquamation, skin, or other organs), generate chemicals in the indoor environment.

Figure 2-1 illustrates some of these primary emission sources and reservoirs that are considered important in the indoor environment. Primary source emission rates are driven by an array of factors, including human activities, building characteristics, physical parameters, environmental conditions, and spatial and temporal factors. Secondary source emission rates are dependent on some of these same factors, with the additional influences of complex and dynamic indoor chemistry (see Chapter 4).

This chapter first aims to provide an overview of the major types of continuous and episodic primary sources and reservoirs of chemicals in the indoor environment and to describe the dominant factors that affect their emission rates. Next, it summarizes the main classes of compounds emitted from primary sources and reservoirs, their typical concentration ranges if they have been reported in indoor air or dust, and key physicochemical properties that dictate their behavior and lifetime indoors within and across these matrices. The chapter continues by introducing various

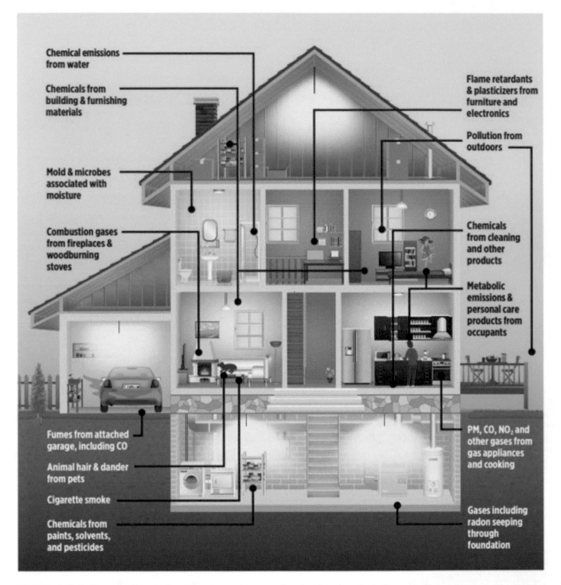

FIGURE 2-1 Chemical emission sources and reservoirs that impact the indoor environment.
NOTE: CO = carbon monoxide; NO$_2$ = nitrogen dioxide; PM = particulate matter.

approaches to chemical inventories, which are important because they enable more accurate chemical exposure estimates, followed by a discussion of the wide range of analytical approaches that can be used to measure the chemical contents of an indoor space. Finally, the chapter concludes by highlighting emerging chemicals of concern indoors, current sampling and measurement capabilities and limitations, and areas of research gaps and future directions. This chapter sets the stage for the committee's subsequent discussions of how these primary emissions are redistributed in the indoor environment (Chapter 3) and chemically transformed into secondary sources (Chapter 4).

MAJOR PRIMARY SOURCES, RESERVOIRS, AND FACTORS THAT INFLUENCE EMISSION RATES

Primary sources and reservoirs of chemicals in the indoor environment are numerous and diverse in terms of their origin, chemical composition, and emission rates (EPA, 2021a). They range from continuous (or sustained) to episodic (or intermittent) and can originate indoors or outdoors. Emission rates can be highly variable and are driven or influenced by multiple factors, including human activities, building characteristics, physical parameters, environmental conditions, and spatial and/or temporal factors.

We define *continuous sources* as those constantly emitting chemicals into indoor environments at rates that may vary due to environmental conditions (e.g., temperature, humidity, and ventilation). When reservoirs (which can both emit and accumulate chemicals, depending on conditions) emit chemicals, they behave like continuous sources. *Episodic* sources are defined as sources that intermittently, not constantly, release chemicals as a result of discrete events (e.g., humans and human activities such as cooking, cleaning, or the use of products). Emission rates of episodic sources generally depend on operational conditions or occupancy-related and behavioral factors.

Factors that Influence Emission Rates

A wide range of variables influence emission rates of chemicals from primary indoor sources and reservoirs discussed in this chapter, from environmental to building- and human-related. This section introduces and defines major categories of these factors, recognizing that these terms (e.g., spatial and temporal) might be used interchangeably across different research and knowledge domains. Therefore, definitions are given below for the purposes of this chapter and in relation to primary source emission rates. This section intentionally excludes any discussion of absolute source strengths, defined as emission rates in mass per time (Ferro et al., 2004), because these are often modulated based on the scale and specific nature of the source or activity emitting chemicals (e.g., the number of people in a room at a certain time, amount and style of cooking). The subsequent two sections reference these factors as they relate to particular continuous or episodic sources and source categories.

Space (or spatial) factors refer to three-dimensional spatial factors that operate at multiple scales of relevance, starting from the microlevel within materials, to the surface area/volume of a room, the position of rooms relative to each other, the floor level, and the position of a residential unit relative to other units (e.g., in multi-unit apartment buildings). Space can also refer to factors such as the location of the building ventilation intake relative to outdoor sources or street-level emissions. Finally, space can refer to the proximity of major outdoor sources of chemicals and spatial gradients in outdoor (or underground) sources of chemicals that may infiltrate or enter the indoor environment.

Time (or temporal) factors refer to diurnal, daily, or seasonal patterns that may influence concentrations of outdoor pollutants (e.g., traffic emissions, photochemistry, and allergens) and their infiltration rate indoors (e.g., the seasonal operation of heating, ventilation, and air-conditioning [HVAC] systems or the opening of windows). Temporal factors may also relate to patterns of human occupancy or behaviors that occur indoors (e.g., diurnal patterns of waking time or cooking).

Building- and ventilation-related factors include different modes of operating buildings and their ventilation systems (e.g., window and door opening, HVAC systems, recirculation) that influence air change rates, pressure gradients, building age, infiltration rates from outdoors, and indoor emission (e.g., HVAC systems as sources of contaminants) and removal rates (e.g., operating fans, filters, and air cleaners). These can also correlate with weather (i.e., environmental factors) and may have temporal patterns or trends. Removal is discussed in more detail in Chapter 5 of this report.

Environmental factors include temperature and temperature gradients, relative humidity, water activity, and light (intensity and wavelength spectrum).

Human occupancy, activities, and behaviors refer to factors including the presence and number of humans in indoor environments and their activities and behaviors, including both passive (e.g., bioeffluents and dust resuspension due to movement) and active (e.g., smoking, cooking, cleaning, or use of products) impacts.

Having defined these terms, the sections below aim to provide a high-level overview of major categories of continuous and episodic primary sources and reservoirs of chemicals, examples of which are listed in Tables 2-1 and 2-2, and the main factors that influence or drive their emission rates. The categories of sources are not meant to be comprehensive or exhaustive. Emphasis is placed on newer sources or source categories of relevance to indoor chemical concentrations and human exposures and on emerging science within each category.

Continuous Sources and Reservoirs

Building materials are continuous sources of gas-phase emissions in indoor environments. Volatile organic compounds (VOCs) and semivolatile organic compounds (SVOCs) are emitted from many commonly used materials, including wood and wood composites, insulation, plastic piping used for water distribution, electrical cables and wiring, adhesives, paint, surface treatments and coatings, carpeting, and vinyl flooring. Primary emission rates for chemicals inherent to the materials or their production processes have been shown to be highest when new, then depend on the age of materials and environmental factors. Continuous emissions from building materials also occur as a result of reaction processes including ozone chemistry, photochemistry (e.g., in lacquers or paints), hydrolysis (e.g., in floorings), or metabolism (e.g., by microbes and fungi). These transformations are not considered primary sources and are discussed in more detail in Chapter 4. Additionally, legacy continuous emissions of concern from building materials include polychlorinated biphenyls (PCBs) in caulking/sealants, lead in paint, and asbestos. Mass-transfer models are used to estimate emission sources of VOCs and SVOCs from building materials under varying conditions, with emissions depending on the quantity of material, diffusion coefficients, air/material partition coefficients, rate constants for adsorption/desorption, and mass-transfer coefficients. Often these parameters are quantified experimentally, but they can also be estimated using modeling approaches.

Over the past few decades, the types of materials used in building construction have changed significantly, which can lead to differences in chemical emissions to the indoor environment. For instance, the market share of single-family homes built with crosslinked polyethylene plastic water pipes increased from 16 percent in 2002 to 63 percent in 2016 (USHUD, 2018). New or updated building codes and standards are sometimes the cause of changes in the chemical composition of building materials, while other changes are motivated by performance improvements to the materials themselves. For example, due to concerns about moisture accumulation and potential for fungal growth, wallboard is now treated with fungicides; however, new research suggests that some fungicides may be migrating out of the wallboard and accumulating in settled dust (Cooper et al., 2020). In addition, there are growing concerns about the use of recycled building materials, because they can be sources of chemicals that have been deliberately removed from production. For example, polyurethane foam from discarded furniture is sometimes recycled into bonded carpet padding. This practice may inadvertently perpetuate exposure to flame retardants that have been phased out from use due to their persistence and toxicity into the carpet pad (Clean and Healthy New York et al., 2015). In addition, some legacy contaminants that were phased out, such as PCBs, chlordane, and chlorpyrifos, are persistent in indoor environments and contribute to prolonged and chronic exposure in older buildings and apartments, which in some cases can lead to social justice issues.

TABLE 2-1 Primary Source Chemicals Present in Air (Gas and Particle) and Dust Phases in the Indoor Environment[a]

Major Category	Chemical Example or Chemical Class	VP Range (Pa, at 25 °C)[b]	BP Range (°C)[b]	$Log_{10}(K_{oa})$ Range[b]	Example Primary Sources	Concentration Range in Indoor Air (Gas and Particle Phase)[c] (ppb, mol/mol$_{volume}$)	Concentration Range in Indoor Dust (ppb, mass/mass)[d]	Concentration Range References[d]
Highly Reactive Inorganics	Ozone (O$_3$)		−112		Outdoor air, electronic air cleaners	10^{-1}–10^2		Nazaroff and Weschler, 2022; Salonen et al., 2018
	Nitrate Radical (NO$_3$)				-	10^{-3}		Arata et al., 2018; Nazaroff and Weschler, 2020
	Hydroxyl Radical				-	10^{-6}–10^{-5}		Carslaw et al., 2017; Young et al., 2019
Other Inorganics	Carbon Monoxide (CO)	$5.5*10^{5,e}$	−192	0.44^e	Outdoor air, combustion	10^2–10^4		Mullen et al., 2016
	Radon (Rn)	$5.2*10^4$	483		Soil gas	10^0–10^3 Bq m^{-3} (10^{-2}–10^1 pCi/L)		Cothern, 1987; Kitto, 2014; Zhukovsky et al., 2018
	Ammonia (NH$_3$)	4	−33.3		Breath, skin, combustion, concrete containing urea, cleaning agents, cooking	10^0–10^3		Ampollini et al., 2019; Nazaroff and Weschler, 2020
	Nitric Acid Nitrous Acid (HNO$_3$, HONO)		111–541^e		Outdoor air, combustion	10^{-2}–10^1		Nazaroff and Weschler, 2020
	Nitrogen Dioxide Nitric Oxide (NO$_2$, NO)		−152–21		Outdoor air, combustion, gas cooking	10^0–10^1		Mullen et al., 2016; Nazaroff and Weschler, 2020; Zhou et al., 2018
	Chloramines (NH$_2$Cl, NHCl$_2$, NCl$_3$)		486–501^e		Treated water, bleach cleaning	10^{-1}–10^1		Ampollini et al., 2019; Mattila et al., 2020
	Hypochlorous Acid (HClO)				Cleaning agents	10^0–10^2		Mattila et al., 2020

continued

TABLE 2-1 Continued

Major Category	Chemical Example or Chemical Class	VP Range (Pa, at 25 °C)[b]	BP Range (°C)[b]	$Log_{10}(K_{oa})$ Range[b]	Example Primary Sources	Concentration Range in Indoor Air (Gas and Particle Phase)[c] (ppb, mol/mol$_{volume}$)	Concentration Range in Indoor Dust (ppb, mass/mass)[d]	Concentration Range References[d]
Organics	Carbonyls ($CH_2O–C_{10}H_{20}O$) (e.g., formaldehyde)	$1.4*10^1–1.7*10^{5,e}$	$-20.1–208$	$5.1–6.1^e$	Wood, combustion, paints, carpets, cooking	$10^0–10^2$		CA Energy Commission, 2009; Hodgson et al., 2000; Hult et al., 2015; Mullen et al., 2016; Nirlo et al., 2014; Salthammer et al., 2010
	Alkanes ($C_5H_{12}–C_{16}H_{34}$) (e.g., hexane)	$1.3*10^{2l}–6.7*10^4$	$40.9–287$	$2.0–7.5^e$	Vinyl flooring, wood, petroleum-based products, natural gas	$10^{-1}–10^1$		CA Energy Commission, 2009; Hodgson and Levin, 2003; Hodgson et al., 2000
	Carboxylic Acids ($CH_2O_2–CH_3(CH_2)_8COOH$) (e.g., formic acid)	$4.9*10^{-2}–5.7*10^3$	$122–269$	$2.62–7.71^e$	Outdoor air, breath, wood, cooking, cleaning agents	$10^{-2}–10^2$		Duncan et al., 2019; Nazaroff and Weschler, 2020
	Alcohols ($C_2H_6O–C_8H_{18}O$) (e.g., ethanol)	$4.9*10^0–7.9*10^3$	$78.4–194$	$3.25–6.03$	Cooking, solvents, wood, microbes	$10^0–10^1$		Hodgson et al., 2000; Licina and Langer, 2021
	Terpenes ($C_{10}H_{16}–C_{10}H_{16}$) (e.g., limonene)	$2.1*10^2–6.3*10^2$	$155–177$	$4.31–4.52^e$	Wood, cleaning agents, cooking, spices, citrus, fragrances	$10^0–10^2$		Hodgson et al., 2000
	Aromatics ($C_6H_6–C_{10}H_8$) (e.g., benzene)	$3.3*10^0–1.3*10^4$	$80–218$	$2.78–5.19$	Carpet, vinyl flooring, petroleum-based products	$10^{-1}–10^1$		Hodgson et al., 2000
	Halogenated Flame Retardants (e.g., PBDEs)	$8.5*10^{-8}–1.1*10^0$	$295–642$	$8.5–11.7^f$	Furniture, electronics, textiles	$<10^{-1}$ ng/m^3	$<10^5$	Kassotis et al., 2020; Venier et al., 2016
	Organophosphate Ester Plasticizers/Flame Retardants (e.g., TCPP, TPhP)	$2.9*10^1–3.2*10^{-6}$	$219–466$	$5.8–15$	Furniture, electronics, mattresses, insulation	$<10^3$ ng/m^3	$<10^5$	Kassotis et al., 2020; Vykoukalova et al., 2017
	Pesticides (e.g., permethrin, fipronil)	$1.3*10^{-3}–4.3*10^{-9}$	$312–487$	$8.4–11.7^f$	Wallboard, home treatments, flea and tick preventatives	$<10^3$ ng/m^3	$10^1–10^5$	Kassotis et al., 2020; Lucattini et al., 2018
	Chlorinated Paraffins (e.g., $C_{10}H_{16}Cl_6$)	$7.3*10^{-2}$	341	8.0	Sealants, paints, adhesives, surface coatings and textiles	$<10^2$ ng/m^3	$10^4–10^6$	Fridén et al., 2011; He et al., 2019

Stain/Water Repellents (e.g., PFAS)[g]	$2.1*10^{3}$– $2.3*10^{-5}$	121–325	3.5–9.8	Carpeting/flooring, cosmetics, paint, upholstery	<10^{1} ng/m^3	10^{1}–10^{3}	Hall et al., 2020; Padilla-Sanchez et al., 2017; Winkens et al., 2017
Phthalates/Phthalate Alternatives (e.g., DEHP)	$5.6*10^{-1}$– $4.7*10^{-5}$	272–422	5.7–11.7[f]	Flooring, electronics, furnishings	<10^{4} ng/m^3	10^{2}–10^{7}	Kassotis et al., 2020; Lucattini et al., 2018; Lunderberg et al., 2019
Combustion Products (e.g., PAHs)	$1.3*10^{1}$– $5.9*10^{-9}$	226–533	5.2–11.7[f]	Smoking, cooking	<10^{4} ng/m^3	10^{1}–10^{4}	Kassotis et al., 2020; Lucattini et al., 2018
Disperse Azo Dyes (e.g., Disperse Blue 373)	$1.5*10^{-7}$– $1.3*10^{-9}$	52 2591	10.2–11.7[f]	Textiles	N/A	10^{1}–10^{3}	Dhungana et al., 2019
Phenols (e.g., triclosan, BPA)	$8.3*10^{0}$– $3.3*10^{-8}$	222–437	6.3–10.5	Countertops, personal care products	10^{-2}–10^{0} ng/m^3	10^{1}–10^{5}	Fan et al., 2019; Laborie et al., 2016; Levasseur et al., 2021
Antioxidants & UV Inhibitors (plastic additives)	$6.9*10^{-9}$– $3.9*10^{-1}$	270–489	7.3–11.7[f]	Electronics, personal care products, lubricants, rubber products	N/A	10^{1}–10^{5}	Tan et al., 2021; Wang et al., 2013
Surfactants (e.g., octylphenol and nonylphenol ethoxylates)	N/A	N/A	N/A	Cleaning products, personal care products, pharmaceuticals, pesticides	10^{-1}–10^{1} ng/m^3	10^{4}–10^{6}	Fan et al., 2019; Laborie et al., 2016
Organotins	N/A	N/A	N/A	PVC, silicone, polyurethane, glass coatings, pesticides	N/A	<10^{3}	Kannan et al., 2010
Siloxanes	$2.3*10^{25}$– $5.1*10^{3}$	96–339	2.1–11.7[f]	Building materials, cosmetics, cookware, electrical devices	<10^{3} ng/m^3	10^{2}–10^{5}	Katz et al., 2021b; Lucattini et al., 2018; Tran et al., 2015

Organics

[a] Chemicals and classes of compounds listed are not exhaustive. Chemical class properties generally represent the straight chain structures of the class, with ranges shown to represent what has been historically measured in indoor environments.

[b] Most chemical class properties are taken from EPA's CompTox Chemicals Dashboard with noted values ([a]) being estimates rather than experimentally derived data.

[c] Gas and particle phase concentrations are not reported independently because many of the cited studies sampled bulk air.

[d] Concentration ranges listed are from referenced literature with household survey and review references prioritized.

[e] Predicted.

[f] The upper boundary is a reflection of the OPERA model used to predict $\log_{10}(K_{oa})$, which has an upper boundary of 11.7.

[g] Values provided reflect the acid/neutral forms.

NOTE: BP = boiling point; BPA = bisphenol A; DEHP = di(2-ethylhexyl)phthalate; PAH = polycyclic aromatic hydrocarbon; PBDE = polybrominated diphenyl ether; PFAS = per- and polyfluoroalkyl substances; PVC = polyvinyl chloride; UV = ultraviolet; VP = vapor pressure; TCPP = tris (1-chloro-2-propyl) phosphate; TPhP = triphenyl phosphate.

TABLE 2-2 Primary Source Chemicals Present in the Particle Phase in Indoor Air

MASS AND NUMBER CONCENTRATIONS OF PM IN VARIOUS SIZE FRACTIONS

Size Distributions and Chemical Components of Aerosols	Example Primary Sources	Concentration Range in Indoor Air	References for Concentration Ranges
Ultrafine particles, or particulate matter with aerodynamic diameter (d_a) <100 nm (UFPs)	Cooking, candle burning, heating oiled surfaces, printers and photocopiers	10^3–10^6 particles/cm^{3,a} 10^3 particles/cm^{3,b}	Bekö et al., 2013; Fuller et al., 2013; Morawska et al., 2017; Slezakova et al., 2018
Particulate matter with aerodynamic diameter (d_a) <2.5 µm (PM$_{2.5}$)	Smoking, incense and candle burning, cooking	10^0–10^2 µg/m^3	Balasubramanian and Lee, 2007; Bekö et al., 2013; Habre et al., 2014; Lanki et al., 2007; Morawska et al., 2017; Patel et al., 2020; Polidori et al., 2006; Ryan et al., 2015 Slezakova et al., 2018; Vardoulakis et al., 2020; Zwoździak et al., 2013
Particulate matter with aerodynamic diameter (d_a) <10 µm (PM$_{10}$)	Dust re-suspension, vacuuming, mechanical abrasion, paint chipping, skin flakes	10^1–10^3 µg/m^3	Chen et al., 2018; Morawska et al., 2017; Slezakova et al., 2018; Vardoulakis et al., 2020; Zwoździak et al., 2013

ELEMENTAL COMPOSITION OF PM$_{2.5}$ GROUPED BY TYPICAL SOURCE

Chemical Component	Example Primary Sources	Concentration Range in Indoor Air	References for Concentration Ranges
Organic Carbon (OC)	Cooking, candles	10^0–10^2 µg/m^3	Brown et al., 2021; Habre et al., 2014; Price et al., 2019
Elemental Carbon (EC)	Infiltration from outdoors, coal, wood, solid fuel combustion, some cooking	10^0 µg/m^{3,c}	Balasubramanian and Lee, 2007; Habre et al., 2014; Lanki et al., 2007; Polidori et al., 2006; Sankhyan et al., 2021
Nitrogen (N)	Nicotine (smoking/ vaping), nitrogenated compounds from bleach cleaning, particulate ammonium (pNH$_4$) particulate nitrate (pNO$_3$)	Nicotine (in PM$_{10}$): 10^2 µg/m^{3,d} pNH$_4$: 10^{-2}–10^1 µg/m^3 pNO$_3$: 10^{-3}–10^2 µg/m^3	Chen et al., 2018; DeCarlo et al., 2018; Johnson et al., 2017; NASEM, 2016; Wong et al., 2017
Sulfur (S)	Infiltration from outdoors, aerosolization from water, mercaptans in natural gas	S: 10^2–10^3 ng/m^3	Habre et al., 2014; Ryan et al., 2015; Zwoździak et al., 2013
Lead (Pb)	Infiltration from outdoors	Pb: 10^0–10^2 ng/m^3	Balasubramanian and Lee, 2007; Habre et al., 2014; Ryan et al., 2015; Zwoździak et al., 2013
Nickel (Ni), Vanadium (V)	Infiltration from outdoors, residual fuel oil burning for residential building heating	Ni: 10^0–10^1 ng/m^3 V: 10^0 ng/m^3	Balasubramanian and Lee, 2007; Habre et al., 2014; Ryan et al., 2015; Zwoździak et al., 2013
Aluminum (Al), Calcium (Ca), Silicon (Si), Iron (Fe)	Resuspended dust, crustal material, soil	Al: 10^1–10^2 ng/m^3 Ca: 10^1–10^3 ng/m^3 Si: 10^1–10^3 ng/m^3 Fe: 10^1–10^3 ng/m^3	Balasubramanian and Lee, 2007; Habre et al., 2014; Ryan et al., 2015; Zwoździak et al., 2013
Potassium (K)	Biomass burning	K: 10^1–10^2 ng/m^3	Habre et al., 2014; Ryan et al., 2015; Zwoździak et al., 2013

TABLE 2-2 Continued

	ELEMENTAL COMPOSITION OF $PM_{2.5}$ GROUPED BY TYPICAL SOURCE		
Chemical Component	Example Primary Sources	Concentration Range in Indoor Air	References for Concentration Ranges
Chlorine (Cl)	Cooking, cleaning with bleach	Cl: 10^1–10^3 ng/m^3	Habre et al., 2014; Mattila et al., 2020; Ryan et al., 2015; Wong et al., 2017; Zwoździak et al., 2013
Zinc (Zn)	Infiltration from outdoors	Zn: 10^1–10^2 ng/m^3	Habre et al., 2014; Ryan et al., 2015; Zwoździak et al., 2013

[a] Secondary data reported in Morawska et al. (2017) based on review of studies using similar UFP measurement instruments in Bekö et al. (2013) and Isaxon et al. (2015).

[b] Median concentrations from 18 homes within 1,500 m of an interstate highway in Massachusetts in Fuller et al. (2013).

[c] Reported as absorbance, a proxy for elemental carbon, using an optical method in Lanki et al. (2007).

[d] Levels reported in PM_{10} indoors at a vaping conference on 2 separate days in Chen et al. (2018).

Chapter 5 discusses ways in which material substitution or improvements can be used to manage indoor chemistry in more detail, and Chapter 6 discusses exposure disparities.

Furnishings and other materials (e.g., furniture, mattresses, rugs, electronics, toys, and other household items) can contain and emit many of the same chemicals as building materials, with similar criteria affecting their emission rates. Furnishings or other household items often contain additional materials not typically found in the building structure itself such as fabrics and textiles, coatings, adhesives, surface treatments, microplastics, and electronics. The introduction of new chemicals in furnishings and other materials may be accompanied by novel emissions that are not yet well characterized or understood. One example of new chemicals found in the indoor environment is the identification of azo dyes in dust samples (Kutarna et al., 2021). Azo dyes are chemicals containing -N=N- bonds; some chemicals in this category are chlorinated and brominated, and many are toxic, carcinogenic, and can bioaccumulate (Berradi et al., 2019). Azo dyes are broadly used in the manufacture of clothing, furniture textiles, and rugs, constituting 70 percent of the world's annual production of synthetic dyes, and have been identified in common children's clothing and house dust (Ferguson and Stapleton, 2017; Overdahl et al., 2021). New research also shows that volatile neutral per- and polyfluoroalkyl substances (PFAS) can be elevated in indoor air in kindergartens and certain retail or commercial spaces like carpeting stores, suggesting that inhalation may be a more important exposure route than previously thought for PFAS and their breakdown products (Morales-McDevitt et al., 2021).

Dust is material that builds up on surfaces that can be mechanically removed (e.g., by wiping, vacuuming, or scrubbing) or resuspended in the air when disturbed. In addition to settled particles, dust can contain fibers (e.g., textiles, and pet and human hair), crustal and abrasion material, and biological material (e.g., bacteria and viruses, pollen, fungi, dust mites, endotoxins, and skin dander). Dust is a reservoir for a variety of metals, VOCs, and SVOCs, acting at times as a sink or a source. Dust generally can be considered a continuous reservoir because it emits chemicals at a rate that is dependent on environmental conditions, especially temperature and relative humidity. Concentrations of chemicals measured in dust can also be used to infer concentrations in other indoor compartments (e.g., air, surface films) if the system is in equilibrium (Parnis et al., 2020). When indoor human and pet movements resuspend dust, it can also be an episodic source of suspended particles, with rates dependent on occupancy, activities, and behaviors.

The migration of chemicals from buildings, materials, and consumer products into dust can mimic the pathways that lead to human exposure. This suggests that the chemical composition of dust can be representative of personal exposure to some chemicals in the indoor environment.

As an example, studies have shown that house dust concentrations of some chemicals, such as flame retardants and plasticizers, are correlated to internal dose for home occupants (Percy et al., 2020; Stapleton et al., 2012). Due to this association, there is increased focus and attention on house dust, including a push by researchers to characterize house dust using non-targeted mass spectrometry approaches to support identification of other chemicals that may be of concern for human exposure (Ouyang et al., 2017; Ulrich et al., 2019).

Biological materials such as microbes, endotoxins, and pests can also be considered continuous indoor sources of VOCs, allergens, and irritants. Their emission rates are largely dependent on environmental factors such as temperature and moisture, with warmer temperatures and wetter conditions leading to increased concentrations of fungi indoors. Emission rates of fungal spores can also be influenced by mechanical ventilation and vibration or disturbance by air currents as well as humidity. The abundance of dust mites in dust correlates with higher indoor humidity levels (D'Amato et al., 2020). Cockroaches, rodents, dust mites, and pets are known sources of allergens especially for individuals with asthma and in lower-income, urban homes (Ahluwalia et al., 2013; Eggleston, 2017). Exposure to dust mites and endotoxins in farming communities is also associated with shaping the innate immune response and conveying protective effects especially in terms of lowering asthma prevalence (Stein et al., 2016). Outdoor plant-based allergen concentrations such as weed, grass, or tree allergens can be driven by seasonal patterns, and these temporal patterns may also vary geographically (e.g., peak concentration season varies in timing, duration, and intensity by region) (Schmidt, 2016). Spatial factors, such as proximity to farming facilities and rural versus urban environments, are associated with greater concentrations of endotoxins in indoor dust (Barnig et al., 2013). Ventilation-related factors, such as window-opening, central air conditioning, and particle filtration, can significantly influence infiltration rates of these biologicals indoors. New analytical techniques allow the quantification of microbial VOC (mVOC) emissions in laboratory settings as a function of microbial respiration, life stage, and substrate (Misztal et al., 2018). Recent efforts have demonstrated mVOC composition from sinks and showers may be more stable than the underlying microbial communities, which can be highly variable in space and time in response to environmental conditions (Adams et al., 2017). The need to quantify the diversity of microbial species that influence indoor air across a variety of indoor environments and materials remains (Bekö et al., 2020; Prussin and Marr, 2015).

Infiltration from interstitial and buffer spaces within a residence, or from neighboring units in multi-unit residential or occupational settings, can be considered a continuous source of chemicals indoors. Emission rates in these spaces depend on temperature gradients between outdoor and indoor surfaces as well as proper sealing and mechanical ventilation. Attached garages (or parking structures in occupational settings) have been shown to introduce hydrocarbons and other pollutants in indoor air. Emission rates also depend on building ventilation and design, air change rates, and the relative position of air intakes to these sources (spatial factors). Wind shear and vertical temperature and pressure gradients create chimney or stack effects within multi-unit apartment buildings, especially high rises, and can also play a role in introducing chemicals indoors from underground or outdoor sources, transferring chemicals within and across units, and accumulating chemicals in elevated floor levels (Man et al., 2019). Connected central ventilation systems and structures can also act as sources of chemicals infiltrating from nearby residences, with secondhand environmental tobacco smoke being a well-known example (Snyder et al., 2016). Not all chemicals in indoor air come from visible spaces. Attics, crawl spaces, and dry wall cavities have been found to contain VOCs emitted from insulation materials, paint, and fungal growth into indoor spaces. Most building materials are porous to some extent, allowing the transport of VOCs through the materials (Xu and Zhang, 2011). Building envelope materials can be sources of chemicals that migrate into the indoor air. Aliphatic alkanes only found in a building envelope air barrier have been measured in the indoor air of a house (Poppendieck et al., 2015). The extent to which VOCs migrate through building materials depends on the material, environmental factors, and the individual chemical. Studies investigating these three factors have been largely limited to wallboard, relative humidity and/or temperature, and a limited number of chemicals (e.g., formaldehyde) (Tran Le et al., 2021).

Outdoor air is a continuous source of particulate matter (PM), ozone, VOCs and SVOCs, and other chemicals. Entry rates depend on natural and mechanical ventilation by opening windows and doors, operating HVAC systems that bring in outdoor air, or infiltrating through small cracks and openings in building structures that, taken together, determine air change rates. Important spatial and temporal factors, including the proximity of the building air intake to outdoor sources of these chemicals (e.g., to street-level car exhaust) and the building's location relative to nearby outdoor sources (e.g., roads and industrial facilities), also influence emission rates. Spatial gradients or patterns can also vary at regional or global scales (e.g., wildfire smoke, ozone, dust storms). Temporal factors influencing rates of chemicals infiltrating from outdoors into indoor environments include diurnal or weekday/weekend patterns (e.g., traffic-related nitrogen oxides [NO_x] and ozone) and seasonal patterns (e.g., ozone or allergens). Climate change–related factors influencing infiltration of outdoor chemicals (e.g., increased wildfires, increased temperatures) could also operate at longer timescales of years and decades (Field, 2010; IOM, 2011; Lee et al., 2017; Liang et al., 2021).

The chemical composition and size distribution of aerosols changes as particles move from outdoors to indoors. During cool outdoor weather, chemicals on outdoor particles moving into warmer indoor spaces volatilize into the gas phase. During warmer outdoor weather, the opposite occurs, and indoor chemicals partition onto aerosols of outdoor origin (Avery et al., 2019b; Cummings et al., 2021). Recent measurements have shown that variation in partitioning onto outdoor aerosols migrating indoors is dependent upon the aerosol water liquid fraction, which is a function of temperature and relative humidity. Neglecting this phenomenon can result in errors on the order of 40 percent to 400 percent when modeling the chemical composition of aerosols (Cummings et al., 2021).

Intrusion of chemical vapors from underground sources is also a continuous source with significant spatial and geographic variability influencing its entry rates. Pesticides, solvents, chemicals from leaky underground storage tanks, contaminated soil and groundwater, and radon gas are some of the sources falling under this category. Examples include VOCs like toluene, trichloroethylene (TCE), perchloroethylene, formaldehyde, and petroleum products (Provoost et al., 2011). Temperature gradients, vertical space, and building design are important factors influencing emission rates (Ma et al., 2020; Unnithan et al., 2021). Stack effects in buildings can also enhance the intrusion rates of these chemicals indoors, depending on building envelope tightness and weatherization (Francisco et al., 2020; Pigg et al., 2018).

Chlorinated solvents, such as TCE, are increasingly recognized as potentially hazardous chemicals that can enter the indoor environment through vapor intrusion (Forand et al., 2012). In 2021, California issued new draft vapor screening guidance in an effort to provide more consistency in evaluating exposure and health risks despite the challenges of significant spatial and temporal variability in subsurface and indoor air concentrations. The lack of data and high degree of variability in space and time of both sources and concentrations in indoor air has been described as a challenge by both the U.S. Environmental Protection Agency (EPA) and California guidance documents and in the literature (Johnston and Gibson, 2015).

Radon gas concentrations are usually highest in underground levels and basements, with greater soil permeability and fractures in foundations of homes increasing intrusion rates indoors (Kellenbenz and Shakya, 2021). Although inconclusive, some studies have hypothesized that increased building weatherization and modern construction could potentially lead to higher radon exposure in North American indoor residential environments, despite wide geographic variability in the geologic distribution of radon and its precursor minerals, like uranium, thorium, and radium

(Stanley et al., 2019). Naturally present radon gas (^{222}Rn) and its decay products, like lead and polonium, are widely recognized as an indoor air concern and a risk factor for lung cancer (Darby et al., 2005; Krewski et al., 2006). Recent work has shown that ^{222}Rn progeny in the form of radioactive nuclides can react with water vapor and gases to form clusters. These clusters attach onto particle surfaces, emit high energy radiation, and get transported into the lungs and human body with inhaled PM, which essentially acts as a delivery vector (Liu et al., 2019). Several recent studies reported that radon gas exposure as a surrogate of gross α-, β-, and γ- particle radioactivity, or directly measured particle radioactivity on filters (Liu et al., 2020), might potentially explain differential toxicity of $PM_{2.5}$ seen in health studies (Blomberg et al., 2019; Huang et al., 2020, 2021; Papatheodorou et al., 2021).

Episodic Sources

Humans and human activities, in terms of occupancy and behaviors, are an important episodic source of indoor chemicals. This category encompasses several classes of compounds in gas- and particle-phase emission processes, given the complexity of human activities and their interactions with the indoor environment. They are considered episodic, since emissions only occur when humans are present or performing certain activities. Types of sources falling under this category include metabolites excreted from skin (e.g., ammonia; see Li et al., 2020) and breath (e.g., carbon dioxide [CO_2], VOCs; see Tang et al., 2015) or other human excretions; desquamation and hair; microbes (bacteria and viruses; see Yang et al., 2021); generation of chemicals through activities such as cooking, smoking (Wallace and Ott, 2011), and burning candles (Afshari and Ekberg, 2005); use or manufacture of illicit drugs (Kuhn et al., 2019); use of disinfectants and cleaning products (Afshari and Ekberg, 2005); use of personal care products (e.g., fragrances and lotions); application of pesticides and insecticides; use of air fresheners; operation of HVAC systems; and the act of vacuuming and moving around that might resuspend or re-entrain settled dust, its components, and larger PM size fractions (e.g., PM_{10}) in indoor spaces.

Next, some of the major sources that fall under this category are discussed, recognizing that humans and human activities as an episodic source is too broad of a topic to cover comprehensively here. For example, **cooking** emits pollutants based on the fuel source, as well as the range of activities that contribute to food preparation, including CO, CO_2, NO/NO_2 (gas combustion or other high temperature combustion), organics, ultrafine particles (UFPs), and fine particles (Bhangar et al., 2011; Patel et al., 2020; Wallace et al., 2004; Wallace and Ott, 2011). Emission rates depend on fuel type (Wallace et al., 2008), cooking method (e.g., frying emits more UFPs than boiling or steaming), food types (e.g., oils or meats), temperature, duration of time, and cooking surface materials. Emission rates can exhibit diurnal or day of the week patterns that mimic typical human time-activity patterns (e.g., meal time peaks). **Consumer products** (e.g., cleaners, disinfectants, pesticide applications on pets or in the home/garden, and room air fresheners) and personal care products introduce chemicals to indoor environments episodically during and following their use. Near-field exposure depends on product formulations, which chemicals are released, how much, where the chemicals reside (e.g., dust or air), and human presence and activity level. Humans can also introduce **"take-home" chemicals** into the indoor environment on clothes—for example, lead, asbestos, and pesticides from occupational settings (Kalweit et al., 2020). A recent study in a German movie theater found that third-hand smoke is an important source of exposure in nonsmoking indoor environments, where VOCs, hazardous air pollutants (e.g., benzene and formaldehyde), and nicotine can off-gas from clothing of moviegoers and partition onto aerosols and dust (Sheu et al., 2020). Temporal factors (time of day and day of the week patterns corresponding to work schedules) and environmental factors (temperature, relative humidity) influence these emission rates. **Home renovations and repairs** can introduce new building materials and cause novel emissions owing to the materials used. Considerations for emissions are similar to building materials as described above for continuous sources. **HVAC operation** may appear like a periodic source impacting indoor air chemical composition. Dust coated heating coils can act as a reservoir of SVOCs, taking up and releasing them as a function of temperature or moisture. Water condensing on cooling coils or other parts of the HVAC system can take up and release soluble gases. Wet surfaces in the HVAC system may also promote biological activity and thus act as a primary source of mVOCs or promote chemical transformations that take up some and release other chemicals. Temporal factors, like time of day, and environmental factors, like weather, influence these emission rates.

The personal cloud effect is a term referring to increased concentrations of chemicals found near or around humans in their immediate surroundings. The personal cloud is the result of several factors that include, but are not limited to, bioeffluents, personal care products, reactions with skin and clothes, and the thermal plume (air moved upward over the torso as adults release about 100 W of heat). Hence, human presence is a direct episodic source of indoor chemicals. While the personal cloud effect is well documented (Licina et al., 2017; Wallace, 2000; Weschler, 2016), several recent studies have characterized emission rates of human activities in indoor environments, including the effects of a crawling infant on dust resuspension (Hyytiäinen et al., 2018; Wu et al., 2021a) and the movement and heat from sleeping resulting in VOCs and SVOCs being released from the bedding environment (Boor et al., 2014, 2017). Recent work also shows that SVOCs adsorbed to the surfaces of cooking pans and burner coils can desorb and form UFPs when heated (Wallace et al., 2008, 2017).

Pets also emit chemicals into the indoor environment, though pets are considered a less significant episodic indoor source than humans. These emissions include biologicals (e.g., metabolites in breath or urine, dander, microbes) emitted directly from pets, chemicals applied to pets (e.g., flea and tick treatments), or chemicals brought indoors from the outdoor environment (e.g., allergens, pollen, metals, pesticides) on pet hair or feet, especially for animals that spend time indoors and outdoors. Pets and human occupants are known sources of ammonia (NH_3) indoors (Ampollini et al., 2019; Ito et al., 1998), with one recent study showing a correlation between presence or number of pets indoors and NH_3 concentrations, especially with warmer temperatures (Uchiyama et al., 2015). Cat litter boxes or other forms of indoor pet waste storage can be important sources. Pets' urine contains urea, which decomposes anaerobically to release ammonia. Very few studies have measured emission rates of NH_3 from pets indoors; however, these are now more feasible with recent developments in on-line ammonia analyzers. Agricultural environments and extreme environments (e.g., hoarding) can also have high concentrations of NH_3.

Water-sourced chemicals released from domestic water use indoors in gas or aerosol form include brominated, chlorinated, and nitrogenated compounds; mineral aerosols; SVOCs; and biologicals. Chlorine and chloramines are disinfectants added during drinking water treatment. They can react with natural organic material to form trihalomethanes, which can be released into indoor air via showers, dishwashers, and washing machines (Gordon et al., 2006). Detergents used in dishwashers can release hypochlorous acid (Dawe et al., 2019) and react with organics to release further chemicals, including chloroform (Olson and Corsi, 2004). In addition, water delivery systems and source waters can become contaminated, delivering chemicals to the indoor environment, especially after forest fires (Omur-Ozbek et al., 2016; Proctor et al., 2020). Depending on geography and source water quality and contamination levels, emission rates can vary significantly between regions. Seasonal trends can also be at play (e.g., concentration of chemicals and dilution rates will depend on seasonal rain, area discharges and runoff, and surrounding land uses, such as agricultural spraying).

Recent studies have reported elevated concentrations of chemicals such as trace metals, elements, and VOCs in water supply systems following wildfires. These chemicals have been detected in water bodies and water delivery systems in the immediate vicinity of wildfire-impacted areas, as well as further downstream, depending on the environmental transport pathways involved (Burton et al., 2016; Tecle and Neary, 2015). Chemicals detected in water systems post-wildfire also differ depending on the fuels involved, with wildland fires releasing lead, iron, cadmium, mercury, and other metals that have potentially deposited in mature trees over decades. Fires in the wildland-urban interface, on the other hand, can involve furniture, building materials, plastics, and other man-made products, releasing chemicals such as pyrolyzed organic carbon, benzene, and other new and potentially unknown chemicals (Bladon et al., 2014; Proctor et al., 2020). In turn, these chemicals can be aerosolized or volatilized after making their way indoors via the drinking water supply, influencing indoor chemistry in ways that are yet to be fully understood.

CLASSES OF COMPOUNDS IN INDOOR ENVIRONMENTS

While some chemicals are well characterized and studied in indoor environments from the perspective of understanding exposure (e.g., radon, ozone, CO, NO_2, lead, and some VOCs), other chemical classes have only recently been identified in the indoor environment (e.g., disperse azo dyes and ultraviolet [UV] stabilizers). Tables 2-1 and 2-2 present a summary of several chemicals and chemical classes that have primary sources and reservoirs in the indoor environment, in the gas, dust (Table 2-1), and aerosol (Table 2-2) phases. This is not an exhaustive or complete list but rather a list to highlight examples of some of the more common or novel chemicals that stem from sources and/or reservoirs and are of concern for the indoor environment.

It is also important to note that some of the chemicals listed in the tables can have both primary and secondary sources, and the concentration ranges listed may reflect both primary and secondary sources. For instance, the primary source of hypochlorous acid is cleaning solutions, which can directly release, or in some cases react with the indoor environment to produce, chloroform, carbon tetrachloride, chloramines, nitrogen trichloride, and chlorine gas (Mattila et al., 2020; Odabasi, 2008; Wong et al., 2017). Surface reactions of nitrogen dioxide (NO_2) are also considered important secondary sources of nitrous acid (HONO) indoors (Collins et al., 2018).

As shown by the information presented in Table 2-1, concentrations of some chemicals are abundant in indoor air, while others are abundant on surfaces and in indoor dust and airborne particles (refer to Table 2-2 for aerosol phase). These differences are attributable to their physico-chemical properties and use patterns. In particular, chemicals with high vapor pressures are often more abundant in the gas phase of indoor air, and chemicals with lower vapor pressures (and thus higher $\log_{10}(K_{oa})$ values) are more abundant in the condensed phase. The wide range in concentrations within many of these chemical classes is evident, even though most of the reported values are from time-averaged sampling of continuous sources. These concentration differences are related to a number of factors and indicate the high degree of variability observed in indoor environments.

Aerosols

An aerosol is defined as a stable suspension of solid or liquid particles in air. Aerosol particles, which are also known as PM, contain a variety of organic and inorganic chemicals (NASEM, 2016). PM also exhibits a wide size distribution in indoor environments, and some of its components, especially SVOCs, may shift from the particle phase to the gas phase and vice versa as they move indoors from outdoor environments or depending on changes in indoor environmental conditions.

While airborne PM occurs indoors in a wide range of sizes, with aerodynamic diameters ranging from a few nanometers (nm) to ~100 micrometers (μm), referred to as Total Suspended Particles, this section will focus on three size fractions of highest relevance to indoor aerosol exposures and associated health risks. For a more detailed discussion of indoor PM chemical composition, size distributions, and health risks, refer to *Health Risks of Indoor Exposure to Particulate Matter: A Workshop Summary* (NASEM, 2016).

These size fractions are defined as follows:

- **UFPs** are most commonly defined as particles with an aerodynamic diameter ≤ 100 nm. However, the definition of UFPs can vary across fields and disciplines, with some referring to particles ≤ 10 nm (e.g., emissions standards) and others to particles ≤ 200 nm (e.g., health disciplines) as UFPs. These are emitted from primary combustion or formed by conversion indoors. They are similarly formed outdoors and can be transported indoors via infiltration and ventilation. Given their negligible mass, UFPs are usually reported in particle number concentrations (particles/cm^3). They span a wide size range and tend to exhibit diffusion-driven behavior in the lungs especially in the smallest size range (<10 nm) (Oberdörster, 2001; Oberdörster and Utell, 2002). Larger UFPs (generally >10 nm) have a much higher surface-area-to-mass ratio compared to larger particles and are thus efficient at transporting chemicals adsorbed onto their surfaces into the alveolar region of the lungs. UFPs are also very efficient at crossing the alveolar epithelial tissue into systemic circulation and reaching other organs (Schraufnagel, 2020). Computational fluid dynamics simulations demonstrate that very small nanoparticles (1-2 nm) will deposit in the olfactory region of the human nose to a larger extent than larger UFPs (>10 nm) (Garcia et al., 2015). Some nanoparticles deposited in the olfactory region during inhalation might undergo direct transport along the olfactory nerve to the olfactory bulb (Elder et al., 2006) raising concern about possible neurological effects (Lucchini et al., 2012).
- **Fine particles (PM$_{2.5}$)**, or particles with an aerodynamic diameter ≤ 2.5 μm, are usually emitted from primary sources like combustion and secondary formation indoors and reported in mass concentrations (μg/m^3). If inhaled, PM$_{2.5}$ deposits in the tracheobronchial and alveolar regions of the lungs and has been associated with a multitude of adverse health outcomes (Dominici et al., 2006; Pope and Dockery, 2006).
- **PM$_{10}$**, or particles ≤ 10 μm in aerodynamic diameter, are typically primary in nature, emitted from mechanical abrasion, natural sources (e.g., sea salt), and dust resuspension and are reported in μg/m^3. Notably, PM$_{10}$ is considered a subset of larger particles often referred to as "super-coarse" which can reach up to 100 μm in aerodynamic diameter. Coarse PM, typically defined as PM between 2.5 and 10 μm in aerodynamic diameter, primarily deposits in the upper nasopharyngeal region of the lungs and is cleared with mucociliary transport.

Organic compounds as a class usually make up a significant portion (~40–70 percent) of the mass concentration of indoor PM$_{2.5}$ (Habre et al., 2014; Polidori et al., 2006; Turpin et al., 2007). Similarly, particulate sulfates, nitrates, ammonium, and water can be present in larger concentrations, as compared to elemental carbon (EC), trace elements, and metals such as nickel, vanadium, and iron, which are usually present at lower concentrations (NASEM, 2016).

The composition of PM will vary based on the source or process generating it (Habre et al., 2014). It is important to note that PM$_{2.5}$ in itself is considered a mixture, and the concentration of species within PM$_{2.5}$ will depend on the major sources contributing to indoor PM$_{2.5}$ concentrations, including those of indoor and outdoor origin, which also vary based on spatial, temporal, and other factors, as described in the next section. For example, concentrations of lead and arsenic in indoor PM$_{2.5}$ might be elevated in residences or schools near battery recycling and lead smelter facilities (Meyer et al., 1999). Similarly, concentrations of EC might be higher inside homes that burn solid fuels (e.g., coal or wood) for heating or cooking. Indoor concentrations of UFPs in homes located within 100 m of a major interstate highway in Massachusetts were higher than in homes

>1,000 m away (Fuller et al., 2013). See Chapter 6 for a discussion of the settings and determinants that correlate with higher exposure to indoor PM.

Table 2-2 contains typical mass (or number, in the case of UFPs) concentrations of indoor aerosols across size fractions and typical concentrations of organic and elemental components of $PM_{2.5}$ as a more well-studied example of indoor aerosols. Elements and metals are grouped together where they are known to originate from a specific source or source category. Where available, the reported concentrations are for daily, 48-hour, or weekly time-averaged or integrated measurements primarily from studies of real indoor environments (versus laboratory or chamber experiments) and do not capture peak concentrations or episodic sources. References from the House Observations of Microbial and Environmental Chemistry (HOMEChem) study, which simulated a realistic indoor environment, are the one exception to this rule. These are also examples meant to illustrate the wide range of variability in indoor concentrations reported in the literature. They include data from residential, office, and school settings, and from the United States and international studies, but they are not meant to be comprehensive or representative of all types of indoor environments, geographies, or time periods.

As with Table 2-1, many of these aerosols and their components have both primary and secondary sources indoors; however, Table 2-2 does not include secondary organic aerosol (SOA). Formation of SOA indoors is covered in more detail in Chapter 4 of this report.

INDOOR CHEMICAL INVENTORIES

EPA's Toxic Substances Control Act (TSCA) Chemical Substance Inventory contains more than 33,000 active (and more than 34,000 inactive) registered chemicals for use in the United States (EPA, n.d.). A recent study attempted to catalogue the regional and national chemical inventories for the first time (Wang et al., 2020). An interesting result from this study was that of the ~350,000 chemicals in the inventory, approximately 50,000 were classified as confidential, and about 20 percent (~70,000) were ambiguously described. Global chemical inventories for plastic production alone exceed 10,000 different chemicals (Wiesinger et al., 2021).

Thousands of chemicals are present in indoor environments that potentially can be released into indoor air. Inventorying chemicals and their concentrations in the indoor environment enables more accurate chemical exposure estimates, including for chemicals whose indoor source may be unknown. Cataloguing all the chemicals present in indoor air can theoretically be addressed by two major approaches: bottom-up inventories, which determine the chemical composition and/or emissions of every indoor object; or top-down inventories, which measure the total chemical composition of the gas, particle, dust, and surface phases.

Bottom-Up Inventories of Chemicals in Indoor Environments

A bottom-up approach determines the chemical composition of or emission from every material, item, or product found indoors, including the wide range of examples found in this chapter thus far. Two drivers currently exist for measuring emissions from building materials in a bottom-up manner. First, there are regulatory limits for the emission of certain chemicals in some countries—for example, EPA's Formaldehyde Emission Standards for Composite Wood Products Act, which limits formaldehyde emission from composite wood products. Second, there are building product certification programs that can be regulatory or voluntary—for example, GREENGUARD, Blue Angel, AgBB, natureplus, and EU Eco Flower. However, these schemes are generally limited to building products, focused on more volatile chemicals, and tested under only one set of chamber parameters (e.g., ASTM International, 2016a), and they can limit reporting to summarizing parameters such as total VOC that lack the detail needed to fully inform indoor chemistry.

Ideally, the potential exposure and health risks could be evaluated in a bottom-up approach prior to use of a new product or chemical by, for example, requesting toxicity testing before production and marketing. Hundreds of chemicals are now submitted annually to EPA for screening-level assessments under the TSCA Premanufacture Notice (EPA, 2021b), some of which are used in building materials and furnishings. These screening-level assessments use manufacturer-supplied data, and often limited data are available to conduct a thorough assessment of risk. With the signing of the Lautenberg Chemical Safety Act in 2016, TSCA was amended to support more detailed risk assessments for chemicals of potential concern. As a consequence, 20 priority chemicals are selected from the TSCA inventory every 3 years to undergo a detailed risk assessment using the best-available science. However, chemicals in circulation prior to the TSCA being passed in 1976 are not necessarily assessed. In addition, many chemicals in the TSCA inventory are classified as confidential business information, often preventing disclosure of risk assessment information. Furthermore, companies are not required to disclose the chemical name or structure in a Safety Data Sheet (SDS), thus making it challenging to know what chemicals are in some materials and products. As a result, chemicals of concern are identified after their use, rather than prior to their use, in the United States.

PCBs are an early case study demonstrating how chemicals can migrate from the original products to become widespread and persistent in the indoor environment. In 1966, PCBs were first discovered in the environment after a Swedish chemist, Soren Jensen, accidentally discovered PCBs when he was analyzing bird tissues for dichloro-diphenyl-trichloroethane (DDT) (Jensen, 1972). PCBs had been on the market since 1929, yet it was not until this accidental discovery 37 years later that scientists became aware of their ubiquitous contamination in the environment. Production of PCBs was banned in the United States in the 1970s, and it was not until the 1990s that production finally ceased in other parts of the world; however, even today, these chemicals are still found in indoor air and dust in homes and schools (Andersen et al., 2020; Bannavti et al., 2021; Herrick et al., 2016; Marek et al., 2017). In addition to PCBs, a more recent discovery of this type was made about flame retardants (Box 2-1). These case examples suggest the need to use bottom-up inventories to understand chemicals present in indoor products and the risks they pose before they become common indoor pollutants.

BOX 2-1
Flame Retardants in Indoor Dust Samples

Firemaster 550 (FM550) is a commercial flame retardant mixture containing both brominated aromatic compounds and nonhalogenated organophosphate esters. FM550 was introduced as a replacement for a commercial mixture of polybrominated diphenyl ethers commonly known as PentaBDE, a common flame retardant used in furniture foam that was phased out beginning in 2002 owing to concerns about its persistence, bioaccumulation, and toxicity. While analyzing house dust samples for PentaBDE using gas chromatography-mass spectrometry in 2006, researchers discovered a peak corresponding to a chemical with four bromine atoms that did not match any previously observed compounds. After further characterization, the chemical was identified as bis(2-ethylhexyl, 2,3,4,5-tetrabromo)phthalate, or BEHTBP (also known as TBPH) (Stapleton et al., 2008). BEHTBP was one of the proprietary compounds in FM550, and was therefore undisclosed to the public, along with 2-ethylhexyl, 2,3,4,5-tetrabromobenzoate (EHTBB; also known as TBB). Today, significant research has focused on the toxicity and health effects of EHTBB and BEHTBP, while they and other flame retardants continue to be commonly detected in indoor air, dust, and other environmental samples.

This migration of chemicals from products into the indoor environment highlights how important it is to understand chemicals present in indoor products, and the risks they pose, before they become a common indoor contaminant. However, this bottom-up approach is often stymied by the lack of information about the composition of the building materials and consumer products that exist in the indoor environment. Even when products list their contents, trace or contaminated product chemical components may not be listed. New composition or emission testing requirements of indoor products could help mitigate health risks.

Top-Down Inventories of Chemicals in Indoor Environments

A top-down approach to determining the presence and abundance of chemicals in the indoor environment requires sampling the gas, particle, dust, and surface phases and conducting a comprehensive chemical analysis. Top-down analyses have two major limitations:

1. Top-down analyses are limited by the representativeness of the environment that is sampled. Every indoor environment contains different materials, products, and people; furthermore, each indoor environment is operated differently with respect to heating, cooling, and ventilation and is affected differently by weather and geography. As discussed earlier in this chapter, all of these factors can impact the chemical identity and intensity of emissions. Indoor spaces will also vary by their function or intended use (e.g., chemicals present in residences, schools, and office buildings can vary based on products placed within them or occupancy and building operation characteristics). This means that effective top-down inventories need to sample large numbers and types of indoor environments over a wide range of conditions to obtain representativeness.

2. Top-down analyses are also limited by the ability to detect specific chemicals at high spatial and temporal resolution in quickly changing or dynamic environments. Historically, targeted analyses (i.e., quantification of known chemicals determined *a priori*) of specific chemicals in the indoor environment were conducted but were limited in number and classes of chemicals due to feasibility. However, recent analytical and instrumental advances have led to greater sensitivity for detecting chemicals at lower concentrations (i.e., lower detection limits), and the increasing use of non-targeted analyses is revealing the presence of chemicals in indoor air, particulates, and dust that were unknown until recently. Box 2-2 discusses one such example.

BOX 2-2
Insights into Real Time Dynamics

The recent application of proton-transfer-reaction mass spectrometer (PTR-MS) instruments to indoor environments has allowed for the analysis of chemicals overlooked with traditional techniques. For instance, using a PTR-MS, decamethylcyclopentasiloxane (D5), an ingredient in personal care products, was found to make up 30 percent of the total volatile organic compund (VOC) mass in a classroom (Tang et al., 2015). In previous gas chromatography (GC) analyses of indoor air, siloxane signals were often ignored, as they were assumed to be GC column artifacts. PTR-MS instruments have also been used to document how VOC emissions from cinema audiences vary with movie genre (Williams et al., 2016). In addition, PTR-MS instruments have shown that cleaning agents and sweat in a gym can combine to produce N-chloraldimines (Finewax et al., 2021).

ANALYTICAL METHODS AND CHALLENGES

Bottom-up and top-down inventories require quantification of chemicals present in consumer products and in gas, particle, dust, and surface phases. Chosen methods for analyzing indoor environments need to be fit for purpose. Fit-for-purpose considerations include sampling cost, detection limit, time resolution, duration, and ease of use. For instance, samples collected to inform average (e.g., daily or weekly) or long-term (e.g., monthly to annually) exposure modeling might include passive or time averaged sampling methods. Samples collected to understand indoor gas-phase chemical reaction and transformations (e.g., after a cleaning event) may require continuous, high temporal resolution (on the order of seconds) analytical techniques. Recent advances in analytical techniques and application of outdoor measurement technologies and methods to indoor air quality have decreased detection limits of known indoor compounds, increased the number of different chemicals detected in the indoor environment, and increased the time resolution of samples. The following sections briefly describe both traditional techniques and new approaches to chemical analysis in the gas, particle, dust, and surface phases.

Trace Gas Sampling

Traditional Techniques

Historically, inventories of airborne indoor chemicals have been accomplished using time-integrated averaging approaches such as canister, sorbent tube, derivatizing agent, passive sampler, and direct measurement (e.g., ozone) methods. The time-integrated averaging methods have been used to quantify indoor concentrations of alkanes, aromatic hydrocarbons, alcohols, ketones, alkenes, ethers, and esters (Hodgson et al., 2000). These established time-integrated averaging indoor air sampling methods have standardized consensus methods for application (ASTM International, 2015, 2016b, 2017, 2021) and are well suited for high-concentration environments, and the samples can typically be collected by personnel with minimal-to-moderate training. In addition, these methods produce time-integrated average concentrations (typically to an hour or more), making them useful for exposure modeling. Finally, individual samples from these methods can be analyzed by commercial laboratories at lower costs than the upfront cost of most direct measurement or continuous analyses with high temporal resolution.

However, canisters, sorption tubes, derivatizing agents, and passive samplers all have limitations. Polar and low-volatility chemicals may have low recovery rates from canisters (Woolfenden, 2010). Sorbent tube retention is dependent upon properties of the sorbent and chemicals being sampled (ASTM International, 2015), and labile chemicals may decompose in the thermal desorption process. Derivatizing agents can have stability and interference issues (ASTM International, 2016b). Passive samplers are prone to air flow and temperature fluctuations (ASTM International, 2017). While particle filters can be used with these techniques, these methods are sometimes used to sample the entire air phase, collecting both particles and gas.

Until recently, direct measurement of gas-phase indoor chemicals has been limited to a subset of individual chemicals (e.g., formaldehyde, ozone, carbon dioxide, nitric oxide, and nitrogen dioxide) for which instrumentation was available. Instruments directly measuring carbon dioxide, like nondispersive infrared sensors, now cost less than $500, allowing widespread use in field campaigns, in real-time building demand control ventilation management, and by the general public. However, instruments directly measuring continuous indoor relevant concentrations of formaldehyde, ozone, nitric oxide, and nitrogen dioxide have remained more costly, limiting use to a few locations in field campaigns.

New Analytical Approaches

Recent funding interest in the indoor environment has led to the application of new analytical equipment and techniques. Continuous, high temporal resolution VOC analysis, such as proton-transfer-reaction mass spectrometry (PTR-MS), has been applied in a limited number of indoor environments (e.g., movie theater, home, gym, classroom, office, and museum) (Brown et al., 2021; Finewax et al., 2021; Liu et al., 2019; Price et al., 2021; Schripp et al., 2014; Tang et al., 2015; Williams et al., 2016; Wu et al., 2021b). PTR-MS instruments have been used to measure aldehydes, aromatic hydrocarbons, amines, alcohols, ketones, phenols, thiols, terpenes, and siloxanes. The use of chemical ionization mass spectrometry (CIMS) in the indoor environment has been expanded beyond protons to use other ion sources, such as acetate (Liu et al., 2017), nitrate (Price et al., 2021), and iodide (Duncan et al., 2019; Finewax et al., 2021; Price et al., 2019). Ammonium could also be another useful ion source for chemicals found in indoor environments. Combining data from multiple ion sources using positive matrix factorization (PMF) can allow identification of source types (e.g., human or building materials) for various gas-phase chemicals (Price et al., 2021). CIMS methods have been used to measure carboxylic acids, inorganic chlorinated compounds, organic nitrogen compounds, and polar VOCs (Duncan et al., 2019; Liu et al., 2017; Price et al., 2021). These instruments can analyze chemicals at speeds relevant to dynamic indoor environments (seconds to minutes). CIMS instruments can be calibrated in novel ways that rely on reaction rates rather than via authentic standards (Farmer, 2019). In addition, these instruments can provide detection limits for 0.1-s duration samples that are equivalent to 1-h sample durations for historically used time-integrated analysis techniques (ASTM International, 2020).

While these instruments' high-resolution time-of-flight mass spectrometers can determine chemical composition, they cannot distinguish structural isomers (Farmer, 2019). However, coupling CIMS instruments with chromatography has the potential to allow quantification of structural isomers with time resolution of minutes rather than hours (Claflin et al., 2021). With proper configuration, these instruments can perform non-targeted (or untargeted) analysis to start to identify the wide range of "unknown" chemicals present in indoor air. However, to date non-targeted analysis has not quantified the thousands of indoor chemicals that would be necessary for a comprehensive down inventory of indoor gas-phase chemicals.

Despite these advantages in quantifying indoor gas-phase concentrations, these instruments are expensive (typically costing much more than $100,000) and require highly trained personnel to operate and interpret the results. These factors prevent the application of these instruments to large numbers of homes or other indoor spaces in field campaigns. Since these instruments have not been used in a significant number of indoor environments, potential interference issues in high-concentration indoor environments are unknown. Facilitating the greater application of these instruments to a wide range of indoor spaces and environmental conditions could make top-down approaches to the inventory of indoor gas-phase constituents more robust and give a greater understanding of the dynamic nature of indoor sources.

Particulate Matter Sampling

Traditional Techniques

Indoor airborne PM has been commonly analyzed using techniques that quantify total mass, particle count, surface area, size distribution, and chemical composition. Commonly used techniques to count particles include optical sensors, condensation particle growth counters, diffusion chargers (Buonanno et al., 2014), and scanning mobility particle sizers (to count and size particles). Whereas reference methods for quantifying PM in outdoor air require active sampling followed by gravimetric analysis, these commonly used counting techniques estimate mass by assuming particle shape and density. Direct optical sensors are most accurate when particle diameters are greater than the light wavelength (~100 nm to 300 nm). Systems that use particle growth can quantify particles down to 1 nm in diameter. These methods do not provide information about chemical composition.

Most commonly, PM needs to be captured on a filter first to determine its chemical constituents. Active or passive sampling methods have been employed to collect particles on filters for gravimetric and compositional analysis, with active size-selective samplers often used to collect 50 percent of particles below a specific cut-point in aerodynamic diameter, assuming spherical shape. The filter is extracted with a solvent, and that extract is then treated with a variety of techniques to purify and isolate chemical classes of concern (e.g., water-soluble metals). The final extract is then often evaporated, reconstituted, and analyzed using a variety of mass spectrometry techniques. Non-destructive methods such as X-ray fluorescence or thermal/optical reflectance/transmittance have also been used to characterize elemental composition and elemental versus organic carbon content based on thermal optical properties, respectively, directly on the filter media. Destructive methods like inductively coupled plasma-mass spectrometry have also been used to characterize chemical composition, and many other methods (e.g., radioactivity and scanning electron microscopy) have been used for morphological characterization and counting. However, these techniques only can capture chemical composition of stable particles. Volatilization (e.g., of low molecular weight organics and nitrates) and reactions (e.g., with ozone and nitrogen dioxide) can occur during sampling and storage prior to analysis (ASTM International, 2013). Chemicals are only analyzed if they are in the size range captured by the sampler.

New Analytical Approaches

One new analytical approach for PM sampling is the use of HVAC filters as a collection media for SVOCs associated with particles. HVAC systems move large volumes of air through filters, creating a lengthy, time-integrated sample of airborne particles. Dust on HVAC filters has been shown to have equivalent or higher concentrations of polybrominated diphenyl ethers (PBDEs), phthalates, and organophosphates than settled dust (Bi et al., 2018; He et al., 2016; Xu et al., 2015).

A wide range of instruments can be applied to determine continuous particle composition in outdoor PM, including mass spectrometry, chromatography, and particle-into-liquid sampling coupled to on- or off-line analysis (Farmer, 2019). To date, however, the limited indoor particle studies have relied on aerosol mass spectrometers (AMS) and thermal desorption aerosol gas chromatography (TAG). AMS instruments can provide elemental composition and oxygen/carbon and hydrogen/carbon ratio information (Farmer, 2019). PMF and other source apportionment models have also been used with chemical composition data to identify and apportion sources that contribute to indoor PM concentrations (Habre et al., 2014; Hasheminassab et al., 2014; Molnár et al., 2014). Using PMF, AMS data have been used to examine how particles change as they transport through the building envelope during varying seasons (Avery et al., 2019a,b) and change after cooking (Katz et al., 2021a), as well as the impact of particles on third-hand smoke transport in indoor air (DeCarlo et al., 2018). Yet, AMS is limited to analysis of non-refractory PM smaller than 1 micron; larger particles, metals, soot, and salts are not analyzed by AMS. TAG has been applied in residential studies to measure the dynamics and evolution of speciated organic chemicals under natural ventilation conditions (Fortenberry et al., 2019). Semivolatile TAG (SV-TAG) has been applied to study residential sources, dynamics, and gas/particle partitioning of SVOCs on an hourly basis (Kristensen et al., 2019), including analysis of phthalate partitioning dynamics (Lunderberg et al., 2019), and the role of surface emissions and re-partitioning of SVOCs to $PM_{2.5}$ (Lunderberg et al., 2020). AMS and SV-TAG have been used together to study emissions and partitioning of siloxanes during cooking events (Katz et al., 2021b). Due to cost and complexity, the application of AMS, TAG, and similar instruments to the indoor environment has been limited. Other continuous particle composition techniques have not been widely applied to the indoor environment (Farmer, 2019). These include mass spectrometry using soft ionization, extractive electrospray techniques, coupling rapid filter collection with CIMS, and single-particle techniques. Like for gas-phase sampling, facilitating the greater application of these instruments and methods to a wide range of indoor spaces and environmental conditions could make top-down approaches to inventorying indoor aerosols more robust.

Settled Dust Sampling

Traditional Techniques

The collection and analysis of dust has been occurring for decades and has been important to understanding this reservoir of indoor chemicals. Identifying and measuring chemical burdens in dust samples has provided insight into potential exposure to occupants, particularly as it relates to estimating exposure from hand-to-mouth activities, which is an important pathway not only for children but also for adults. Dust is often collected using a vacuum, although some studies have reported using brooms and brushes to collect settled dust in high-traffic areas. In some cases, clean sampling films are left in a horizontal position (e.g., on a bookshelf or countertop) to passively collect airborne dust that settles over time. The mass accumulating on these films can be quantified, and the films can be extracted and analyzed to characterize their chemical composition. Additionally, dust is often extracted and analyzed in a similar fashion to PM on filters.

Unfortunately, there is no standardization of sampling methods for dust, and the various methods currently used can lead to uncertainty in measurements and comparability among studies. For example, differences in the types of vacuums used by researchers vary considerably and are not standardized. EPA has recommended standardized methods and equipment such as the High Volume Surface Sampler; however, this instrument is heavy and expensive compared to commercial vacuums. Researchers have a tendency to use dust collected either by participants using their home vacuum cleaner or with a smaller vacuum or a handheld device that can be used by researchers from home to home. Differences in vacuums can potentially lead to differences in the size and type of dust particles collected and thus differences in the chemical concentrations reported in dust. Dust often needs to be sieved to remove large particles before chemical analysis; however, researchers typically use different sized sieves that result in different sized particles included in the final analyses and further contribute to differences in measurements across studies.

In addition, differences have been found based on the location (room) in the house where the dust is collected, or between settled dust on the floor and elevated dust (e.g., on a bookshelf). Factors that often influence chemical measurements of SVOCs in dust samples were recently highlighted in a meta-analysis in which flame retardants were used as a case example (Al-Omran et al., 2021). Box 2-3 discusses challenges related to assessing exposure to chemicals in dust.

New Analytical Approaches

Moving forward, it will be important to more thoroughly identify and characterize chemicals present in dust samples. Non-targeted approaches that utilize high-resolution mass spectrometry to identify unknown chemical features or responses are now being used to characterize dust samples (Ouyang et al., 2017; Rager et al., 2016; Rostkowski et al., 2019). For example, a recent study used a non-targeted method to identify brominated azo disperse dyes in house dust (Peng et al., 2016). These methods provide more insight into the complex and diverse suite of chemicals that are present in indoor environments, yet they are not currently at a point where every chemical in the thousands in a mass spectrum of an indoor air or dust sample can be identified—let alone quantified. Chemical space is vast, and the number of chemicals that are on the market and their potential

---

BOX 2-3
Using Tracers to Enhance Dust Exposure Assessment

Linking chemical concentrations in dust samples with exposure and health risks in people has several challenges. The differences in collection methods can lead to difficulties in comparing data from one study to another and in understanding how to best relate these measurements to human exposure. Currently, exposure is estimated using a relatively simple assumption that people ingest a specific mass of dust daily, which varies depending on the age of the person (EPA, 2017). These dust intake fractions are then multiplied by the concentration of a chemical in dust to estimate exposure. However, exposure to dust, and therefore exposure to chemicals associated with dust, is not accurately reflected in such a simple framework. A number of variables and factors will influence these exposure estimates, such as time spent in various micro-environments, body weight, and behaviors. The U.S. Environmental Protection Agency has recognized this limitation and is seeking more research to identify chemical tracers of dust that can improve our understanding of dust ingestion rates and their variability. Research is underway, but so far there has been little success in identifying a good tracer owing to the multimedia nature and partitioning behavior of most chemicals (Abrahamsson et al., 2021). Further discussion on exposure modeling, including from dust, can be found in Chapter 6.

---

degradation products span tens of thousands of chemicals (EPA, n.d.). Investments in chemoinformatic approaches and non-targeted analyses may improve our ability to fully characterize complex chemical mixtures in indoor dust in the future.

Surface Sampling

Traditional Techniques

Surfaces in indoor environments, including windows, countertops, and walls, accumulate organic films and act as sinks, reaction sites, and reservoirs for SVOCs (Weschler and Nazaroff, 2008). The Occupational Safety and Health Administration's Guidelines for Surface Sampling Methods are available and widely utilized (OSHA, 2007). Surfaces have traditionally been sampled by taking a clean wipe, often made of cotton or a similar material, and wiping down a specific surface area. Surface wipe samples can then be extracted and analyzed in a similar fashion to PM on filters and may experience similar issues like loss of volatile chemical species (Liu et al., 2003).

Growth of organic films on indoor surfaces has been modeled theoretically (Weschler and Nazaroff, 2017). In addition, many studies have been done to understand how specific chemicals react with different surfaces, including ozone oxidation reactions of organic chemicals present on surfaces (Shu and Morrison, 2011; Weschler and Nazaroff, 2017). Indoor surface films are also studied by placing clean substrates or coupons (made of materials commonly found indoors, such as glass, metal, or tile) indoors and allowing films to grow, followed by analysis in the laboratory by surface chemistry or extraction techniques.

New Analytical Approaches

A variety of new surface sampling techniques have been applied in the past decade. Some analytical advancements have focused on improving the chemical characterization of surface wipes to evaluate chemical residues using spectrophotometer techniques (Deming and Ziemann, 2020). Surface chemistry is also being probed by applying portable surface reactors to organic chemical reactions on surfaces (Algrim et al., 2020), flux chambers for emissions from surface films on coupons (Adams et al., 2017) or surfaces (Wu et al., 2016), atomic force microscopy coupled to infrared spectroscopy to measure film coverage and functional group analysis (Or et al., 2018), and analysis of glass collection plates using Fourier Transform Ion Cyclotron Resonance Mass Spectrometry and off-line AMS (O'Brien et al., 2021). Application of more standard, advanced, and emerging techniques for surface chemistry analysis to relevant indoor surfaces holds significant promise for advancing understanding of surface films and their emissions and chemical interactions.

CONCLUSIONS

Chemical sources in the indoor environment are numerous and varied. Potential human exposure to these chemicals is dynamic owing to the complexities inherent not only in these sources but also in the environment and human behaviors. Chemical emissions and abundances are linked to both chronic and episodic sources and are modified by behavioral, environmental, and physical factors. Building materials, consumer products, infiltration of outdoor air, and human behavior (particularly cooking) can strongly influence the chemicals detected in the indoor environment. Dust and indoor surfaces also serve as a sink and reservoir of chemicals that are slowly released from primary sources, and they play a role in understanding human exposure, which is discussed further in Chapter 6. New research also suggests that clothing mediates our exposure to chemicals in the indoor environment, acting as a sink for chemicals present in indoor air (Licina et al., 2019).

Over the past few decades, new materials containing unique chemicals have emerged for use in the indoor environment—for example, new types of building insulation and greater use of electronics and smart devices in indoor environments, particularly in homes. These devices and materials can also be sources of chemicals to the indoor environment, such as plastics, plasticizers, antioxidants, UV stabilizers, and flame retardants. Recycling of materials could lead to increased exposure to hazardous chemicals that have been phased out due to concerns about human exposure. Increased use of cleaning and disinfection agents, particularly during the COVID-19 pandemic, is leading to an increase in the levels of some chemicals, such as quaternary ammonium compounds (Zheng et al., 2020), in the indoor environment. Little information is available on the health effects or distribution of these chemicals in the indoor environment. Over time, some chemicals that were emitted by primary sources in the indoor environment have been phased out of use owing to concerns about their elevated exposure and/or toxicity (e.g., PBDEs, benzyl butyl phthalate). These unfortunate chemical substitutions are often replaced with chemicals that have less data available on their emissions, exposure, and potential hazards.

A wide range of analytical techniques are currently being used to identify new chemicals, in both bottom-up and top-down approaches, that may be released into the indoor environment, but these approaches are costly and time-consuming. The lack of transparency in chemical use (i.e., confidential business information) and challenges in identifying chemical sources in the indoor environment will continue to be major obstacles to chemical management and risk evaluation.

RESEARCH NEEDS

Given its findings about the current state of the science, the committee has identified priority research areas to help drive future advances in understanding primary sources and reservoirs in indoor environments:

- **Prioritize acquisition of actionable data and research to link sources with exposures and understand impacts of mixtures on health.** Research is needed that provides greater resolution on spatial and temporal trends in chemical emissions indoors, and how these vary for both chronic and episodic sources. Specific needs include improved understanding of the combined influence of ventilation rates, humidity, and temperature on chemical emissions indoors, as well as the interaction of the indoor and outdoor environment and its influence on indoor sources and exposures, especially for SVOCs. Better understanding of sources would provide more actionable information to control sources and, consequently, exposures. In addition, the indoor environment, where humans spend the majority of their time, contains a highly complex chemical mixture found in multiple phases and originating from many sources.

- **Increase transparency in chemical applications/use in building materials to minimize time and effort needed to establish evidence of exposure and health risks.** New or improved analytical methods have helped to identify new chemicals in indoor environments and their transformation products. However, every year new materials are introduced to the market that may become new sources (new types of flooring, new surface treatments for water and grease repellency, etc.). More research is needed to evaluate chemical emissions, and thus potential human exposure, from new materials before they are broadly introduced to the market. Greater transparency would expedite the process of establishing evidence for potential harmful exposures.

- **Improve analytical methods and non-targeted approaches to support discovery.** In the past, a number of chemicals of human health concern have been identified in indoor settings using targeted or screening methods. Use of non-targeted methods has the potential to identify chemicals of concern earlier. Thus, the committee recommends that greater emphasis is placed on improving non-targeted approaches that utilize high-resolution mass spectrometry to identify previously unknown chemicals and characterize the complex chemical mixtures present in the indoor environment.

- **Develop and maintain harmonized chemical information databases.** Information databases now exist for a broad range of volatile chemical products and other consumer products (e.g., from SDSs and ingredient labels, and the Chemical and Products Database [CPDat]). These can support the use of non-targeted methods that are key to improving our understanding. However, maintaining and updating these databases requires resources. Those who develop and maintain databases (i.e., regulatory bodies, scientific research groups, or consumer and health protection agencies) would benefit from harmonization of terminology, tracking of data provenance, consistent identification of standard product categories, and consistent processes for data updates.

- **Expand research further into nonresidential settings and underrepresented countries and contexts.** The vast majority of the information presented in this chapter refers to research conducted in residential settings and in high-income countries. However, people spend time in other indoor environments, which may have different sources and emission rates. For example, less information is available on chemical sources in schools, hospitals, government buildings, retail stores, restaurants, and occupational and recreational buildings (hair and nail salons, rock-climbing gyms, trampoline parks, indoor pools, etc.) across a wide range of socioeconomic strata (Mandin et al., 2017). In addition, most research is conducted in high-income countries or in low-income countries with households that rely on solid fuels for indoor cooking and heating (typically termed "household air pollution" by the World Health Organization). Populations living in low- to middle-income countries likely experience widely variable chemistries indoors due to differences in regional activities, human behaviors, climates, fuels used in cooking and heating, and the use of different materials in building construction. Similarly, significant variability exists among developed countries based on social and economic factors (e.g., public housing, lower-income residences, or disadvantaged neighborhoods). Researchers could consider fostering community-based partnerships to increase the amount of information available and better incorporate local knowledge and context when investigating these understudied environments.

REFERENCES

Abrahamsson, D. P., J. R. Sobus, E. M. Ulrich, K. Isaacs, C. Moschet, T. M. Young, D. H. Bennett, and N. S. Tulve. 2021. A quest to identify suitable organic tracers for estimating children's dust ingestion rates. *Journal of Exposure Science & Environmental Epidemiology* 31(1):70–81. https://doi.org/10.1038/s41370-020-0244-0.

Adams, R. I., D. S. Lymperopoulou, P. K. Misztal, R. De Cassia Pessotti, S. W. Behie, Y. Tian, A. H. Goldstein, S. E. Lindow, W. W. Nazaroff, J. W. Taylor, M. F. Traxler, and T. D. Bruns. 2017. Microbes and associated soluble and volatile chemicals on periodically wet household surfaces. *Microbiome* 5(1):128. https://doi.org/10.1186/s40168-017-0347-6.

Afshari, A. U. M., and L. E. Ekberg. 2005. Characterization of indoor sources of fine and ultrafine particles: A study conducted in a full-scale chamber. *Indoor Air* 15(2):141–150. https://doi.org/10.1111/j.1600-0668.2005.00332.x.

Ahluwalia, S. K., R. D. Peng, P. N. Breysse, G. B. Diette, J. Curtin-Brosnan, C. Aloe, and E. C. Matsui. 2013. Mouse allergen is the major allergen of public health relevance in Baltimore City. *Journal of Allergy and Clinical Immunology* 132(4):830–835.e2. https://doi.org/10.1016/j.jaci.2013.05.005.

Algrim, L. B., D. Pagonis, J. A. de Gouw, J. L. Jimenez, and P. J. Ziemann. 2020. Measurements and modeling of absorptive partitioning of volatile organic compounds to painted surfaces. *Indoor Air* 30(4):745–756. https://doi.org/10.1111/ina.12654.

Al-Omran, L. S., S. Harrad, and M. Abou-Elwafa Abdallah. 2021. A meta-analysis of factors influencing concentrations of brominated flame retardants and organophosphate esters in indoor dust. *Environmental Pollution* 285:117262. https://doi.org/10.1016/j.envpol.2021.117262.

Ampollini, L., E. F. Katz, S. Bourne, Y. Tian, A. Novoselac, A. H. Goldstein, G. Lucic, M. S. Waring, and P. F. DeCarlo. 2019. Observations and contributions of real-time indoor ammonia concentrations during HOMEChem. *Environmental Science & Technology* 53(15):8591–8598. https://doi.org/10.1021/acs.est.9b02157.

Andersen, H. V., L. Gunnarsen, L. E. Knudsen, and M. Frederiksen. 2020. PCB in air, dust and surface wipes in 73 Danish homes. *International Journal of Hygiene and Environmental Health* 229:113429. https://doi.org/10.1016/j.ijheh.2019.113429.

Arata, C., K. J. Zarzana, P. K. Misztal, Y. Liu, S. S. Brown, W. W. Nazaroff, and A. H. Goldstein. 2018. Measurement of NO_3 and N_2O_5 in a residential kitchen. *Environmental Science & Technology Letters* 5(10):595–599. https://doi.org/10.1021/acs.estlett.8b00415.

ASTM International. 2013. *ASTM D6209-13 Standard Test Method for Determination of Gaseous and Particulate Polycyclic Aromatic Hydrocarbons in Ambient Air (Collection on Sorbent-Backed Filters with Gas Chromatographic/Mass Spectrometric Analysis)*. https://doi.org/10.1520/D6209-13.

ASTM International. 2015. *ASTM D6196-15 Standard Practice for Choosing Sorbents, Sampling Parameters and Thermal Desorption Analytical Conditions for Monitoring Volatile Organic Chemicals in Air*. https://doi.org/10.1520/D6196-15.

ASTM International. 2016a. *ASTM D6007-14 Standard Test Method for Determining Formaldehyde Concentrations in Air from Wood Products Using a Small-Scale Chamber*.

ASTM International. 2016b. *ASTM D5197-16 Standard Test Method for Determination of Formaldehyde and Other Carbonyl Compounds in Air (Active Sampler Methodology)*. https://doi.org/10.1520/D5197-16.

ASTM International. 2017. *ASTM D6306-17 Standard Guide for Placement and Use of Diffusive Samplers for Gaseous Pollutants in Indoor Air*.

ASTM International. 2020. *ASTM WK71196 New Guide for Measurement Techniques for Formaldehyde in Air*.

ASTM International. 2021. *ASTM D5466-21: Standard Test Method for Determination of Volatile Organic Compounds in Atmospheres (Canister Sampling, Mass Spectrometry Analysis Methodology)*.

Avery, A. M., M. S. Waring, and P. F. DeCarlo. 2019a. Human occupant contribution to secondary aerosol mass in the indoor environment. *Environmental Science: Processes & Impacts* 21(8):1301–1312. https://doi.org/10.1039/C9EM00097F.

Avery, A. M., M. S. Waring, and P. F. DeCarlo. 2019b. Seasonal variation in aerosol composition and concentration upon transport from the outdoor to indoor environment. *Environmental Science: Processes & Impacts* 21(3):528–547. https://doi.org/10.1039/C8EM00471D.

Balasubramanian, R., and S. S. Lee. 2007. Characteristics of indoor aerosols in residential homes in urban locations: A case study in Singapore. *Journal of the Air & Waste Management Association* 57(8):981–990. https://doi.org/10.3155/1047-3289.57.8.981.

Bannavti, M. K., J. C. Jahnke, R. F. Marek, C. L. Just, and K. C. Hornbuckle. 2021. Room-to-room variability of airborne polychlorinated biphenyls in schools and the application of air sampling for targeted source evaluation. *Environmental Science & Technology* 55(14):9460–9468. https://doi.org/10.1021/acs.est.0c08149.

Barnig, C., G. Reboux, S. Roussel, A. Casset, C. Sohy, J.-C. Dalphin, and F. de Blay. 2013. Indoor dust and air concentrations of endotoxin in urban and rural environments. *Letters in Applied Microbiology* 56(3):161–167. https://doi.org/10.1111/lam.12024.

Bekö, G., N. Carslaw, P. Fauser, V. Kauneliene, S. Nehr, G. Phillips, D. Saraga, C. Schoemaecker, A. Wierzbicka, and X. Querol. 2020. The past, present, and future of indoor air chemistry. *Indoor Air* 30(3):373–376. https://doi.org/10.1111/ina.12634.

Bekö, G., C. J. Weschler, A. Wierzbicka, D. G. Karottki, J. Toftum, S. Loft, and G. Clausen. 2013. Ultrafine particles: Exposure and source apportionment in 56 Danish homes. *Environmental Science & Technology* 47(18):10240–10248.

Berradi, M., R. Hsissou, M. Khudhair, M. Assouag, O. Cherkaoui, A. El Bachiri, and A. El Harfi. 2019. Textile finishing dyes and their impact on aquatic environs. *Heliyon* 5(11):e02711. https://doi.org/10.1016/j.heliyon.2019.e02711.

Bhangar, S., N. A. Mullen, S. V. Hering, N. M. Kreisberg, and W. W. Nazaroff. 2011. Ultrafine particle concentrations and exposures in seven residences in northern California. *Indoor Air* 21(2):132–144. https://doi.org/10.1111/j.1600-0668.2010.00689.x.

Bi, C., J. P. Maestre, H. Li, G. Zhang, R. Givehchi, A. Mahdavi, K. A. Kinney, J. Siegel, S. D. Horner, and Y. Xu. 2018. Phthalates and organophosphates in settled dust and HVAC filter dust of U.S. low-income homes: Association with season, building characteristics, and childhood asthma. *Environment International* 121:916–930. https://doi.org/10.1016/j.envint.2018.09.013.

Bladon, K. D., M. B. Emelko, U. Silins, and M. Stone. 2014. Wildfire and the future of water supply. *Environmental Science & Technology* 48(16):8936–8943. https://doi.org/10.1021/es500130g.

Blomberg, A. J., B. A. Coull, I. Jhun, C. L. Z. Vieira, A. Zanobetti, E. Garshick, J. Schwartz, and P. Koutrakis. 2019. Effect modification of ambient particle mortality by radon: A time series analysis in 108 U.S. cities. *Journal of the Air & Waste Management Association* 69(3):266–276. https://doi.org/10.1080/10962247.2018.1523071.

Boor, B. E., H. Järnström, A. Novoselac, and Y. Xu. 2014. Infant exposure to emissions of volatile organic compounds from crib mattresses. *Environmental Science & Technology* 48(6):3541–3549. https://doi.org/10.1021/es405625q.

Boor, B. E., M. P. Spilak, J. Laverge, A. Novoselac, and Y. Xu. 2017. Human exposure to indoor air pollutants in sleep microenvironments: A literature review. *Building and Environment* 125:528–555. https://doi.org/10.1016/j.buildenv.2017.08.050.

Brown, W. L., D. A. Day, H. Stark, D. Pagonis, J. E. Krechmer, X. Liu, D. J. Price, E. F. Katz, P. F. DeCarlo, C. G. Masoud, D. S. Wang, L. Hildebrandt Ruiz, C. Arata, D. M. Lunderberg, A. H. Goldstein, D. K. Farmer, M. E. Vance, and J. L. Jimenez. 2021. Real-time organic aerosol chemical speciation in the indoor environment using extractive electrospray ionization mass spectrometry. *Indoor Air* 31(1):141–155. https://doi.org/10.1111/ina.12721.

Buonanno, G., R. E. Jayaratne, L. Morawska, and L. Stabile. 2014. Metrological performances of a diffusion charger particle counter for personal monitoring. *Aerosol and Air Quality Research* 14(1):156–167. https://doi.org/10.4209/aaqr.2013.05.0152.

Burton, C. A., T. M. Hoefen, G. S. Plumlee, K. L. Baumberger, A. R. Backlin, E. Gallegos, and R. N. Fisher. 2016. Trace elements in stormflow, ash, and burned soil following the 2009 station fire in Southern California. *PLoS ONE* 11(5):e0153372. https://doi.org/10.1371/journal.pone.0153372.

CA Energy Commission. 2009. Ventilation and indoor air quality in new homes. https://ww2.arb.ca.gov/sites/default/files/classic//research/apr/past/04-310.pdf.

Carslaw, N., L. Fletcher, D. Heard, T. Ingham, and H. Walker. 2017. Significant OH production under surface cleaning and air cleaning conditions: Impact on indoor air quality. *Indoor Air* 27(6):1091–1100. https://doi.org/10.1111/ina.12394.

Chen, R., A. Aherrera, C. Isichei, P. Olmedo, S. Jarmul, J. E. Cohen, A. Navas-Acien, and A. M. Rule. 2018. Assessment of indoor air quality at an electronic cigarette (vaping) convention. *Journal of Exposure Science & Environmental Epidemiology* 28(6):522–529. https://doi.org/10.1038/s41370-017-0005-x.

Claflin, M. S., D. Pagonis, Z. Finewax, A. V. Handschy, D. A. Day, W. L. Brown, J. T. Jayne, D. R. Worsnop, J. L. Jimenez, P. J. Ziemann, J. de Gouw, and B. M. Lerner. 2021. An in situ gas chromatograph with automatic detector switching between PTR- and EI-TOF-MS: isomer-resolved measurements of indoor air. *Atmos. Meas. Tech.* 14(1):133–152. https://doi.org/10.5194/amt-14-133-2021.

Clean and Healthy New York, Clean Water Action, and Conservation Minnesota. 2015. *Flame Retardants in Furniture, Foam, Floors.* https://static1.squarespace.com/static/5fa197d3a325783e2e6f7b89/t/6008a10ce402840fe7a01c90/1611178271769/Flame+Retardants+report+2015.

Collins, D. B., R. F. Hems, S. Zhou, C. Wang, E. Grignon, M. Alavy, J. A. Siegel, and J. P. D. Abbatt. 2018. Evidence for gas-surface equilibrium control of indoor nitrous acid. *Environmental Science & Technology* 52(21):12419–12427. https://doi.org/10.1021/acs.est.8b04512.

Cooper, E. M., R. Rushing, K. Hoffman, A. L. Phillips, S. C. Hammel, M. J. Zylka, and H. M. Stapleton. 2020. Strobilurin fungicides in house dust: Is wallboard a source? *Journal of Exposure Science & Environmental Epidemiology* 30(2):247–252.

Cothern, C. R. 1987. *Environmental Radon.* Edited by C. R. Cothern and J. E. Smith, Jr. New York: Plenum Press.

Cummings, B. E., A. M. Avery, P. F. DeCarlo, and M. S. Waring. 2021. Improving predictions of indoor aerosol concentrations of outdoor origin by considering the phase change of semivolatile material driven by temperature and mass-loading gradients. *Environmental Science & Technology* 55(13):9000–9011. https://doi.org/10.1021/acs.est.1c00417.

D'Amato, G., O. P. M. Ortega, I. Annesi-Maesano, and M. D'Amato. 2020. Prevention of allergic asthma with allergen avoidance measures and the role of exposure. *Current Allergy and Asthma Reports* 20(3):8. https://doi.org/10.1007/s11882-020-0901-3.

Darby, S., D. Hill, A. Auvinen, J. M. Barros-Dios, H. Baysson, F. Bochicchio, H. Deo, R. Falk, F. Forastiere, M. Hakama, I. Heid, L. Kreienbrock, M. Kreuzer, F. Lagarde, I. Mäkeläinen, C. Muirhead, W. Oberaigner, G. Pershagen, A. Ruano-Ravina, E. Ruosteenoja, A. S. Rosario, M. Tirmarche, L. Tomásek, E. Whitley, H.-E. Wichmann, and R. Doll. 2005. Radon in homes and risk of lung cancer: Collaborative analysis of individual data from 13 European case-control studies. *BMJ* 330:223. https://doi.org/10.1136/bmj.38308.477650.63.

Dawe, K. E. R., T. C. Furlani, S. F. Kowal, T. F. Kahan, T. C. VandenBoer, and C. J. Young. 2019. Formation and emission of hydrogen chloride in indoor air. *Indoor Air* 29(1):70–78. https://doi.org/10.1111/ina.12509.

DeCarlo, P. F., A. M. Avery, and M. S. Waring. 2018. Thirdhand smoke uptake to aerosol particles in the indoor environment. *Science Advances* 4(5):eaap8368. https://doi.org/10.1126/sciadv.aap8368.

Deming, B. L., and P. J. Ziemann. 2020. Quantification of alkenes on indoor surfaces and implications for chemical sources and sinks. *Indoor Air* 30(5):914–924. https://doi.org/10.1111/ina.12662.

Dhungana, B., H. Peng, S. Kutarna, G. Umbuzeiro, S. Shrestha, J. Liu, P. D. Jones, B. Subedi, J. P. Giesy, and G. P. Cobb. 2019. Abundances and concentrations of brominated azo dyes detected in indoor dust. *Environmental Pollution* 252:784–793. https://doi.org/10.1016/j.envpol.2019.05.153.

Dominici, F., R. D. Peng, M. L. Bell, L. Pham, A. McDermott, S. L. Zeger, and J. M. Samet. 2006. Fine particulate air pollution and hospital admission for cardiovascular and respiratory diseases. *JAMA* 295(10):1127–1134. https://doi.org/10.1001/jama.295.10.1127.

Duncan, S. M., S. Tomaz, G. Morrison, M. Webb, J. Atkin, J. D. Surratt, and B. J. Turpin. 2019. Dynamics of residential water soluble organic gases: Insights into sources and sinks. *Environmental Science & Technology* 53(4):1812–1821. https://doi.org/10.1021/acs.est.8b05852.

Eggleston, P. A. 2017. Cockroach allergy and urban asthma. *Journal of Allergy and Clinical Immunology* 140(2):389–390. https://doi.org/10.1016/j.jaci.2017.04.033.

Elder, A., R. Gelein, V. Silva, T. Feikert, L. Opanashuk, J. Carter, R. Potter, A. Maynard, Y. Ito, J. Finkelstein, and G. Oberdörster. 2006. Translocation of inhaled ultrafine manganese oxide particles to the central nervous system. *Environmental Health Perspectives* 114(8):1172–1178. https://doi.org/10.1289/ehp.9030.

EPA (U.S. Environmental Protection Agency). n.d. Toxic Substances Control Act (TSCA) Chemical Substance Inventory. Accessed January 19, 2022. https://www.epa.gov/tsca-inventory.

EPA. 2017. Chapter 5: Soil and dust ingestion. *Exposure Factors Handbook.* https://www.epa.gov/expobox/exposure-factors-handbook-chapter-5.

EPA. 2021a. *Controlling Pollutants and Sources: Indoor Air Quality Design Tools for Schools.* https://www.epa.gov/iaq-schools/controlling-pollutants-and-sources-indoor-air-quality-design-tools-schools.

EPA. 2021b. *Premanufacture Notices (PMNs) and Significant New Use Notices (SNUNs) Table.* https://www.epa.gov/reviewing-new-chemicals-under-toxic-substances-control-act-tsca/premanufacture-notices-pmns-and.

Fan, X., C. Kubwabo, F. Wu, and P. E. Rasmussen. 2019. Analysis of bisphenol a, alkylphenols, and alkylphenol ethoxylates in NIST SRM 2585 and indoor house dust by gas chromatography-tandem mass spectrometry (GC/MS/MS). *Journal of AOAC International* 102(1):246–254. https://doi.org/10.5740/jaoacint.18-0071.

Farmer, D. K. 2019. Analytical challenges and opportunities for indoor air chemistry field studies. *Analytical Chemistry* 91(6):3761–3767. https://doi.org/10.1021/acs.analchem.9b00277.

Ferguson, P. L., and H. M. Stapleton. 2017. Comment on "Mutagenic azo dyes, rather than flame retardants, are the predominant brominated compounds in house dust." *Environmental Science & Technology* 51(6):3588–3590. https://doi.org/10.1021/acs.est.7b00372.

Ferro, A. R., R. J. Kopperud, and L. M. Hildemann. 2004. Source strengths for indoor human activities that resuspend particulate matter. *Environmental Science & Technology* 38(6):1759–1764. https://doi.org/10.1021/es0263893.

Field, W. R. 2010. *Climate change and indoor air quality.* Contractor Report prepared for U.S. Environmental Protection Agency, Office of Radiation and Indoor Air. https://www.epa.gov/sites/default/files/2014-08/documents/field_climate_change_iaq.pdf.

Finewax, Z., D. Pagonis, M. S. Claflin, A. V. Handschy, W. L. Brown, O. Jenks, B. A. Nault, D. A. Day, B. M. Lerner, J. L. Jimenez, P. J. Ziemann, and J. A. de Gouw. 2021. Quantification and source characterization of volatile organic compounds from exercising and application of chlorine-based cleaning products in a university athletic center. *Indoor Air* 31(5):1323–1339. https://doi.org/10.1111/ina.12781.

Forand, S. P., E. L. Lewis-Michl, and M. I. Gomez. 2012. Adverse birth outcomes and maternal exposure to trichloroethylene and tetrachloroethylene through soil vapor intrusion in New York State. *Environmental Health Perspectives* 120(4):616–621. https://doi.org/10.1289/ehp.1103884.

Fortenberry, C., M. Walker, A. Dang, A. Loka, G. Date, K. Cysneiros de Carvalho, G. Morrison, and B. Williams. 2019. Analysis of indoor particles and gases and their evolution with natural ventilation. *Indoor Air* 29(5):761–779. https://doi.org/10.1111/ina.12584.

Francisco, P. W., S. Gloss, J. Wilson, J. Rose, Y. Sun, S. L. Dixon, J. Breysse, E. Tohn, and D. E. Jacobs. 2020. Radon and moisture impacts from interventions integrated with housing energy retrofits. *Indoor Air* 30(1):147–155. https://doi.org/10.1111/ina.12616.

Fridén, U. E., M. S. McLachlan, and U. Berger. 2011. Chlorinated paraffins in indoor air and dust: concentrations, congener patterns, and human exposure. *Environment International* 37(7):1169–1174. https://doi.org/10.1016/j.envint.2011.04.002.

Fuller, C. H., D. Brugge, P. L. Williams, M. A. Mittleman, K. Lane, J. L. Durant, and J. D. Spengler. 2013. Indoor and outdoor measurements of particle number concentration in near-highway homes. *Journal of Exposure Science & Environmental Epidemiology* 23(5):506–512. https://doi.org/10.1038/jes.2012.116.

Garcia, G. J., J. D. Schroeter, and J. S. Kimbell. 2015. Olfactory deposition of inhaled nanoparticles in humans. *Inhalation Toxicology* 27(8):394–403. https://doi.org/10.3109/08958378.2015.1066904.

Gordon, S. M., M. C. Brinkman, D. L. Ashley, B. C. Blount, C. Lyu, J. Masters, and P. C. Singer. 2006. Changes in breath tri-halomethane levels resulting from household water-use activities. *Environmental Health Perspectives* 114(4):514–521. https://doi.org/10.1289/ehp.8171.

Habre, R., B. Coull, E. Moshier, J. Godbold, A. Grunin, A. Nath, W. Castro, N. Schachter, A. Rohr, M. Kattan, J. Spengler, and P. Koutrakis. 2014. Sources of indoor air pollution in New York City residences of asthmatic children. *Journal of Exposure Science & Environmental Epidemiology* 24(3):269 278. https://doi.org/10.1038/jes.2013.74.

Hall, S. M., S. Patton, M. Petreas, S. Zhang, A. L. Phillips, K. Hoffman, and H. M. Stapleton. 2020. Per- and Polyfluoroalkyl Substances in Dust Collected from Residential Homes and Fire Stations in North America. *Environmental Science & Technology* 54(22):14558–14567. https://doi.org/10.1021/acs.est.0c04869.

Hasheminassab, S., N. Daher, M. M. Shafer, J. J. Schauer, R. J. Delfino, and C. Sioutas. 2014. Chemical characterization and source apportionment of indoor and outdoor fine particulate matter ($PM_{2.5}$) in retirement communities of the Los Angeles Basin. *The Science of the Total Environment* 490:528–537. https://doi.org/10.1016/j.scitotenv.2014.05.044.

He, R., Y. Li, P. Xiang, C. Li, C. Zhou, S. Zhang, X. Cui, and L. Q. Ma. 2016. Organophosphorus flame retardants and phthalate esters in indoor dust from different microenvironments: Bioaccessibility and risk assessment. *Chemosphere* 150:528–535. https://doi.org/10.1016/j.chemosphere.2015.10.087.

He, C., S. H. Brandsma, H. Jiang, J. W. O'Brien, L. M. van Mourik, A. P. Banks, X. Wang, P. K. Thai, and J. F. Mueller. 2019. Chlorinated paraffins in indoor dust from Australia: Levels, congener patterns and preliminary assessment of human exposure. *Science of the Total Environment* 682:318–323. https://doi.org/10.1016/j.scitotenv.2019.05.170.

Herrick, R. F., J. H. Stewart, and J. G. Allen. 2016. Review of PCBs in US schools: A brief history, an estimate of the number of impacted schools, and an approach for evaluating indoor air samples. *Environmental Science and Pollution Research* 23(3):1975–1985. https://doi.org/10.1007/s11356-015-4574-8.

Hodgson, A. T., and H. Levin. 2003. Volatile organic compounds in indoor air: A review of concentrations measured in North America since 1990. Lawrence Berkeley National Laboratory Report LBNL-51715.

Hodgson, A. T., A. F. Rudd, D. Beal, and S. Chandra. 2000. Volatile organic compound concentrations and emission rates in new manufactured and site-built houses. *Indoor Air* 10(3):178-192. https://doi.org/10.1034/j.1600-0668.2000.010003178.x.

Huang, S., E. Garshick, C. L. Z. Vieira, S. T. Grady, J. D. Schwartz, B. A. Coull, J. E. Hart, F. Laden, and P. Koutrakis. 2020. Short-term exposures to particulate matter gamma radiation activities and biomarkers of systemic inflammation and endo-thelial activation in COPD patients. *Environmental Research* 180:108841. https://doi.org/10.1016/j.envres.2019.108841.

Huang, S., P. Koutrakis, S. T. Grady, C. L. Z. Vieira, J. D. Schwartz, B. A. Coull, J. E. Hart, F. Laden, J. Zhang, and E. Garshick. 2021. Effects of particulate matter gamma radiation on oxidative stress biomarkers in COPD patients. *Journal of Exposure Science & Environmental Epidemiology* 31(4):727–735. https://doi.org/10.1038/s41370-020-0204-8.

Hult, E. L., H. Willem, P. N. Price, T. Hotchi, M. L. Russell, and B. C. Singer. 2015. Formaldehyde and acetaldehyde exposure mitigation in US residences: In-home measurements of ventilation control and source control. *Indoor Air* 25(5):523–535. https://doi.org/10.1111/ina.12160.

Hyytiäinen, H. K., B. Jayaprakash, P. V. Kirjavainen, S. E. Saari, R. Holopainen, J. Keskinen, K. Hämeri, A. Hyvärinen, B. E. Boor, and M. Täubel. 2018. Crawling-induced floor dust resuspension affects the microbiota of the infant breath-ing zone. *Microbiome* 6(1):25. https://doi.org/10.1186/s40168-018-0405-8.

IOM (Institute of Medicine). 2011. *Climate Change, the Indoor Environment, and Health.* Washington, DC: The National Academies Press.

Isaxon, C., A. Gudmundsson, E. Z. Nordin, L. Lönnblad, A. Dahl, G. Wieslander, M. Bohgard, and A. Wierzbicka. 2015. Contribution of indoor-generated particles to residential exposure. *Atmospheric Environment* 106:458–466. https://doi.org/10.1016/j.atmosenv.2014.07.053.

Ito, K., C. C. Chasteen, H.-K. Chung, S. K. Poruthoor, Z. Genfa, and P. K. Dasgupta. 1998. A continuous monitoring system for strong acidity in aerosols. *Analytical Chemistry* 70(14):2839–2847. https://doi.org/10.1021/ac980135b.

Jensen, S. 1972. The PCB story. *Ambio* 1(4): 123–131. http://www.jstor.org/stable/4311963.

Johnson, A. M., M. S. Waring, and P. F. DeCarlo. 2017. Real-time transformation of outdoor aerosol components upon trans-port indoors measured with aerosol mass spectrometry. *Indoor Air* 27(1):230–240. https://doi.org/10.1111/ina.12299.

Johnston, J., and J. MacDonald Gibson. 2015. Indoor air contamination from hazardous waste sites: Improving the evidence base for decision-making. *International Journal of Environmental Research and Public Health* 12(12):15040–15057. https://doi.org/10.3390/ijerph121214960.

Kalweit, A., R. F. Herrick, M. A. Flynn, J. D. Spengler, J. K. Berko, Jr, J. I. Levy, and D. M. Ceballos. 2020. Eliminating take-home exposures: Recognizing the role of occupational health and safety in broader community health. *Annals of Work Exposures and Health* 64(3):236-249. https://doi.org/10.1093/annweh/wxaa006.

Kannan, K., S. Takahashi, N. Fujiwara, H. Mizukawa, and S. Tanabe. 2010. Organotin compounds, including butyltins and octyltins, in house dust from Albany, New York, USA. *Archives of Environmental Contamination and Toxicology* 58(4):901-907. https://doi.org/10.1007/s00244-010-9513-6.

Kassotis, C. D., N. J. Herkert, S. C. Hammel, K. Hoffman, Q. Xia, S. W. Kullman, J. A. Sosa, and H. M. Stapleton. 2020. Thyroid receptor antagonism of chemicals extracted from personal silicone wristbands within a papillary thyroid cancer pilot study. *Environmental science & technology* 54(23):15296–15312. https://doi.org/10.1021/acs.est.0c05972.

Katz, E. F., H. Guo, P. Campuzano-Jost, D. A. Day, W. L. Brown, E. Boedicker, M. Pothier, D. M. Lunderberg, S. Patel, K. Patel, P. L. Hayes, A. Avery, L. Hildebrandt Ruiz, A. H. Goldstein, M. E. Vance, D. K. Farmer, J. L. Jimenez, and P. F. DeCarlo. 2021a. Quantification of cooking organic aerosol in the indoor environment using aerodyne aerosol mass spectrometers. *Aerosol Science and Technology* 55(10):1099–1114. https://doi.org/10.1080/02786826.2021.1931013.

Katz, E. F., D. M. Lunderberg, W. L. Brown, D. A. Day, J. L. Jimenez, W. W. Nazaroff, A. H. Goldstein, and P. F. DeCarlo. 2021b. Large emissions of low-volatility siloxanes during residential oven use. *Environmental Science & Technology Letters* 8(7):519–524. https://doi.org/10.1021/acs.estlett.1c00433.

Kellenbenz, K. R., and K. M. Shakya. 2021. Spatial and temporal variations in indoor radon concentrations in Pennsylvania, USA from 1988 to 2018. *Journal of Environmental Radioactivity* 233:106594. https://doi.org/10.1016/j.jenvrad.2021.106594.

Kitto, M. 2014. Radon testing in schools in New York State: A 20-year summary. *Journal of Environmental Radioactivity* 137:213–216. https://doi.org/10.1016/j.jenvrad.2014.07.020.

Krewski, D., J. H. Lubin, J. M. Zielinski, M. Alavanja, V. S. Catalan, R. W. Field, J. B. Klotz, E. G. Létourneau, C. F. Lynch, J. L. Lyon, D. P. Sandler, J. B. Schoenberg, D. J. Steck, J. A. Stolwijk, C. Weinberg, and H. B. Wilcox. 2006. A combined analysis of North American case-control studies of residential radon and lung cancer. *Journal of Toxicology and Environmental Health, Part A* 69(7-8):533–597. https://doi.org/10.1080/15287390500260945.

Kristensen, K., D. M. Lunderberg, Y. Liu, P. K. Misztal, Y. Tian, C. Arata, W. W. Nazaroff, and A. H. Goldstein. 2019. Sources and dynamics of semivolatile organic compounds in a single-family residence in northern California. *Indoor Air* 29(4):645–655. https://doi.org/10.1111/ina.12561.

Kuhn, E. J., G. S. Walker, H. Whiley, J. Wright, and K. E. Ross. 2019. Household contamination with methamphetamine: Knowledge and uncertainties. *International Journal of Environmental Research and Public Health* 16(23):4676. https://doi.org/10.3390/ijerph16234676.

Kutarna, S., S. Tang, X. Hu, and H. Peng. 2021. Enhanced nontarget screening algorithm reveals highly abundant chlorinated azo dye compounds in house dust. *Environmental Science & Technology* 55(8):4729–4739. https://doi.org/10.1021/acs.est.0c06382.

Laborie, S., E. Moreau-Guigon, F. Alliot, A. Desportes, L. Oziol, and M. Chevreuil. 2016. A new analytical protocol for the determination of 62 endocrine-disrupting compounds in indoor air. *Talanta* 147:132–141. https://doi.org/10.1016/j.talanta.2015.09.028.

Lanki, T., A. Ahokas, S. Alm, N. A. H. Janssen, G. Hoek, J. J. De Hartog, B. Brunekreef, and J. Pekkanen. 2007. Determinants of personal and indoor $PM_{2.5}$ and absorbance among elderly subjects with coronary heart disease. *Journal of Exposure Science & Environmental Epidemiology* 17(2):124–133. https://doi.org/10.1038/sj.jes.7500470.

Lee, W.-C., L. Shen, P. J. Catalano, L. J. Mickley, and P. Koutrakis. 2017. Effects of future temperature change on $PM_{2.5}$ infiltration in the Greater Boston area. *Atmospheric Environment* 150:98–105. https://doi.org/10.1016/j.atmosenv.2016.11.027.

Levasseur, J. L., S. C. Hammel, K. Hoffman, A. L. Phillips, S. Zhang, X. Ye, A. M. Calafat, T. F. Webster, and H. M. Stapleton. 2021. Young children's exposure to phenols in the home: Associations between house dust, hand wipes, silicone wristbands, and urinary biomarkers. *Environment International* 147:106317. https://doi.org/10.1016/j.envint.2020.106317.

Li, M., C. J. Weschler, G. Beko, P. Wargocki, G. Lucic, and J. Williams. 2020. Human ammonia emission rates under various indoor environmental conditions. *Environmental Science & Technology* 54(9):5419–5428. https://doi.org/10.1021/acs.est.0c00094.

Liang, D., W.-C. Lee, J. Liao, J. Lawrence, J. M. Wolfson, S. T. Ebelt, C.-M. Kang, P. Koutrakis, and J. A. Sarnat. 2021. Estimating climate change-related impacts on outdoor air pollution infiltration. *Environmental Research* 196:110923. https://doi.org/10.1016/j.envres.2021.110923.

Licina, D., and S. Langer. 2021. Indoor air quality investigation before and after relocation to WELL-certified office buildings. *Building and Environment* 204:108182. https://doi.org/10.1016/j.buildenv.2021.108182.

Licina, D., G. C. Morrison, G. Beko, C. J. Weschler, and W. W. Nazaroff. 2019. Clothing-mediated exposures to chemicals and particles. *Environmental Science & Technology* 53:5559–5575.

Licina, D., Y. Tian, and W. W. Nazaroff. 2017. Emission rates and the personal cloud effect associated with particle release from the perihuman environment. *Indoor Air* 27:79–802. https://doi.org/10.1111/ina.12365.

Liu, M., C.-M. Kang, J. M. Wolfson, L. Li, B. Coull, J. Schwartz, and P. Koutrakis. 2020. Measurements of gross α- and β-activities of archived $PM_{2.5}$ and PM_{10} Teflon filter samples. *Environmental Science & Technology* 54(19):11780–11788. https://doi.org/10.1021/acs.est.0c02284.

Liu, Q.-T., R. Chen, B. E. McCarry, M. L. Diamond, and B. Bahavar. 2003. Characterization of polar organic compounds in the organic film on indoor and outdoor glass windows. *Environmental Science & Technology* 37(11):2340–2349. https://doi.org/10.1021/es020848i.

Liu, S., S. L. Thompson, H. Stark, P. J. Ziemann, and J. L. Jimenez. 2017. Gas-phase carboxylic acids in a university classroom: Abundance, variability, and sources. *Environmental Science & Technology* 51(10):5454–5463. https://doi.org/10.1021/acs.est.7b01358.

Liu, Y., P. K. Misztal, J. Xiong, Y. Tian, C. Arata, R. J. Weber, W. W. Nazaroff, A. H. Goldstein. 2019. Characterizing sources and emissions of volatile organic compounds in a northern California residence using space- and time-resolved measurements. *Indoor Air* 29(4):630–644. https://doi.org/10.1111/ina.12562.

Lucattini, L., G. Poma, A. Covaci, J. de Boer, M. H. Lamoree, and P. E. G. Leonards. 2018. A review of semi-volatile organic compounds (SVOCs) in the indoor environment: occurrence in consumer products, indoor air and dust. *Chemosphere* 201:466–482. https://doi.org/10.1016/j.chemosphere.2018.02.161.

Lucchini, R. G., D. C. Dorman, A. Elder, and B. Veronesi. 2012. Neurological impacts from inhalation of pollutants and the nose-brain connection. *Neurotoxicology* 33(4):838–841. https://doi.org/10.1016/j.neuro.2011.12.001.

Lunderberg, D. M., K. Kristensen, Y. Liu, P. K. Misztal, Y. Tian, C. Arata, R. Wernis, N. Kreisberg, W. W. Nazaroff, and A. H. Goldstein. 2019. Characterizing airborne phthalate concentrations and dynamics in a normally occupied residence. *Environmental Science & Technology* 53(13):7337–7346. https://doi.org/10.1021/acs.est.9b02123.

Lunderberg, D. M., K. Kristensen, Y. Tian, C. Arata, P. K. Misztal, Y. Liu, N. Kreisberg, E. F. Katz, P. F. DeCarlo, S. Patel, M. E. Vance, W. W. Nazaroff, and A. H. Goldstein. 2020. Surface emissions modulate indoor SVOC concentrations through volatility-dependent partitioning. *Environmental Science & Technology* 54(11):6751–6760. https://doi.org/10.1021/acs.est.0c00966.

Ma, J., T. McHugh, L. Beckley, M. Lahvis, G. DeVaull, and L. Jiang. 2020. Vapor intrusion investigations and decision-making: A critical review. *Environmental Science & Technology* 54(12):7050–7069. https://doi.org/10.1021/acs.est.0c00225.

Man, X., Y. Lu, G. Li, Y. Wang, and J. Liu. 2019. A study on the stack effect of a super high-rise residential building in a severe cold region in China. *Indoor and Built Environment* 29(2):255–269. https://doi.org/10.1177/1420326X19856045.

Mandin, C., M. Trantallidi, A. Cattaneo, N. Canha, V. G. Mihucz, T. Szigeti, R. Mabilia, E. Perreca, A. Spinazzè, S. Fossati, Y. De Kluizenaar, E. Cornelissen, I. Sakellaris, D. Saraga, O. Hänninen, E. De Oliveira Fernandes, G. Ventura, P. Wolkoff, P. Carrer, and J. Bartzis. 2017. Assessment of indoor air quality in office buildings across Europe - The OFFICAIR study. *Science of the Total Environment* 579:169–178.

Marek, R. F., P. S. Thorne, N. J. Herkert, A. M. Awad, and K. C. Hornbuckle. 2017. Airborne PCBs and OH-PCBs inside and outside urban and rural U.S. schools. *Environmental Science & Technology* 51(14):7853–7860. https://doi.org/10.1021/acs.est.7b01910.

Mattila, J. M., P. S. J. Lakey, M. Shiraiwa, C. Wang, J. P. D. Abbatt, C. Arata, A. H. Goldstein, L. Ampollini, E. F. Katz, P. F. DeCarlo, S. Zhou, T. F. Kahan, F. J. Cardoso-Saldaña, L. H. Ruiz, A. Abeleira, E. K. Boedicker, M. E. Vance, and D. K. Farmer. 2020. Multiphase chemistry controls inorganic chlorinated and nitrogenated compounds in indoor air during bleach cleaning. *Environmental Science & Technology* 54(3):1730–1739. https://doi.org/10.1021/acs.est.9b05767.

Meyer, I., J. Heinrich, and U. Lippold. 1999. Factors affecting lead, cadmium, and arsenic levels in house dust in a smelter town in Eastern Germany. *Environmental Research* 81(1):32–44. https://doi.org/10.1006/enrs.1998.3950.

Misztal, P. K., D. S. Lymperopoulou, R. I. Adams, R. A. Scott, S. E. Lindow, T. Bruns, J. W. Taylor, J. Uehling, G. Bonito, R. Vilgalys, and A. H. Goldstein. 2018. Emission factors of microbial volatile organic compounds from environmental bacteria and fungi. *Environmental Science & Technology* 52(15):8272–8282. https://doi.org/10.1021/acs.est.8b00806.

Molnár, P., S. Johannesson, and U. Quass. 2014. Source apportionment of $PM_{2.5}$ using positive matrix factorization (PMF) and PMF with factor selection. *Aerosol and Air Quality Research* 14(3):725–733. https://doi.org/10.4209/aaqr.2013.11.0335.

Morales-McDevitt, M. E., J. Becanova, A. Blum, T. A. Bruton, S. Vojta, M. Woodward, and R. Lohmann. 2021. The air that we breathe: Neutral and volatile PFAS in indoor air. *Environmental Science & Technology Letters* 8(10):897–902. https://doi.org/10.1021/acs.estlett.1c00481.

Morawska, L., G. A. Ayoko, G. N. Bae, G. Buonanno, C. Y. H. Chao, S. Clifford, S. C. Fu, O. Hänninen, C. He, C. Isaxon, M. Mazaheri, T. Salthammer, M. S. Waring, and A. Wierzbicka. 2017. Airborne particles in indoor environment of homes, schools, offices and aged care facilities: The main routes of exposure. *Environment International* 108:75–83. https://doi.org/10.1016/j.envint.2017.07.025.

Mullen, N. A., J. Li, M. L. Russell, M. Spears, B. D. Less, and B. C. Singer. 2016. Results of the California Healthy Homes Indoor Air Quality Study of 2011–2013: Impact of natural gas appliances on air pollutant concentrations. *Indoor Air* 26(2):231–245. https://doi.org/10.1111/ina.12190.

NASEM (National Academies of Sciences, Engineering, and Medicine). 2016. *Health Risks of Indoor Exposure to Particulate Matter: Workshop Summary*. Washington, DC: The National Academies Press.

Nazaroff, W. W., and C. J. Weschler. 2020. Indoor acids and bases. *Indoor Air* 30(4):559–644. https://doi.org/10.1111/ina.12670.

Nazaroff, W. W., and C. J. Weschler. 2022. Indoor ozone: Concentrations and influencing factors. *Indoor Air* 32:e12942.

Nirlo, E. L., N. Crain, R. L. Corsi, and J. A. Siegel. 2014. Volatile organic compounds in fourteen U.S. retail stores. *Indoor Air* 24(5):484–494. https://doi.org/10.1111/ina.12101.

Oberdörster, G. 2001. Pulmonary effects of inhaled ultrafine particles. *International Archives of Occupational and Environmental Health* 74(1):1–8. https://pubmed.ncbi.nlm.nih.gov/11196075/.

Oberdörster, G., and M. J. Utell. 2002. Ultrafine particles in the urban air: To the respiratory tract—and beyond? *Environmental Health Perspectives* 110(8):A440–441. https://pubmed.ncbi.nlm.nih.gov/12153769/.

O'Brien, R. E., Y. Li, K. J. Kiland, E. F. Katz, V. W. Or, E. Legaard, E. Q. Walhout, C. Thrasher, V. H. Grassian, P. F. DeCarlo, A. K. Bertram, and M. Shiraiwa. 2021. Emerging investigator series: Chemical and physical properties of organic mixtures on indoor surfaces during HOMEChem. *Environmental Science: Processes & Impacts* 23(4):559–568. https://doi.org/10.1039/D1EM00060H.

Odabasi, M. 2008. Halogenated volatile organic compounds from the use of chlorine-bleach-containing household products. *Environmental Science & Technology* 42(5):1445–1451. https://doi.org/10.1021/es702355u.

Olson, D. A., and R. L. Corsi. 2004. In-home formation and emissions of trihalomethanes: The role of residential dishwashers. *Journal of Exposure Science & Environmental Epidemiology* 14(2):109–119. https://doi.org/10.1038/sj.jea.7500295.

Omur-Ozbek, P., D. Akalp, and A. Whelton. 2016. Tap water and indoor air contamination due to an unintentional chemical spill in source water. In *Proceedings of International Structural Engineering and Construction: Interaction between Theory and Practice in Civil Engineering and Construction* 3(1), edited by R. Komurlu, A. P. Gurgun, A. Singh, and S. Yazdani. Fargo, ND: ISEC. https://www.isec-society.org/ISEC_PRESS/EURO_MED_SEC_01/html/AW-11.xml. Accessed May 13, 2022.

Or, V. W., M. R. Alves, M. Wade, S. Schwab, R. L. Corsi, and V. H. Grassian. 2018. Crystal clear? Microspectroscopic imaging and physicochemical characterization of indoor depositions on window glass. *Environmental Science & Technology Letters* 5(8):514–519. https://doi.org/10.1021/acs.estlett.8b00355.

OSHA (Occupational Safety and Health Administration). 2007. *Evaluation Guidelines for Surface Sampling Methods. T-006-01-0104-M*. Salt Lake City, UT: OSHA Salt Lake Technical Center.

Ouyang, X., J. M. Weiss, J. de Boer, M. H. Lamoree, and P. E. G. Leonards. 2017. Non-target analysis of household dust and laundry dryer lint using comprehensive two-dimensional liquid chromatography coupled with time-of-flight mass spectrometry. *Chemosphere* 166:431–437. https://doi.org/10.1016/j.chemosphere.2016.09.107.

Overdahl, K. E., D. Gooden, B. Bobay, G. J. Getzinger, H. M. Stapleton, and P. L. Ferguson. 2021. Characterizing azobenzene disperse dyes in commercial mixtures and children's polyester clothing. *Environmental Pollution* 287:117299. https://doi.org/10.1016/j.envpol.2021.117299.

Padilla-Sánchez, J. A., E. Papadopoulou, S. Poothong, and L. S. Haug. 2017. Investigation of the Best Approach for Assessing Human Exposure to Poly- and Perfluoroalkyl Substances through Indoor Air. *Environmental Science & Technology* 51(21):12836–12843. https://doi.org/10.1021/acs.est.7b03516.

Papatheodorou, S., W. Yao, C. L. Z. Vieira, L. Li, B. J. Wylie, J. Schwartz, and P. Koutrakis. 2021. Residential radon exposure and hypertensive disorders of pregnancy in Massachusetts, USA: A cohort study. *Environment International* 146:106285. https://doi.org/10.1016/j.envint.2020.106285.

Parnis, J. M., T. Taskovic, A. K. D. Celsie, and D. Mackay. 2020. Indoor dust/air partitioning: Evidence for kinetic delay in equilibrium for low-volatility SVOCs. *Environmental Science & Technology* 54(11):6723–6729. https://doi.org/10.1021/acs.est.0c00632.

Patel, S., S. Sankhyan, E. K. Boedicker, P. F. DeCarlo, D. K. Farmer, A. H. Goldstein, E. F. Katz, W. W. Nazaroff, Y. Tian, J. Vanhanen, and M. E. Vance. 2020. Indoor particulate matter during HOMEChem: Concentrations, size distributions, and exposures. *Environmental Science & Technology* 54(12):7107–7116. https://doi.org/10.1021/acs.est.0c00740.

Peng, H., D. M. V. Saunders, J. Sun, P. D. Jones, C. K. C. Wong, H. Liu, and J. P. Giesy. 2016. Mutagenic azo dyes, rather than flame retardants, are the predominant brominated compounds in house dust. *Environmental Science & Technology* 50(23):12669–12677. https://doi.org/10.1021/acs.est.6b03954.

Percy, Z., A. M. Vuong, A. M. Ospina, A. M. Calafat, M. J. La Guardia, Y. Xu, R. C. Hale, K. N. Dietrich, C. Xie, B. P. Lanphear, J. M. Braun, K. M. Cecil, K. Yolton, and A. Chen. 2020. Organophosphate esters in a cohort of pregnant women: Variability and predictors of exposure. *Environmental Research* 184:109255. https://doi.org/10.1016/j.envres.2020.109255.

Pigg, S., D. Cautley, and P. W. Francisco. 2018. Impacts of weatherization on indoor air quality: A field study of 514 homes. *Indoor Air* 28(2):307–317. https://doi.org/10.1111/ina.12438.

Polidori, A., B. Turpin, Q. Y. Meng, J. H. Lee, C. Weisel, M. Morandi, S. Colome, T. Stock, A. Winer, J. Zhang, J. Kwon, S. Alimokhtari, D. Shendell, J. Jones, C. Farrar, and S. Maberti. 2006. Fine organic particulate matter dominates indoor-generated PM$_{2.5}$ in RIOPA homes. *Journal of Exposure Science & Environmental Epidemiology* 16(4):321–331. https://doi.org/10.1038/sj.jes.7500476.

Pope, C. A., and D. W. Dockery. 2006. Health effects of fine particulate air pollution: Lines that connect. *Journal of the Air & Waste Management Association* 56(6):709–742. https://doi.org/10.1080/10473289.2006.10464485.

Poppendieck, D. G., L. C. Ng, A. K. Persily, and A. T. Hodgson. 2015. Long term air quality monitoring in a net-zero energy residence designed with low emitting interior products. *Building and Environment* 94:33-42. https://doi.org/10.1016/j.buildenv.2015.07.001.

Price, D. J., D. A. Day, D. Pagonis, H. Stark, L. B. Algrim, A. V. Handschy, S. Liu, J. E. Krechmer, S. L. Miller, J. F. Hunter, J. A. de Gouw, P. J. Ziemann, and J. L. Jimenez. 2019. Budgets of organic carbon composition and oxidation in indoor air. *Environmental Science & Technology* 53(22):13053–13063. https://doi.org/10.1021/acs.est.9b04689.

Price, D. J., D. A. Day, D. Pagonis, H. Stark, A. V. Handschy, L. B. Algrim, S. L. Miller, J. A. de Gouw, P. J. Ziemann, and J. L. Jimenez. 2021. Sources of gas-phase species in an art museum from comprehensive real-time measurements. *ACS Earth and Space Chemistry* 5(9):2252–2267. https://doi.org/10.1021/acsearthspacechem.1c00229.

Proctor, C. R., J. Lee, D. Yu, A. D. Shah, and A. J. Whelton. 2020. Wildfire caused widespread drinking water distribution network contamination. *AWWA Water Science* 2(4):e1183. https://doi.org/10.1002/aws2.1183.

Provoost, J., R. Ottoy, L. Reijnders, J. Bronders, I. V. Keer, F. Swartjes, D. Wilczek, and D. Poelmans. 2011. Henry's equilibrium partitioning between ground water and soil air: Predictions versus observations. *Journal of Environmental Protection* 2(7):9, https://doi.org/10.4236/jep.2011.27099

Prussin, A. J., and L. C. Marr. 2015. Sources of airborne microorganisms in the built environment. *Microbiome* 3(1):78. https://doi.org/10.1186/s40168-015-0144-z.

Rager, J. E., M. J. Strynar, S. Liang, R. L. McMahen, A. M. Richard, C. M. Grulke, J. F. Wambaugh, K. K. Isaacs, R. Judson, A. J. Williams, and J. R. Sobus. 2016. Linking high resolution mass spectrometry data with exposure and toxicity forecasts to advance high-throughput environmental monitoring. *Environment International* 88:269–280. https://doi.org/10.1016/j.envint.2015.12.008.

Rostkowski, P., P. Haglund, R. Aalizadeh, N. Alygizakis, N. Thomaidis, J. B. Arandes, P. B. Nizzetto, P. Booij, H. Budzinski, P. Brunswick, A. Covaci, C. Gallampois, S. Grosse, R. Hindle, I. Ipolyi, K. Jobst, S. L. Kaserzon, P. Leonards, F. Lestremau, T. Letzel, J. Magnér, H. Matsukami, C. Moschet, P. Oswald, M. Plassmann, J. Slobodnik, and C. Yang. 2019. The strength in numbers: Comprehensive characterization of house dust using complementary mass spectrometric techniques. *Analytical and Bioanalytical Chemistry* 411(10):1957–1977. https://doi.org/10.1007/s00216-019-01615-6.

Ryan, P. H., C. Brokamp, Z.-H. Fan, and M. B. Rao. 2015. Analysis of personal and home characteristics associated with the elemental composition of PM$_{2.5}$ in indoor, outdoor, and personal air in the RIOPA study. *Research Report (Health Effects Institute)* 185:3–40.

Salonen, H., T. Salthammer, and L. Morawska. 2018. Human exposure to ozone in school and office indoor environments. *Environment International* 119:503–514. https://doi.org/10.1016/j.envint.2018.07.012.

Salthammer, T., S. Mentese, and R. Marutzky. 2010. Formaldehyde in the indoor environment. *Chemical Reviews* 110(4):2536–2572. https://doi.org/10.1021/cr800399g.

Sankhyan, S., S. Patel, E. F. Katz, P. F. DeCarlo, D. K. Farmer, W. W. Nazaroff, and M. E. Vance. 2021. Indoor black carbon and brown carbon concentrations from cooking and outdoor penetration: Insights from the HOMEChem study. *Environmental Science: Processes & Impacts* 23(10):1476–1487. https://doi.org/10.1039/D1EM00283J.

Schmidt, C. W. 2016. Pollen overload: Seasonal allergies in a changing climate. *Environmental Health Perspectives* 124(4):A70–A75. https://doi.org/10.1289/ehp.124-A70.

Schraufnagel, D. E. 2020. The health effects of ultrafine particles. *Experimental & Molecular Medicine* 52:311–317. https://doi.org/10.1038/s12276-020-0403-3.

Schripp, T., S. Etienne, C. Fauck, F. Fuhrmann, L. Märk, and T. Salthammer. 2014. Application of proton-transfer-reaction-mass-spectrometry for Indoor Air Quality research. *Indoor Air* 24(2):178–189. https://doi.org/10.1111/ina.12061.

Sheu, R., C. Stönner, J. C. Ditto, T. Klüpfel, J. Williams, and D. R. Gentner. 2020. Human transport of thirdhand tobacco smoke: A prominent source of hazardous air pollutants into indoor nonsmoking environments. *Science Advances* 6(10):eaay4109. https://doi.org/10.1126/sciadv.aay4109.

Shu, S., and G. C. Morrison. 2011. Surface reaction rate and probability of ozone and alpha-terpineol on glass, polyvinyl chloride, and latex paint surfaces. *Environmental Science & Technology* 45(10):4285–4292. https://doi.org/10.1021/es200194e.

Slezakova, K., C. Peixoto, M. Oliveira, C. Delerue-Matos, M. d. C. Pereira, and S. Morais. 2018. Indoor particulate pollution in fitness centres with emphasis on ultrafine particles. *Environmental Pollution* 233:180–193. https://doi.org/10.1016/j.envpol.2017.10.050.

Snyder, K., J. H. Vick, and B. A. King. 2016. Smoke-free multiunit housing: A review of the scientific literature. *Tobacco Control* 25(1):9–20. https://doi.org/10.1136/tobaccocontrol-2014-051849.

Stanley, F. K. T., J. L. Irvine, W. R. Jacques, S. R. Salgia, D. G. Innes, B. D. Winquist, D. Torr, D. R. Brenner, and A. A. Goodarzi. 2019. Radon exposure is rising steadily within the modern North American residential environment, and is increasingly uniform across seasons. *Scientific Reports* 9(1):18472. https://doi.org/10.1038/s41598-019-54891-8.

Stapleton, H. M., J. G. Allen, S. M. Kelly, A. Konstantinov, S. Klosterhaus, D. Watkins, M. D. McClean, and T. F. Webster. 2008. Alternate and new brominated flame retardants detected in U.S. house dust. *Environmental Science & Technology* 42(18):6910–6916. https://doi.org/10.1021/es801070p.

Stapleton, H. M., S. Eagle, A. Sjödin, and T. F. Webster. 2012. Serum PBDEs in a North Carolina toddler cohort: Associations with handwipes, house dust, and socioeconomic variables. *Environmental Health Perspectives* 120(7):1049–1054. https://doi.org/10.1289/ehp.1104802.

Stein, M. M., C. L. Hrusch, C. Gozdz, C. Igartua, V. Pivniouk, S. E. Murray, J. G. Ledford, M. Marques dos Santos, R. L. Anderson, N. Metwali, J. W. Neilson, R. M. Maier, J. A. Gilbert, M. Holbreich, P. S. Thorne, F. D. Martinez,

E. von Mutius, D. Vercelli, C. Ober, and A. I. Sperling. 2016. Innate immunity and asthma risk in Amish and Hutterite farm children. *New England Journal of Medicine* 375(5):411–421. https://doi.org/10.1056/NEJMoa1508749.

Tan, H., L. Yang, Y. Huang, L. Tao, and D. Chen. 2021. "Novel" Synthetic Antioxidants in House Dust from Multiple Locations in the Asia-Pacific Region and the United States. *Environmental Science & Technology* 55(13):8675–8682. https://doi.org/10.1021/acs.est.1c00195.

Tang, X., P. K. Misztal, W. W. Nazaroff, and A. H. Goldstein. 2015. Siloxanes are the most abundant volatile organic compound emitted from engineering students in a classroom. *Environmental Science & Technology Letters* 2(11):303–307. https://doi.org/10.1021/acs.estlett.5b00256.

Tecle, A., and D. G. Neary. 2015. Water quality impacts of forest fires. *Journal of Pollution Effects and Control* 3:1–7. https://doi.org/10.4172/2375-4397.1000140.

Tran, T. M., K. O. Abualnaja, A. G. Asimakopoulos, A. Covaci, B. Gevao, B. Johnson-Restrepo, T. A. Kumosani, G. Malarvannan, T. B. Minh, H.-B. Moon, H. Nakata, R. K. Sinha, and K. Kannan. 2015. A survey of cyclic and linear siloxanes in indoor dust and their implications for human exposures in twelve countries. *Environment International* 78:39–44. https://doi.org/10.1016/j.envint.2015.02.011.

Tran Le, A. D., J. S. Zhang, and Z. Liu. 2021. Impact of humidity on formaldehyde and moisture buffering capacity of porous building material. *Journal of Building Engineering* 36:102114. https://doi.org/10.1016/j.jobe.2020.102114.

Turpin, B., C. Weisel, M. Morandi, S. Colome, T. Stock, S. Eisenreich, and B. Buckley. 2007. Relationships of Indoor, Outdoor, and Personal Air (RIOPA): Part II. Analyses of concentrations of particulate matter species. *Research Report (Health Effects Institute)* 130:1–77.

Uchiyama, S., T. Tomizawa, A. Tokoro, M. Aoki, M. Hishiki, T. Yamada, R. Tanaka, H. Sakamoto, T. Yoshida, K. Bekki, Y. Inaba, H. Nakagome, and N. Kunugita. 2015. Gaseous chemical compounds in indoor and outdoor air of 602 houses throughout Japan in winter and summer. *Environmental Research* 137:364–372. https://doi.org/10.1016/j.envres.2014.12.005.

Ulrich, E. M., J. R. Sobus, C. M. Grulke, A. M. Richard, S. R. Newton, M. J. Strynar, K. Mansouri, and A. J. Williams. 2019. EPA's non-targeted analysis collaborative trial (ENTACT): Genesis, design, and initial findings. *Analytical and Bioanalytical Chemistry* 411(4):853–866. https://doi.org/10.1007/s00216-018-1435-6.

Unnithan, A., D. N. Bekele, S. Chadalavada, and R. Naidu. 2021. Insights into vapour intrusion phenomena: Current outlook and preferential pathway scenario. *Science of the Total Environment* 796:148885. https://doi.org/10.1016/j.scitotenv.2021.148885.

USHUD. 2018. *Water Piping in the Home.* https://www.huduser.gov/portal/pdredge/pdr-edge-trending-091018.html. Accessed May 13, 2022.

Vardoulakis, S., E. Giagloglou, S. Steinle, A. Davis, A. Sleeuwenhoek, K. Galea, K. Dixon, and J. Crawford. 2020. Indoor exposure to selected air pollutants in the home environment: A systematic review. *International Journal of Environmental Research and Public Health* 17(23):8972. https://doi.org/10.3390/ijerph17238972.

Venier, M., O. Audy, Š. Vojta, J. Bečanová, K. Romanak, L. Melymuk, M. Krátká, P. Kukučka, J. Okeme, A. Saini, M. L. Diamond, and J. Klánová. 2016. Brominated flame retardants in the indoor environment - Comparative study of indoor contamination from three countries. *Environ Int* 94:150–160. https://doi.org/10.1016/j.envint.2016.04.029.

Vykoukalová, M., M. Venier, Š. Vojta, L. Melymuk, J. Bečanová, K. Romanak, R. Prokeš, J. O. Okeme, A. Saini, M. L. Diamond, and J. Klánová. 2017. Organophosphate esters flame retardants in the indoor environment. *Environ Int* 106:97–104. https://doi.org/10.1016/j.envint.2017.05.020.

Wallace, L. 2000. Correlations of personal exposure to particles with outdoor air measurements: A review of recent studies. *Aerosol Science and Technology* 32(1):15–25. https://doi.org/10.1080/027868200303894.

Wallace, L., and W. Ott. 2011. Personal exposure to ultrafine particles. *Journal of Exposure Science & Environmental Epidemiology* 21(1):20–30. https://doi.org/10.1038/jes.2009.59.

Wallace, L., F. Wang, C. Howard-Reed, and A. Persily. 2008. Contribution of gas and electric stoves to residential ultrafine particle concentration between 2 and 64 nm: Size distributions and emission and coagulation rates. *Environmental Science & Technology* 42(23):8641–8647. https://doi.org/10.1021/es801402v.

Wallace, L. A., S. J. Emmerich, and C. Howard-Reed. 2004. Source strengths of ultrafine and fine particles due to cooking with a gas stove. *Environmental Science & Technology* 38(8):2304–2311. https://doi.org/10.1021/es0306260.

Wallace, L. A., W. R. Ott, C. J. Weschler, and A. C. K. Lai. 2017. Desorption of SVOCs from heated surfaces in the form of ultrafine particles. *Environmental Science & Technology* 51(3):1140–1146. https://doi.org/10.1021/acs.est.6b03248.

Wang, L., A. G. Asimakopoulos, H.-B. Moon, H. Nakata, and K. Kannan. 2013. Benzotriazole, benzothiazole, and benzophenone compounds in indoor dust from the United States and East Asian countries. *Environmental Science & Technology* 47(9):4752–4759. https://doi.org/10.1021/es305000d.

Wang, Z., G. W. Walker, D. C. G. Muir, and K. Nagatani-Yoshida. 2020. Toward a global understanding of chemical pollution: A first comprehensive analysis of national and regional chemical inventories. *Environmental Science & Technology* 54(5):2575–2584. https://doi.org/10.1021/acs.est.9b06379.

Weschler, C. J. 2016. Roles of the human occupant in indoor chemistry. *Indoor Air* 26:6–24.

Weschler, C. J., and W. W. Nazaroff. 2008. Semivolatile organic compounds in indoor environments. *Atmospheric Environment* 42(40): 9018–9040. https://doi.org/10.1016/j.atmosenv.2008.09.052.

Weschler, C. J., and W. W. Nazaroff. 2017. Growth of organic films on indoor surfaces. *Indoor Air* 27(6):1101–1112. https://doi.org/10.1111/ina.12396.

Wiesinger, H., Z. Wang, and S. Hellweg. 2021. Deep dive into plastic monomers, additives, and processing aids. *Environmental Science & Technology* 55(13):9339–9351. https://doi.org/10.1021/acs.est.1c00976.

Williams, J., C. Stönner, J. Wicker, N. Krauter, B. Derstroff, E. Bourtsoukidis, T. Klüpfel, and S. Kramer. 2016. Cinema audiences reproducibly vary the chemical composition of air during films, by broadcasting scene specific emissions on breath. *Scientific Reports* 6(1):25464. https://doi.org/10.1038/srep25464.

Winkens, K., J. Koponen, J. Schuster, M. Shoeib, R. Vestergren, U. Berger, A. M. Karvonen, J. Pekkanen, H. Kiviranta, and I. T. Cousins. 2017. Perfluoroalkyl acids and their precursors in indoor air sampled in children's bedrooms. *Environmental Pollution* 222:423–432. https://doi.org/10.1016/j.envpol.2016.12.010.

Wong, J. P. S., N. Carslaw, R. Zhao, S. Zhou, and J. P. D. Abbott. 2017. Observations and impacts of bleach washing on indoor chlorine chemistry. *Indoor Air* 27(6):1082–1090. https://doi.org/10.1111/ina.12402.

Woolfenden, E. 2010. Sorbent-based sampling methods for volatile and semi-volatile organic compounds in air: Part 1: Sorbent-based air monitoring options. *Journal of Chromatography A* 1217(16):2674–2684. https://doi.org/10.1016/j.chroma.2009.12.042.

Wu, T., M. Fu, M. Valkonen, M. Täubel, Y. Xu, and B. E. Boor. 2021a. Particle resuspension dynamics in the infant near-floor microenvironment. *Environmental Science & Technology* 55(3):1864–1875. https://doi.org/10.1021/acs.est.0c06157.

Wu, T., A. Tasoglou, H. Huber, P. S. Stevens, and B. E. Boor. 2021b. Influence of mechanical ventilation systems and human occupancy on time-resolved source rates of volatile skin oil ozonolysis products in a LEED-certified office building. *Environmental Science & Technology* 55:16477–16488.

Wu, Y., S. S. Cox, Y. Xu, Y. Liang, D. Won, X. Liu, P. A. Clausen, L. Rosell, J. L. Benning, and Y. Zhang. 2016. A reference method for measuring emissions of SVOCs in small chambers. *Building and Environment* 95:126–132. https://doi.org/10.1016/j.buildenv.2015.08.025.

Xu, J., and J. S. Zhang. 2011. An experimental study of relative humidity effect on VOCs' effective diffusion coefficient and partition coefficient in a porous medium. *Building and Environment* 46(9):1785–1796. https://doi.org/10.1016/j.buildenv.2011.02.007.

Xu, Y., Y. Liang, J. R. Urquidi, and J. A. Siegel. 2015. Semi-volatile organic compounds in heating, ventilation, and air-conditioning filter dust in retail stores. *Indoor Air* 25(1):79–92. https://doi.org/10.1111/ina.12123.

Yang, S., G. Beko, P. Wargocki, J. Williams, and D. Licina. 2021. Human emissions of size-resolved fluorescent-aerosol particles: Influence of personal and environmental factors. *Environmental Science & Technology* 55(1):509–518. https://doi.org/10.1021/acs.est.0c06304.

Young, C. J., S. Zhou, J. A. Siegel, and T. F. Kahan. 2019. Illuminating the dark side of indoor oxidants. *Environmental Science: Processes & Impacts* 21(8):1229–1239. https://doi.org/10.1039/C9EM00111E.

Zheng, G., G. M. Filippelli, and A. Salamova. 2020. Increased indoor exposure to commonly used disinfectants during the COVID-19 pandemic. *Environmental Science & Technology Letters* 7(10):760–765. https://doi.org/10.1021/acs.estlett.0c00587.

Zhou, S., C. J. Young, T. C. VandenBoer, S. F. Kowal, and T. F. Kahan. 2018. Time-resolved measurements of nitric oxide, nitrogen dioxide, and nitrous acid in an occupied New York home. *Environmental Science & Technology* 52(15):8355–8364. https://doi.org/10.1021/acs.est.8b01792.

Zhukovsky, M., A. Vasilyev, A. Onishchenko, and I. Yarmoshenko. 2018. Review of indoor radon concentrations in schools and kindergartens. *Radiation Protection Dosimetry* 181(1):6–10. https://doi.org/10.1093/rpd/ncy092.

Zwoździak, A., I. Sówka, B. Krupińska, J. Zwoździak, and A. Nych. 2013. Infiltration or indoor sources as determinants of the elemental composition of particulate matter inside a school in Wrocław, Poland? *Building and Environment* 66:173.

3

Partitioning of Chemicals in Indoor Environments

Partitioning of chemicals plays an important role in indoor chemistry, indoor air quality, and occupant exposure to chemicals. Partitioning refers to the distribution of chemicals among phases or compartments. At the state of thermodynamic equilibrium, partitioning determines the concentration of a chemical in air, on surfaces, and elsewhere. Partitioning can also refer to the process of molecular exchange between phases. Chemicals in indoor environments are often not at equilibrium, and net transfer can occur between phases. Therefore, partitioning refers to both the thermodynamic state of chemicals distributed among phases in a system and the processes that transfer chemicals among phases, generally with a net tendency to approach equilibrium. Partitioning can also influence the rate of primary emissions (see Chapter 2) of some chemicals. Because of the very high surface-area-to-volume ratio of indoor environments, partitioning influences the manner, extent, and duration over which occupants are exposed to contaminants. The distribution of chemicals among phases and compartments also influences efforts to remove chemicals from indoor environments and mitigate exposure. While chemical transformations can subsequently occur, as discussed in Chapter 4, this chapter will focus on the role of partitioning of chemicals in indoor environments as it relates to indoor chemistry and indoor air quality.

Chemicals can have a greater affinity for one reservoir than another, and this is characterized by a "partition coefficient" defined later in this chapter. For example, nicotine has a high affinity for indoor material surfaces; therefore, there tends to be more nicotine on surfaces than in the air. The sizes of the reservoirs are also important. By combining partition coefficients with the sizes of reservoirs, one can determine where most of the mass of a chemical contaminant will be at equilibrium. However, partitioning does not occur instantaneously, and it can take time, sometimes years, for reservoirs to approach equilibrium with their neighbors owing to slow rates of molecular transport. As will be discussed in this chapter, challenges remain in understanding partitioning from the molecular to whole building scales and how partitioning influences exposure and chemistry. These challenges include limited information on the detailed composition of building materials across the building stock; complexity of spatial and temporal variations in environmental conditions; and poor understanding of the complexity of surfaces and indoor materials on molecular and nanometer (nm) length scales (Ault et al., 2020; Liu et al., 2020), as well as the molecular interactions which drive the partitioning of a chemical from the gas to other phases (Abbatt and Wang, 2020; Ault et al., 2020).

This chapter first describes building reservoirs and surfaces as well as how contaminants interact with surfaces and bulk phases. Then the "partition coefficient" is defined along with a discussion of partitioning equilibria, chemical transport limitations on equilibria, and how environmental conditions influence partitioning. Specific reservoirs are discussed with a focus on the most recent studies and discoveries, and then the chapter concludes with a description of future research needs.

INDOOR ENVIRONMENTAL RESERVOIRS AND SURFACES

The indoor environment is often characterized by its different reservoirs. For the purposes of this report, reservoirs are defined as any surface or volume present in indoor environments to which chemicals can partition. Practically, when defining reservoirs for the purposes of talking about partitioning behavior, it is useful to combine materials of the same type into a reservoir. For example, it might be useful to define textiles as a reservoir or to distinguish textile reservoirs by type of fiber. Other useful reservoirs include paint films; the surfaces of glass, wood, dust, and indoor air; and even the volume of air surrounding a building. Important points to consider in the partitioning of chemicals are the fact that the physicochemical properties of these different materials are quite varied and the surfaces (see Box 3-1) of these materials, including their surface area, are often poorly defined. However, the properties of these materials are important to understand, as surface reservoirs of chemicals have recently been shown to be a dominant source of chemicals and potentially of chemical exposure (Wang et al., 2020b).

From a surface chemistry perspective, indoor environments provide a large variety of surfaces that a gas-phase molecule present in air can partition into (Figure 3-1). The available indoor surface area for this partitioning is many times greater than outdoors. The flat projected area is approximately 3 m^2 for every cubic meter of room volume, for a room with contents such as furnishings (Manuja et al., 2019); much more area is available by accounting for the internal surface area of rough and porous materials.

Partitioning to the surface can be more or less important than the bulk, depending on the nature of the material and chemical. Surfaces can be impermeable, such as window glass, metal, and granite counters, or they can be highly permeable and porous, such as fabrics used in home furnishings, carpets, textiles used for clothing, and even painted walls.

For nonporous, solid surfaces, deposition of particles and partitioning of chemicals result in the buildup of thin films of material with thicknesses on the order of nanometers to micrometers (Eichler et al., 2019b; Liu et al., 2020; Weschler and Nazaroff, 2017). For porous materials or

---

BOX 3-1
Defining the Surface Region

In this report, the term *surface* describes the top-most (or outer-most) region of any solid or liquid material accessible to chemicals present in air. This region includes the interface between air and the material as well as some depth of material below that interface that also interacts in a meaningful way with the gas phase over relevant timescales. This definition is consistent with the way surfaces are implicitly defined in most indoor chemistry literature as it takes into account *internal surface area* due to the porous nature of most materials, but it differs from the way *surface* is typically defined as the interface between two phases. The report distinguishes the surface reservoir of a material from its *bulk* because partitioning phenomena can be quite different and occur over very different timescales, even for the same material.

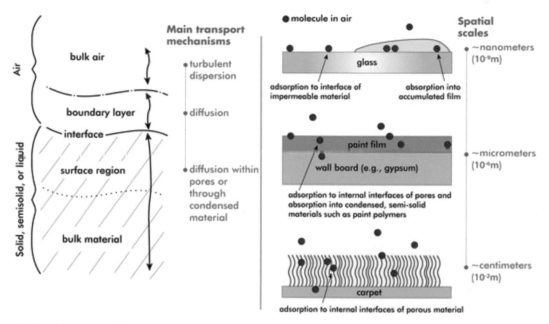

FIGURE 3-1 Several distinct regions are associated with a surface (left). Different types of surfaces lead to different partitioning behaviors with a variety of relevant spatial and timescales (right).

aqueous reservoirs, surfaces are the "gateway" into the bulk. Surfaces can control the ability of a chemical to partition from one phase into another. The surface is dynamic—in other words, changing over time. Surfaces can be cleaned, refurbished, or even repainted. Surfaces can soil through the buildup of organic thin films and dust particles, and the deposition of aerosols. In this region, thin films of organics can become surfaces themselves. Thus, there are complexities in understanding these surfaces, from clean to dirty, that are part of indoor environments.

PARTITIONING AMONG RESERVOIRS AND PHASES FOUND IN INDOOR ENVIRONMENTS

There are different phases and reservoirs in which gas-phase chemicals can partition (Figure 3-2). Gas-phase chemicals can partition to particles suspended in air (aerosol particles), impermeable solid surfaces like glass, thin films, and dust particles present on surfaces, permeable materials like wood, and water reservoirs (condensed aqueous phases; see Chapter 4). Partitioning is often regarded as an equilibrium process whereby the equilibrium in different phases (or different reservoirs) for a chemical is dictated by size of the reservoir and the nature and strength of the molecular interactions that occur. Thermodynamics govern these processes: the net transfer of molecules from one phase to another is due to a difference in chemical potential; thus, gradients drive this process until the chemical potential in each phase approaches equilibrium and is equal. This is similar to temperature differences or gradients that reach zero as they approach equilibrium. The equilibrium processes involving chemicals in the gas phase with other phases can be written generally as follows: M (phase 1) $\rightleftarrows$ M (phase 2), where, for example, phase 1 is the gas phase (air) and phase 2 is a solid phase (e.g., a wood surface). These equilibria can be written for a variety of processes involving different reservoirs and phases. If equilibrium is established, then all reservoirs and phases are in equilibrium with one another.

These equilibria can be characterized by partition coefficients. While there are numerous ways to define partition coefficients, here, two are considered: K_{area} and K_{vol}. The area-specific partition

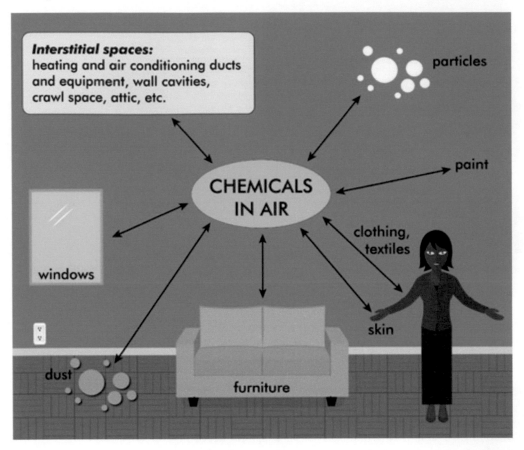

FIGURE 3-2 Chemicals can partition among a number of different reservoirs in the indoor environment. SOURCE: Modified from Weschler and Nazaroff (2008).

coefficient, K_{area}, is the equilibrium ratio of the areal concentration on a surface, C_s (in μg/m^2), to the gas-phase (air) concentration, C_g (in μg/m^3), or $K_{area} = C_s/C_g$; its units are then m^3/m^2 or simply m. This parameter is useful when describing partitioning to surfaces like glass or metal. The volume-specific partition coefficient, K_{vol}, refers to partitioning to the bulk of a material and is defined as the equilibrium ratio of the volume concentration in that material, C_v (in μg/m^3), to the concentration in air, or $K_{vol} = C_v/C_g$; its units are then m^3/m^3, but it is often shown without units. Partition coefficients are specific to the contaminant and the material it partitions to, but they can be influenced by environmental conditions, such as relative humidity (as discussed later in this chapter).

Gas- to aqueous-phase partitioning can be described by Henry's law, $K_H = [A]/C_{A,g}$, where $C_{A,g}$ is the gas-phase concentration of species A, [A] is its concentration in the aqueous phase, and K_H is Henry's law constant. Note that the Henry's law constant and the water-air partition coefficient, K_{wa}, are equivalent after accounting for the gas constant and temperature (Weschler and Nazaroff, 2008). For partitioning into organic materials, including organic films, which are prevalent on surfaces, the K_{oa} value for a molecular species (i.e., the ratio of the amount of a chemical in octanol compared to air) is often chosen to approximate the equilibrium state. There have been several approaches formulated toward understanding chemical partitioning, especially of semivolatile organic compounds (SVOCs) (Wang et al., 2020b; Weschler and Nazaroff, 2008). Additional considerations of indoor dust/air partitioning models have also been described (Parnis et al., 2020).

SIZE AND CAPACITY OF DIFFERENT INDOOR RESERVOIRS

Buildings vary greatly in size, design, types of materials, and environment. To help put partitioning in perspective, this section uses the example of a typical U.S. residence with a floor area of 140 m^2 and volume of 360 m^3 (living space) (U.S. Census, 2021). The internal, exposed surface area is ~1,150 m^2, with painted walls and ceiling, flooring, windows, and furnishings comprising the bulk of that area. Most of the volume is air, followed by the building shell and walls, then furnishings. An accounting of areas and volumes of all indoor materials is beyond the scope of this report, but Figure 3-3 provides a few for scale; here, values are estimated from multiple sources (Hodgson et al., 2005; Manuja et al., 2019, Nazaroff and Weschler, 2020).

An estimate of the amount of a chemical present in a specific reservoir relative to indoor air can be obtained by multiplying an area-specific partition coefficient (K_{area}) by the surface-area-to-volume ratio of that material indoors. For example, glass and metal are responsible for approximately 0.1 m^2 of exposed surface area per cubic meter of residence volume. Therefore, using a value of K_{area} =1,000 m^3/m^2 for the partition coefficient of a typical SVOC with glass and steel (see the later section in this chapter on Impermeable Surfaces), ~100 times more contaminant mass will be present on these surfaces than in the air volume of the house. The mass of SVOCs can be orders of magnitude higher in the bulk of materials like paint, wood, and textile than at their outermost surface region. In general, the amount that can partition to interfacial reservoirs and organic films is much lower than the amount that can partition into bulk materials. Indoor reservoirs can accumulate very large amounts of contaminants relative to the amount in air.

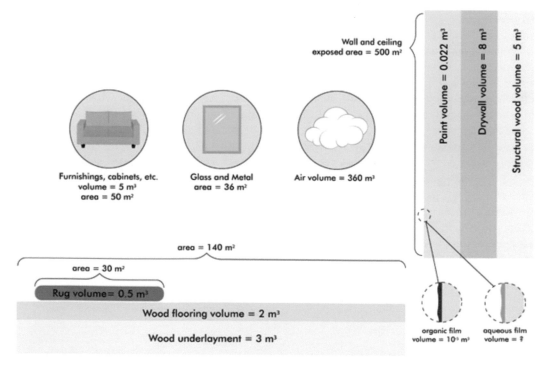

FIGURE 3-3 Approximate areas and volumes of select indoor reservoirs in a typical U.S. residence with a volume = 360 m^3. Areas shown are defined as the simple, geometrically projected areas that can be determined using simple tools, such as a tape measure. This excludes the area associated with internal porosity and therefore represents a lower-bound value for all accessible surface area. Note that organic and aqueous films can exist on all indoor surfaces, not only on paint as shown in the figure.

PARTITIONING THERMODYNAMICS: EFFECTS OF TEMPERATURE AND RELATIVE HUMIDITY

The strength of the intermolecular interactions between a sorbing molecule and the chemicals in the condensed-phase reservoir play a very strong role in determining the degree to which gas-to-surface partitioning occurs. These interactions involve van der Waal's, dipole-dipole, dipole-induced dipole, and H-bonding forces, whereas covalent bonds arise when chemical transformations occur (see Chapter 4). Together with entropic factors, the collective strength of these intermolecular forces—referred to as the sorption enthalpy—controls the magnitude of the equilibrium partition coefficient defined above.

An important factor to consider is the temperature dependence of the partition coefficient, which is controlled by the sorption enthalpy (i.e., larger sorption enthalpies are related to a stronger temperature dependence of the partitioning). Partitioning of SVOCs to indoor materials has been characterized as a function of temperature (Wei et al., 2018). The enthalpy varies by chemical and the reservoir to which it partitions, but it is typically ~100 kJ/mol for SVOCs (Liang and Xu, 2014). In this case, a 5 °C increase in the temperature (e.g., from 20 °C to 25 °C) results in a 2-fold change in the partition coefficient and thus a 2-fold increase in the concentration in air relative to the surface. For example, a recent field study showed a 1.25- to 2-fold increase in indoor SVOC concentrations when ambient temperature increased from 14 °C to 21 °C, due to the emissions from indoor surfaces (Kristensen et al., 2019).

Many factors, including relative humidity, impact a chemical's partitioning behavior. Water vapor can partition to surfaces, and it has been shown that the amount of water associated with surfaces increases as relative humidity increases and as temperature decreases. The multiple roles of adsorbed water in the partitioning and chemistry of a variety of different surfaces have been previously discussed (Rubasinghege and Grassian, 2013). Water on surfaces can both increase and decrease the uptake of chemicals on surfaces. When increasing the amounts of water on the surface results in a decrease in the sorption of other chemicals, this suggests that there is competition between sorbed water (i.e., on surfaces or inside porous materials) and sorbed chemicals. As the amount of sorbed water on indoor surfaces increases with increasing relative humidity, the sorbed chemicals are displaced from the surface into the gas phase. The extent of this displacement depends on the relative energies of the specific molecular interactions that occur for water compared to the sorbed chemical (Frank et al., 2020; Huang et al., 2021). Several recent studies have shown that, for some organic compounds such as limonene, increasing relative humidity leads to a decrease in the amount of limonene adsorbed on silica (SiO_2) particle surfaces and the presence of adsorbed water on the surface (Frank et al., 2020). However, for more oxygenated, polar chemicals, such as dihydromyrcenol, linalool, and α-terpineol, there is no displacement and little adsorbed water on the SiO_2 particle surface at high relative humidity (Huang et al., 2021).

Studies on more complex indoor surface materials show that increasing relative humidity affects partitioning behavior in different ways depending on the chemical and the material. In some cases, increasing relative humidity has no effect, in other cases it decreases the amount of organics associated with the material, and in a few cases increasing relative humidity is shown to increase the amount of chemical sorbed in the material (Won et al., 2001). Most importantly, recent studies have made clear that a range of behaviors on the impacts of water on surfaces and materials can be found in indoor environments as it relates to the chemical partitioning of organic compounds (Wang et al., 2020b).

PARTITIONING DYNAMICS, TIMESCALES, AND LIMITATIONS ON THE EQUILIBRIUM CONCEPT

Although thermodynamics drive these equilibria, transport limitations and chemical kinetics play an important role in indoor partitioning of chemicals, as high physical and energy barriers for some processes result in nonequilibrium conditions. The amount of any chemical present within reservoirs changes over time with changes in the conditions; these can occur over timescales of minutes to years. For example, SVOCs present in a new piece of furniture may require decades to approach equilibrium with the surrounding air and surfaces. Mechanisms that slow transfer of molecules from one reservoir to another include diffusion across a thin layer of slow-moving air adjacent to all surfaces (i.e., boundary layer) and diffusion into building materials and furnishings. These "mass transfer limitations" are central to understanding the dynamic development of contaminant concentrations in indoor reservoirs and will be highlighted several times in this report; however, because these mechanisms are well described in other sources, this report will not separately discuss mass transfer phenomena. The equilibrium partitioning concept is also somewhat limited, as it only describes scenarios where reactions do not occur. If an organic compound undergoes oxidation or other chemical reactions on the surface, it may or may not re-equilibrate back into the gas phase on relevant timescales.

The combination of reservoir capacity and transport limitations can strongly influence dynamic indoor concentrations. For example, surfaces and porous materials can become large reservoirs of many chemicals. Although air can be "cleaned" in minutes by indoor/outdoor air exchange (e.g., opening windows) or high ventilation rates (e.g., heating, ventilation, and air-conditioning [HVAC] systems and filtration), reservoirs release these chemicals back into indoor air (Uhde et al., 2019; Wang et al., 2020b), reducing the effectiveness of control efforts. This leads to continued exposure of these contaminants with only intermittent decreases under certain conditions. Similarly, low-volatility chemicals can remain on surfaces for durations much longer than timescales of ventilation; this can provide enough time for slow chemical reactions to occur. Eichler et al. (2019a) provide insights into these highly variable surfaces and a mechanistic framework in which to model SVOCs that are emitted from surfaces and materials indoors.

CURRENT SCIENCE ON PARTITIONING OF CHEMICALS IN INDOOR ENVIRONMENTS

Impermeable Surfaces

Impermeable indoor surfaces, to which chemicals can adsorb but not penetrate, include windows and mirrors, glazed tiles, ceramic-top stoves, polished stone counters, stainless steel, copper pipes, and enameled appliances. High-density plastics may sometimes be considered effectively impermeable to large organic chemicals. Indoor air is in direct contact with many such exposed surfaces, providing for rapid uptake and release of chemicals, including reactive species that may not otherwise penetrate below permeable material surfaces. Uptake to such indoor surfaces has been studied since it was recognized that adsorption to the surfaces of steel or glass chambers altered the dynamic air concentration of volatile organic compounds (VOCs) from building materials being studied for their emissions (Tichenor et al., 1991). In a review of SVOC partitioning indoors, Wei et al. (2018) reported that partition coefficients for impermeable surfaces ranged from 6 m to 7,500 m; the wide range is due to the range of temperature, relative humidity, and materials studied. Wu et al. (2017) observed that molecular-scale surface area was of central importance in the partitioning of phthalate esters to metal, glass, and acrylic.

Molecular interactions at the interface can be unique to the primarily inorganic materials that comprise impermeable indoor surfaces. For example, several authors have observed higher than anticipated partitioning of terpenes to SiO_2, the primary component of glass (Chase et al., 2017; Fang et al., 2019b; Liu et al., 2019). Molecular dynamic simulations are consistent with spectroscopic measurements of limonene adsorption to SiO_2, demonstrating the potential of computational methods to quantify molecular-level thermodynamics that have building-scale implications (Fang et al., 2019a). Partitioning behavior of terpenes is important because double bonds and other substituents in terpenes can influence their orientation on the surface (Fang et al., 2019a; Ho et al., 2016) and the rate of surface chemistry (see Chapter 4; Ham and Wells, 2009; Shu and Morrison, 2011; Stokes et al., 2008, 2009).

Organic Thin Films

Film development is an extension of molecular adsorption onto pristine surfaces. As more molecules continue to adsorb to a surface, they can form "islands" or eventually films with thicknesses measured in nanometers. Deposition of particles can also contribute to surface films. Because the formation, composition, and aging of films is analogous to secondary organic aerosols, they have been described as secondary organic films (Ault et al., 2020). Clear evidence has been found for the heterogeneity of glass surfaces in indoor environments using new microspectroscopic methods to analyze surface films and deposited particles on glass surfaces. In particular, Or et al. (2018) developed atomic force microscopy coupled to photothermal infrared spectroscopy to analyze surfaces placed in different indoor environments (e.g., kitchen, garage, office, and copier room). These studies show the variability of these surfaces on nanometer to micrometer length scales. When glass is placed in a kitchen for 6 months, organic films are shown to contain several different types of organics, including fatty acids and oxidized organic compounds, such as nonanal.

Organic films were first identified and characterized in indoor spaces by Liu et al. (2003). The composition is dominated by plasticizers, mono- and dicarboxylic acids, and "cooking organic aerosol" components (Lim and Abbatt, 2020; Liu et al., 2003; O'Brien et al., 2021) but also includes many other SVOCs as discussed below. Weschler and Nazaroff (2017) proposed that indoor organic films develop from accumulation of existing indoor SVOCs. SVOCs with lower octanol-air coefficients (K_{oa}) rapidly equilibrate with an existing thin film and maintain a constant concentration, even as the film grows as the result of absorption of chemicals with higher K_{oa}. This process continues, and the film growth slows as chemicals of higher molecular weight reach equilibrium with the film. Eichler et al. (2019a) extended this model to include the formation of the first layer of the film, and Lakey et al. (2021) further extended that model to the influence of overlying turbulence, film viscosity, and chemical transformations. In experiments intended to follow film development on surfaces, Wallace et al. (2017) heated surfaces that had been deployed indoors and measured the number concentration of ultrafine particles that were generated as the result of evaporation and nucleation in air; they observed that the particle emission rate (proportional to film thickness) increased with exposure up to ~100 days of exposure. Lim and Abbatt (2020) deployed glass capillaries in various locations in a home and measured film growth and composition using direct analysis in real time mass spectrometry (DART-MS). They observed initial growth rates consistent with predictions and other measurements, at least 0.05 nm/day, with a highly homogeneous film composition from location to location throughout a residence. This is consistent with growth of the films by SVOC uptake. Taken as a whole, thin organic films eventually develop on all indoor surfaces, and this has been suggested as one reason why ozone decay rates among buildings are surprisingly similar despite the presence of very different kinds of materials (Weschler and Nazaroff, 2017). Based on a review of equilibrium measurements of compounds found on surfaces and air, Weschler and Nazaroff (2012) inferred that average organic surface film thicknesses range from 3 to 30 nm, which is consistent with direct measurements of organic matter and predictive models, although these films may not be of uniform thickness.

Importantly for this chapter, these films provide volume for absorption of chemicals. Surface films accumulate the many SVOCs that have been identified indoors including phthalates, flame retardants, polycyclic aromatic hydrocarbons, perfluorinated alkyl substances (PFAS), and others (Bennett et al., 2015; Butt et al., 2004a,b; Duigu et al., 2009; Gewurtz et al., 2009; Huo et al., 2016; Liu et al., 2019; Melymuk et al., 2016; Pan et al., 2012; Venier et al., 2016). For most of these SVOCs, their presence in films dwarfs the amount present in the air (Weschler and Nazaroff, 2008), but the amount present in films is dwarfed by bulk reservoirs of permeable materials (see the next section).

Films that develop on exposed surfaces can be highly responsive to changes: there is little mass-transfer resistance for flux to and from bulk indoor air, and in the absence of a dramatic increase in viscosity, diffusion is fast (Lakey et al., 2021; O'Brien et al., 2021). This means that changes in indoor environmental conditions will result in a rapid response from these exposed reservoirs. For example, increasing ventilation by opening a window is usually expected to improve air quality by diluting chemicals that have indoor sources; however, in response to the disruption of equilibrium, near-instantaneous increased emissions of SVOCs stored in surface reservoirs will recharge the air phase, limiting the ability of a temporary increase in ventilation to decrease levels. Consistently high ventilation can eventually deplete these reservoirs, but the time required to lower indoor air concentrations can range from days to years.

Although films have the potential to respond rapidly, low-volatility chemicals take time to equilibrate with these films. Over timescales of weeks, chemicals with $\log_{10}(K_{oa})$ <11 are likely to equilibrate with 10 nm-thick organic surface films, while those with $\log_{10}(K_{oa})$ >14 will take years to approach equilibrium (Lim and Abbatt, 2020; Venier et al., 2016; Weschler and Nazaroff, 2008).

Permeable Materials

Permeable materials comprise the majority of surface reservoirs that are present indoors (Hodgson et al., 2005; Manuja et al., 2019). The term "permeable" here refers to materials that allow chemicals to penetrate below their interface by diffusing into the bulk of the material or via openings or pores that allow access to the internal surface area. Wood, textiles, concrete, plaster, drywall, plastics, and coatings have large surface areas for molecular adsorption and/or large volumes for absorption. Manuja et al. (2019) reported that painted walls and coated wood comprised the majority of exposed surface area in the 22 rooms they studied, in line with the findings of an earlier report (Hodgson et al., 2005). The capacity of some permeable materials relevant to the indoor environment to take up and release chemicals has been measured for VOCs (Algrim et al., 2020; Meininghaus and Uhde, 2002; Won et al., 2001) and SVOCs (Eftekhari and Morrison, 2018; Morrison et al., 2015a; Saini et al., 2016; Wei et al., 2018).

A notable recent advance has been in the application of real-time methods to observe the dynamics of VOC absorption into paint. Algrim et al. (2020) quantified partition and diffusion coefficients for carboxylic acids and related these parameters to volatility. Their framing of the results mirrors the way atmospheric chemists now parameterize partitioning into aerosols. These parameters can then be used in models to predict the time-dependent air concentrations of VOCs as influenced by their interactions with painted surfaces.

In-vitro evidence that clothing influences dermal uptake of indoor SVOCs (Bekö et al., 2018; Morrison et al., 2016, 2017a) has resulted in increased interest in partitioning to textiles (Licina et al., 2019). Because textiles can act as a sink, clothing can help reduce dermal uptake of chemicals at least until the materials become saturated; at that time, they become a net source and may even enhance dermal uptake of some chemicals (Cao et al., 2016; Morrison et al., 2016, 2017b). Measurements of partitioning of flame retardants (Saini et al., 2016; Venier et al., 2016), plasticizers (Cao et al., 2016; Morrison et al., 2015b, 2017b; Saini et al., 2016), and other chemicals (Morrison et al., 2015a) demonstrate that the textile reservoir is large. Textiles act not only as strong chemical sinks when first introduced to homes but also as long-term sources of exposure to those same chemicals. When equilibrated with a typical home, a single pair of cotton jeans has been demonstrated to absorb more phthalates, flame retardants, and other species than are present in the entire air volume of a home (Cao et al., 2016; Morrison et al., 2015a; Saini et al., 2016).

Airborne Particles

SVOCs of indoor origin have long been observed to partition to indoor particles (Weschler, 1980). Advanced measurements of aerosol composition have recently revealed a wealth of time-averaged and real-time compositional information. Several groups have deployed aerosol mass spectrometers, thermal desorption aerosol gas chromatography (TAG), and extractive electro-spray ionization mass spectrometry systems to collect and generate detailed, real-time (or roughly hourly depending on the system), compositional information for aerosol particles (Avery et al., 2019; Brown et al., 2021; DeCarlo et al., 2018; Fortenberry et al., 2019; Lunderberg et al., 2020). Consistently across these studies, indoor sourced SVOCs are observed partitioning to particles that originate outdoors or are generated indoors during cooking or other activities. These include phthalates, siloxanes, paint additives, and components of third-hand smoke. Also evident is that partitioning to aerosol particles shifts the equilibrium between air and building surfaces, causing an increase in emission rates of some compounds.

The total volume and mass of airborne particles in buildings is generally quite small relative to the surrounding air. One cubic meter of air weighs approximately 1.2 kg, within which there is usually less than 0.1 mm^3 of particles present (or less than 100 μg). Despite the relatively small volume of particles, partitioning ensures that some SVOCs can be present at a greater mass in particles than in air (Salthammer and Goss, 2019). Inhalation of airborne particles can therefore dominate total inhalation exposures despite making up a small fraction of inhaled mass or volume. However, this is influenced by particle concentration and chemical partition coefficients, which are themselves strongly influenced by environmental conditions (Zhou et al., 2021). Over a wide range of volatilities, most SVOCs are expected to equilibrate with particles in 1 hour or less (Liu et al., 2014).

Particle Deposition to Surfaces

Besides the partitioning of discrete chemical species, particle deposition also plays a role in the transfer of chemical mass from air to surfaces in the indoor environment. For example, the deposition of aerosols can contribute to organic film formation or mineral dust deposition containing reactive metals from indoor-outdoor air exchange. Deposited salts can contribute to material corrosion (Sinclair et al., 1990). Deposition also acts to reduce inhalation exposure; the concentration of particulate matter of outdoor origin is generally lower indoors than outdoors (Chen and Zhao, 2011). Particle deposition rates depend on particle size and density as well as indoor air mixing (Lai and Nazaroff, 2000; Nazaroff and Cass, 1987; Thatcher et al., 2002). The orientation of surfaces is an important factor in the total mass deposited (Nazaroff and Cass, 1987). Horizontal, upward-facing surfaces accumulate more particles than vertical surfaces, which contribute to organic films on surfaces (Deming and Ziemann, 2020). Walking indoors can resuspend some deposited mass, transferring it back into the air (Boor et al., 2013; Wang et al., 2020a).

Advanced methods are now being applied to probe the physical and chemical characteristics of single deposited particles to indoor surfaces. Microspectrochemical techniques that combine imaging and measurements of chemical speciation can improve our understanding of particle deposition. Or et al. (2018) revealed both organic particles and films on the surface of glass slides placed vertically in a kitchen for 6 months. Glass slides in other indoor environments, including a garage, a copier room, and an office, also showed particle deposition. The greatest number of particles were less than 500 nm in diameter, with the copier room having the greatest number of particles deposited on the glass surface after 6 months. Results from the House Observations of Microbial and Environmental Chemistry (HOMEChem) study showed that glass surfaces exposed to different kitchen activities again resulted in particle-covered surfaces, many of which were <100 nm in size, but that these activities led to different particle coverages (Or et al., 2020). Most interesting is that physical deposition models (Lai and Nazaroff, 2000) underestimated particle deposition rates by an order of magnitude during the stir-fry experiments, suggesting that not all processes or conditions are being captured by the models. This model-measurement discrepancy is consistent with real-time particle measurements taken during HOMEChem that showed that traditional models underestimate deposition rates. Boedicker et al. (2021) used these measurements to propose using an alternate model for particle deposition to better account for interception processes to irregular collection surfaces. Physical measurements of individual particles (e.g., viscosity, phase, shape, and charge) contribute to our understanding of film formation, formation of hydrophobic or hydrophilic "islands," and resuspension. This highlights the importance of probing the physicochemical properties and interactions between particles and common indoor surfaces that influence deposition in more detail.

Dust

Deposited particles eventually accumulate, collectively forming dust. Because dust can be a major source of exposure to chemicals, especially for small children, it has been studied extensively. As discussed in Chapter 2, thousands of chemicals, including flame retardants, pesticides, plasticizers, and chemical stabilizers, have been quantified in the dust of homes, offices, schools, and daycare centers (Lucattini et al., 2018). As such, this report includes dust as a distinct and significant reservoir for indoor chemicals.

Weschler and Nazaroff (2010) evaluated paired air and dust concentration measurements reported in the existing literature. They observed that the concentration in dust could be predicted reasonably well using the octanol-air partition coefficient, K_{oa}. Deviations from such predictions may be due to temperature effects, kinetic limitations in achieving equilibrium with air, uncertainties in the values of thermodynamic parameters (like K_{oa}), and the presence of particles of abraded materials (e.g., plastics) that contribute very large quantities of contaminants to dust. More recent studies confirm that many chemicals partition from air to dust in a generally predictable manner (Melymuk et al., 2016; Watkins et al., 2013), with the notable exception of high molecular weight species that take much longer to equilibrate (Melymuk et al., 2016; Parnis et al., 2020; Weschler and Nazaroff, 2010). Dust that deposits on a surface can more rapidly absorb a contaminant from that surface than from the air if that surface is a primary contaminant source (Rauert and Harrad, 2015; Schripp et al., 2010). This can overcome some mass-transfer limitations for equilibration but also can result in a higher concentration in the dust than would be expected based on the air concentration alone.

Condensed Water

Water is present in homes in many forms, including bulk water, water adsorbed to surfaces as thin films (see Chapter 4), perhaps leading to visible condensation, and water absorbed into materials. Because of the high surface-area-to-volume ratio of indoor spaces, the amount of condensed water available for partitioning is orders of magnitude higher than outdoors (Duncan et al., 2019). The amount of water available for partitioning per volume of air in an indoor space has been estimated to range from less than 10^{-6} l/m^3 (for water adsorbed to indoor surfaces) up to 0.1 l/m^3 (for bulk water present in a toilet) (Nazaroff and Weschler, 2020).

Water-soluble organic gases partition predictably into condensed water, but partitioning can be influenced by the pH of the water (Nazaroff and Weschler, 2020). The pH, in turn, is influenced by the presence of salts and absorbed gases such as ammonia, carbon dioxide, and nitric acid. People are an important indoor source of carbon dioxide, ammonia (Li et al., 2020), and organic acids; therefore, the presence of occupants likely influences partitioning of acids and bases to indoor materials and condensed water (Nazaroff and Weschler, 2020).

While the research community is far from understanding the full extent to which water influences indoor air concentrations of contaminants, some recent studies are suggestive. Duncan et al. (2019) observed that the air concentration of formic and acetic acids dropped when the air conditioner was operating in a test home. The concentrations rose again when the air conditioner was off. They interpreted this behavior as partitioning of these acids into the air-conditioning condensate that forms on cooling coils. Similar behavior has been observed for ammonia (Ampollini et al., 2019) and nitrous acid (Wang et al., 2020b) in the HOMEChem study. The HOMEChem study also revealed that the air concentration of acids rose rapidly upon closing windows during an airing-out perturbation experiment. This evidence of very large, highly accessible reservoirs of acids suggested that water or other polar materials on surfaces play an important role in storing acids (Collins et al., 2018a; Wang et al., 2020b).

Wang et al. (2020b) described partitioning among air, organic, and aqueous phases by relating the equilibrium proportion of a chemical distributed among phases as a function of the octanol-air partition coefficient and the water-air partition coefficient (or Henry's law coefficient). They showed how pH of an aqueous phase controls the phase distribution of organic acids, with increasing pH driving acids from air or the organic phase into the aqueous phase. Experiments conducted at HOMEChem demonstrated that moderately strong acids (e.g., formic acid and HONO) present in indoor surfaces could be re-partitioned to the gas phase via acidification of the aqueous surface reservoirs, via application of acetic acid in a vinegar cleaner (Wang et al., 2020b).

Microorganisms and biofilms present on surfaces can be treated as a condensed phase composed primarily of water with a small amount of organic matter representing microorganisms. As an approximation, chemicals will partition based on the octanol-air partition and Henry's law coefficients, proportional to the relative amount of organic matter and water. Microbial metabolism could shift conditions away from equilibrium, but the extent depends on relative rates of metabolism and chemical transport to and away from the microorganism.

Occupants

Occupants are an important source of carbon dioxide (CO_2), ammonia (NH_3), and numerous VOCs, as discussed in Chapter 2, and can act as sinks for chemicals that influence indoor concentrations and lead to chemical exposure. They also transfer skin and skin oils to surfaces by direct contact and desquamation. Unlike most other indoor reservoirs, occupants come and go, resulting in daily dynamics in source/sink behavior. The thin film of lipids (i.e., skin oils) found on the surface of skin can act much like cooking oils in their partitioning behavior. Weschler and Nazaroff (2012) compared measurements of SVOCs in skin lipids from the Children's Total Exposure to Persistent Pesticides and Other Persistent Organic Pollutants study to values predicted based on air concentrations and octanol-air partition coefficients. They found good agreement, implying that SVOCs in skin lipids were roughly equilibrated with indoor air, allowing partitioning to be predicted from standard thermodynamic parameters. They combined this finding with existing models of transdermal uptake of chemicals to show that air-to-skin partitioning and net uptake could be an important component of chemical exposure in buildings. Experiments with human subjects have now verified this exposure pathway for phthalates (Morrison et al., 2016; Weschler et al., 2015) and nicotine (Bekö et al., 2018). Partitioning of ionizable compounds, like nicotine and methamphetamine, to skin lipids has not been well characterized and may be strongly influenced by the water content, acid concentration, and pH (Bekö et al., 2017; Morrison et al., 2015a; Parker and Morrison, 2016). Given the large capacity for absorbing SVOCs, a recent publication suggests that skin lipids on the hands can act as important agents of chemical transport among indoor surfaces (Diamond et al., 2021). Partitioning to clothing and subsequent exposure is also important, as discussed earlier in this chapter. Box 3-2 discusses third-hand smoke as one example of the complexity of partitioning.

BOX 3-2
Third-Hand Smoke: An Example of the Complexity of Partitioning

Partitioning of contaminants among reservoirs in indoor environments can be complex, dynamic, and interdependent. This is especially true for organic contaminants that partition to organic materials but are polar and even ionizable, which means they also have an affinity for water reservoirs. Countless nitrogen-containing compounds associated with cigarette smoke fall into this category. Cigarette smoke partitioning to surfaces has been of interest for many years, because the presence of residual nicotine has been used as a marker for childhood exposure to cigarette smoke (Van Loy et al., 1998). Over time, smoke residues can also be altered chemically (third-hand smoke or THS) to produce even more hazardous chemicals, such as nitrosamines (Sleiman et al., 2010).

Recent studies have shown how partitioning among indoor reservoirs influences occupant exposure to THS, even in nonsmoking environments. DeCarlo et al. (2018) observed that 29 percent of indoor sub-micron aerosol mass was composed of reduced nitrogen compounds associated with THS. They suggest that enhanced partitioning into an acidic aqueous component of the aerosol explains such high proportions in aerosols. They also observed strong seasonal differences that are explained by aqueous partitioning; drying of aerosols by hot and dry heating, ventilation, and air-conditioning (HVAC) air during the winter; and deliquescence of aerosols in cool, wet HVAC air in summer. This evidence also highlights the possibility that drying and deliquescence of salts in particles (in air or deposited on surfaces) could be of broad importance for partitioning and chemistry (Collins et al., 2018b). By measuring volatile markers of THS in a movie theater in real time, Sheu et al. (2020) showed that the major source of THS compounds was off-gassing from the skin, clothing, and other surfaces belonging to moviegoers. Like DeCarlo et al. (2018), they also observed that nicotine and related compounds partitioned from people to aerosols, contributing significantly to aerosol mass. While partitioning to the aqueous phase is important for aerosols, it may not be important for other surfaces and materials. For example, Bekö et al. (2017, 2018) observed dermal uptake of nicotine from air for the first time and concluded that partitioning to the primarily organic component of skin lipids was more important than aqueous partitioning to skin surface reservoirs. If present in an aqueous skin reservoir at a pH of 5 to 6, most of the nicotine would be protonated and slow to migrate through the skin. As pH can be important for partitioning of THS compounds, the presence of ammonia and CO_2 can alter partitioning equilibrium (Nazaroff and Weschler, 2020; Ongwandee and Sawanyapanich, 2012).

MODELS FOR PARTITIONING BEHAVIOR

Although partitioning occurs via both adsorption and absorption mechanisms to many indoor reservoirs with a vast array of chemical and morphological complexity, models of human exposure to indoor chemicals typically rely on simplified representations of these interactions that use partition coefficients. In particular, such exposure models require predictions of the distribution of the chemicals across indoor spaces in order to assess different exposure mechanisms. For example, a recent model describes how K_{oa} and K_{wa} are used to assess the degree of inhalation, dermal, and dietary and non-dietary ingestion exposures that arise for many organic chemicals in the indoor environment as a function of the fundamental properties of the molecules (Li et al., 2019). See Chapter 6 for a more detailed discussion of exposure assessment and modeling.

Advanced techniques to estimate partition coefficients are an active area of research. This is especially important given the tens of thousands of commercial chemicals released indoors, because it is not realistic to experimentally measure partitioning parameters for each of these chemicals to each reservoir. These methods include advanced polyparameter linear free energy relationships for different materials (Endo and Goss, 2014) and quantum chemical methods that have been applied to thousands of chemicals (Wang et al., 2017). Furthermore, at the individual molecule level, a highly detailed understanding is emerging of how important indoor chemicals (e.g., terpenes and some oxygenated VOCs) interact with select surface components (e.g., silica) through coupled experimental-theoretical approaches (Fang et al., 2019a,b; Huang et al., 2021). It will be valuable to compare the performance of estimated parameters currently used in exposure models with results from these detailed studies. Because of the implications for indoor exposure and health, partitioning models for surfactant species, such as PFAS, need further development.

CONCLUSIONS

Chemical contaminants move from one reservoir to another with a net tendency toward approaching, but not necessarily always attaining, an equilibrium condition. The nature of the equilibrium condition may change with time as environmental parameters such as temperature and relative humidity evolve. This partitioning process distributes chemicals from their initial sources throughout indoor spaces, to air, building materials, furnishings, dust, and so forth. These reservoirs buffer the air concentrations of chemicals, reducing the short-term effectiveness of controls by ventilation or filtration, compared to long-term effectiveness. Because the indoor partitioning capacity is so large, many molecules that are considered to be entirely volatile in the outdoor environment exhibit SVOC behavior indoors (Wang et al., 2020b). Partitioning of indoor chemicals to aerosols increases inhalation exposure, while partitioning to dust and surfaces increases ingestion exposure, especially for toddlers. Partitioning can also result in chemicals moving to locations and conditions favorable for chemical transformations.

Despite a rapidly growing base of knowledge about indoor partitioning, important gaps remain. The materials that are present in buildings, or comprise buildings, are not physically or chemically well characterized. Partition coefficients have been measured only for very few pairs of chemical contaminants and materials. Models to predict thermodynamic parameters exist, but their application to real indoor materials has not been widely demonstrated. Furthermore, models have not yet successfully been applied to many chemical classes important in indoor environments, such as surfactants. The extent to which environmental and other building factors, occupant activities, and control systems influence partitioning and exposure remain to be explored. Acting on the following recommendations could help bridge these gaps.

RESEARCH NEEDS

Given its findings about the current state of the science, the committee has identified priority research areas to help drive future advances in chemical partitioning relevant to indoor environments:

- **Expand equilibrium and nonequilibrium (dynamic and steady-state) partitioning studies to include a larger variety of materials present in buildings.** Detailed studies of partitioning of SVOCs to both impermeable and permeable materials, such as glass, clothing, and paint, have illustrated the sophistication of currently available laboratory techniques that allow for measurement of both uptake kinetics and equilibrium partitioning to representative indoor materials. These studies could be extended to additional indoor materials—wood, concrete, home furnishings, carpet, clothing, etc.—as well as those found in interstitial spaces to develop a more comprehensive understanding of the capacity and chemical affinities of different surface reservoirs for a wide range of sorbing chemicals. This is important to determine lifetime exposure levels and the overall mass flow of chemicals through the built environment. Better understanding of diffusion timescales through materials would inform whether their full volume is available for partitioning and the concentration gradients that may exist with them.
- **Examine the influence of environmental conditions and occupant activities on equilibrium and nonequilibrium partitioning and the influence of partitioning on contamination management.** The effects of seasonal and daily temperature changes and relative humidity are important, including studies of extreme conditions, for example, associated with

air-conditioning systems. Regional differences in building types and building climate control are important to assess, given their connections to total contaminant exposure. Building occupants can add, remove, or modify reservoirs and change conditions that influence partitioning phenomena. Management of indoor environments does not typically account for partitioning phenomena but may improve with such consideration. For example, the effectiveness of ventilation strategies could improve if their timing accounted for partitioning dynamics.

- **Develop a molecular-level understanding of partitioning among indoor reservoirs.** This may arise from use of advanced physical chemistry techniques and models. In particular, soiled surfaces, airborne particles, dust, and surface coatings are important. Laboratory-scale partitioning experiments could be complemented by experiments in genuine indoor experiments, such as when chemicals are added in a controlled manner to an indoor space or as mixtures of chemicals to investigate synergistic effects. Spectrochemical analysis, imaging, and depth profiling could provide important insights into molecular interactions and spatial gradients in indoor materials.

- **Improve predictive models of equilibrium and nonequilibrium partitioning and compare them with observations from laboratory experiments and real-world, occupied buildings.** Multimedia partitioning models that assess human exposure typically use a limited number of partitioning volumes, such as air, polar surface reservoirs (e.g., water), and mildly polar reservoirs (e.g., *n*-octanol). It is important to establish how well this simplified approach represents the full chemical complexity of the system, when assessed against the additional uncertainties associated with source fluxes, human behavior and diversity, and reactive loss rates. Advanced computational methods, including quantum chemical methods and molecular dynamic simulations, can be used to determine partitioning coefficients. Relevant model systems may play a role in providing important insights and a more detailed understanding of the more complex home environment. These studies can also probe temperature, relative humidity, and other important factors that impact chemical partitioning.

- **Identify key species, materials, and partitioning phenomena that strongly influence exposure.** Given that most chemicals are predominantly partitioned to indoor surfaces, it is important for exposure modeling to consider carefully the distribution of chemicals among these surface reservoirs, the gas phase, and aerosol particles. Compositional changes occurring due to partitioning to airborne particles remains an important research topic. For SVOCs, the gas phase acts as a conduit between large surface reservoirs and aerosol particles, which can be inhaled and increase exposure. While this phenomenon has been illustrated by third-hand tobacco smoke, the degree and timescales for other contaminants residing in surface reservoirs to partition to aerosol particles need to be determined. Particles and other contaminants are complex and have many sources, and indoor environments themselves are diverse; the very few predictive models that are now including such complexity have not been well vetted for real environments and will continue to need more comprehensive primary input data (e.g., source terms), a better fundamental understanding of partitioning, and evaluation in real indoor environments.

REFERENCES

Abbatt, J. P. D., and C. Wang. 2020. The atmospheric chemistry of indoor environments. *Environmental Science: Processes & Impacts* 22(1): 25–48. https://doi.org/10.1039/c9em00386j.

Algrim, L. B., D. Pagonis, J. A. de Gouw, J. L. Jimenez, and P. J. Ziemann. 2020. Measurements and modeling of absorptive partitioning of volatile organic compounds to painted surfaces. *Indoor Air* 30(4): 745–756. https://doi.org/10.1111/ina.12654.

Ampollini, L., E. F. Katz, S. Bourne, Y. Tian, A. Novoselac, A. H. Goldstein, G. Lucic, M. S. Waring, and P. F. DeCarlo. 2019. Observations and contributions of real-time indoor ammonia concentrations during HOMEChem. *Environmental Science & Technology* 53(15): 8591–8598. https://doi.org/10.1021/acs.est.9b02157.

Ault, A. P., V. H. Grassian, N. Carslaw, D. B. Collins, H. Destaillats, D. J. Donaldson, D. K. Farmer, J. L. Jimenez, V. F. McNeill, G. C. Morrison, R. E. O'Brien, M. Shiraiwa, M. E. Vance, J. R. Wells, and W. Xiong. 2020. Indoor surface chemistry: Developing a molecular picture of reactions on indoor interfaces. *Chem* 6(12): 3203–3218. https://doi. org/10.1016/j.chempr.2020.08.023.

Avery, A. M., M. S. Waring, and P. F. DeCarlo. 2019. Human occupant contribution to secondary aerosol mass in the indoor environment. *Environmental Science: Processes & Impacts* 21(8): 1301–1312. https://doi.org/10.1039/c9em00097f.

Bekö, G., G. Morrison, C. J. Weschler, H. M. Koch, C. Palmke, T. Salthammer, T. Schripp, A. Eftekhari, J. Toftum, and G. Clausen. 2018. Dermal uptake of nicotine from air and clothing: Experimental verification. *Indoor Air* 28(2): 247–257. https://doi.org/10.1111/ina.12437.

Bekö, G., G. Morrison, C. J. Weschler, H. M. Koch, C. Palmke, T. Salthammer, T. Schripp, J. Toftum, and G. Clausen. 2017. Measurements of dermal uptake of nicotine directly from air and clothing. *Indoor Air* 27(2): 427–433. https:// doi.org/10.1111/ina.12327.

Bennett, D. H., R. E. Moran, X. M. Wu, N. S. Tulve, M. S. Clifton, M. Colon, W. Weathers, A. Sjodin, R. Jones, and I. Hertz-Picciotto. 2015. Polybrominated diphenyl ether (PBDE) concentrations and resulting exposure in homes in California: Relationships among passive air, surface wipe and dust concentrations, and temporal variability. *Indoor Air* 25(2): 220–229. https://doi.org/10.1111/ina.12130.

Boedicker, E. K., E. W. Emerson, G. R. McMeeking, S. Patel, M. E. Vance, and D. K. Farmer. 2021. Fates and spatial variations of accumulation mode particles in a multi-zone indoor environment during the HOMEChem campaign. *Environmental Science: Processes & Impacts* 23(7): 1029–1039. https://doi.org/10.1039/d1em00087j.

Boor, B. E., J. A. Siegel, and A. Novoselac. 2013. Monolayer and multilayer particle deposits on hard surfaces: Literature review and implications for particle resuspension in the indoor environment. *Aerosol Science and Technology* 47(8): 831–847. https://doi.org/10.1080/02786826.2013.794928.

Brown, W. L., D. A. Day, H. Stark, D. Pagonis, J. E. Krechmer, X. Liu, D. J. Price, E. F. Katz, P. F. DeCarlo, C. G. Masoud, D. S. Wang, L. Hildebrandt Ruiz, C. Arata, D. M. Lunderberg, A. H. Goldstein, D. K. Farmer, M. E. Vance, and J. L. Jimenez. 2021. Real-time organic aerosol chemical speciation in the indoor environment using extractive electrospray ionization mass spectrometry. *Indoor Air* 31(1): 141–155. https://doi.org/10.1111/ina.12721.

Butt, C. M., M. L. Diamond, J. Truong, M. G. Ikonomou, P. A. Helm, and G. A. Stern. 2004a. Semivolatile organic compounds in window films from lower Manhattan after the September 11th World Trade Center attacks. *Environmental Science & Technology* 38(13): 3514–3524. https://doi.org/10.1021/es0498282.

Butt, C. M., M. L. Diamond, J. Truong, M. G. Ikonomou, and A. F. ter Schure. 2004b. Spatial distribution of polybrominated diphenyl ethers in southern Ontario as measured in indoor and outdoor window organic films. *Environmental Science & Technology* 38(3): 724–731. https://doi.org/10.1021/es034670r.

Cao, J., C. J. Weschler, J. Luo, and Y. Zhang. 2016. C_m-history method, a novel approach to simultaneously measure source and sink parameters important for estimating indoor exposures to phthalates. *Environmental Science & Technology* 50(2): 825–834. https://doi.org/10.1021/acs.est.5b04404.

Chase, H. M., J. Ho, M. A. Upshur, R. J. Thomson, V. S. Batista, and F. M. Geiger. 2017. Unanticipated stickiness of alpha-Pinene. *Journal of Physical Chemistry A* 121(17): 3239–3246. https://doi.org/10.1021/acs.jpca.6b12653.

Chen, C., and B. Zhao. 2011. Review of relationship between indoor and outdoor particles: I/O ratio, infiltration factor and penetration factor. *Atmospheric Environment* 45(2): 275–288. https://doi.org/10.1016/j.atmosenv.2010.09.048.

Collins, D. B., R. F. Hems, S. Zhou, C. Wang, E. Grignon, M. Alavy, J. A. Siegel, and J. P. D. Abbatt. 2018a. Evidence for gas-surface equilibrium control of indoor nitrous acid. *Environmental Science & Technology* 52(21): 12419–12427. https://doi.org/10.1021/acs.est.8b04512.

Collins, D. B., C. Wang, and J. P. D. Abbatt. 2018b. Selective uptake of third-hand tobacco smoke components to inorganic and organic aerosol particles. *Environmental Science & Technology* 52(22): 13195–13201. https://doi.org/10.1021/ acs.est.8b03880.

DeCarlo, P. F., A. M. Avery, and M. S. Waring. 2018. Thirdhand smoke uptake to aerosol particles in the indoor environment. *Science Advances* 4(5): eaap8368. https://doi.org/10.1126/sciadv.aap8368.

Deming, B. L., and P. J. Ziemann. 2020. Quantification of alkenes on indoor surfaces and implications for chemical sources and sinks. *Indoor Air* 30(5): 914–924. https://doi.org/10.1111/ina.12662.

Diamond, M. L., J. O. Okeme, and L. Melymuk. 2021. Hands as agents of chemical transport in the indoor environment. *Environmental Science & Technology Letters* 8(4): 326–332. https://doi.org/10.1021/acs.estlett.0c01006.

Duigu, J. R., G. A. Ayoko, and S. Kokot. 2009. The relationship between building characteristics and the chemical composition of surface films found on glass windows in Brisbane, Australia. *Building and Environment* 44(11): 2228–2235. https://doi.org/10.1016/j.buildenv.2009.02.019.

Duncan, S. M., S. Tomaz, G. Morrison, M. Webb, J. Atkin, J. D. Surratt, and B. J. Turpin. 2019. Dynamics of residential water-soluble organic gases: Insights into sources and sinks. *Environmental Science & Technology* 53(4): 1812–1821. https://doi.org/10.1021/acs.est.8b05852.

Eftekhari, A., and G. C. Morrison. 2018. A high throughput method for measuring cloth-air equilibrium distribution ratios for SVOCs present in indoor environments. *Talanta* 183:250–257. https://doi.org/10.1016/j.talanta.2018.02.061.

Eichler, C. M. A., J. Cao, G. Isaacman-VanWertz, and J. C. Little. 2019a. Modeling the formation and growth of organic films on indoor surfaces. *Indoor Air* 29(1): 17–29. https://doi.org/10.1111/ina.12518.

Eichler, C. M. A., E. A. Cohen Hubal, and J. C. Little. 2019b. Assessing human exposure to chemicals in materials, products and articles: The international risk management landscape for phthalates. *Environmental Science & Technology* 53(23): 13583–13597. https://doi.org/10.1021/acs.est.9b03794.

Endo, S., and K. U. Goss. 2014. Applications of polyparameter linear free energy relationships in environmental chemistry. *Environmental Science & Technology* 48(21): 12477–12491. https://doi.org/10.1021/es503369t.

Fang, Y., P. S. J. Lakey, S. Riahi, A. T. McDonald, M. Shrestha, D. J. Tobias, M. Shiraiwa, and V. H. Grassian. 2019a. A molecular picture of surface interactions of organic compounds on prevalent indoor surfaces: Limonene adsorption on SiO$_2$. *Chemical Science* 10(10): 2906–2914. https://doi.org/10.1039/c8sc05560b.

Fang, Y., S. Riahi, A. T. McDonald, M. Shrestha, D. J. Tobias, and V. H. Grassian. 2019b. What is the driving force behind the adsorption of hydrophilic surfaces? *Journal of Physical Chemistry Letters* 10(3): 468–473. https://doi.org/10.1021/acs.jpclett.8b03484.

Fortenberry, C., M. Walker, A. Dang, A. Loka, G. Date, K. Cysneiros de Carvalho, G. Morrison, and B. Williams. 2019. Analysis of indoor particles and gases and their evolution with natural ventilation. *Indoor Air* 29(5): 761–779. https://doi.org/10.1111/ina.12584.

Frank, E. S., H. Fan, M. Shrestha, S. Riahi, D. J. Tobias, and V. H. Grassian. 2020. Impact of adsorbed water on the interaction of limonene with hydroxylated SiO$_2$: Implications of π-hydrogen bonding for surfaces in humid environments. *Journal of Physical Chemistry A* 124(50): 10592–10599. https://doi.org/10.1021/acs.jpca.0c08600.

Gewurtz, S. B., S. P. Bhavsar, P. W. Crozier, M. L. Diamond, P. A. Helm, C. H. Marvin, and E. J. Reiner. 2009. Perfluoroalkyl contaminants in window film: Indoor/outdoor, urban/rural, and winter/summer contamination and assessment of carpet as a possible source. *Environmental Science & Technology* 43(19): 7317–7323. https://doi.org/10.1021/es9002718.

Ham, J. E., and J. R. Wells. 2009. Surface chemistry of dihydromyrcenol (2,6-dimethyl-7-octen-2-ol) with ozone on silanized glass, glass, and vinyl flooring tiles. *Atmospheric Environment* 43(26): 4023–4032. https://doi.org/10.1016/j.atmosenv.2009.05.007.

Ho, J., B. T. Psciuk, H. M. Chase, B. Rudshteyn, M. A. Upshur, L. Fu, R. J. Thomson, H.-F. Wang, F. M. Geiger, and V. S. Batista. 2016. Sum frequency generation spectroscopy and molecular dynamics simulations reveal a rotationally fluid adsorption state of α-pinene on silica. *The Journal of Physical Chemistry C* 120(23): 12578–12589. https://doi.org/10.1021/acs.jpcc.6b03158.

Hodgson, A. T., K. Y. Ming, and B. C. Singer. 2005. *Quantifying Object and Material Surface Areas in Residences*. Lawrence Berkeley National Laboratory Technical Report LBNL-56786. https://doi.org/10.2172/861239.

Huang, L., E. S. Frank, M. Shrestha, S. Riahi, D. J. Tobias, and V. H. Grassian. 2021. Heterogeneous interactions of prevalent indoor oxygenated organic compounds on hydroxylated SiO$_2$ surfaces. *Environmental Science & Technology* 55(10): 6623–6630. https://doi.org/10.1021/acs.est.1c00067.

Huo, C. Y., L. Y. Liu, Z. F. Zhang, W. L. Ma, W. W. Song, H. L. Li, W. L. Li, K. Kannan, Y. K. Wu, Y. M. Han, Z. X. Peng, and Y. F. Li. 2016. Phthalate esters in indoor window films in a Northeastern Chinese urban center: Film growth and implications for human exposure. *Environmental Science & Technology* 50(14): 7743–7751. https://doi.org/10.1021/acs.est.5b06371.

Kristensen, K., D. M. Lunderberg, Y. Liu, P. K. Misztal, Y. Tian, C. Arata, W. W. Nazaroff, and A. H. Goldstein. 2019. Sources and dynamics of semivolatile organic compounds in a single-family residence in northern California. *Indoor Air* 29(4): 645–655. https://doi.org/10.1111/ina.12561.

Lai, A. C., and W. W. Nazaroff. 2000. Modeling indoor particle deposition from turbulent flow onto smooth surfaces. *Journal of Aerosol Science* 31(4): 463–476. https://doi.org/10.1016/S0021-8502(99)00536-4.

Lakey, P. S. J., C. M. A. Eichler, C. Wang, J. C. Little, and M. Shiraiwa. 2021. Kinetic multi-layer model of film formation, growth, and chemistry (KM-FILM): Boundary layer processes, multi-layer adsorption, bulk diffusion, and heterogeneous reactions. *Indoor Air* 31(6): 2070–2083. https://doi.org/10.1111/ina.12854.

Li, L., J. A. Arnot, and F. Wania. 2019. How are humans exposed to organic chemicals released to indoor air? *Environmental Science & Technology* 53(19):11276–11284. https://doi.org/10.1021/acs.est.9b02036.

Li, M., C. J. Weschler, G. Beko, P. Wargocki, G. Lucic, and J. Williams. 2020. Human ammonia emission rates under various indoor environmental conditions. *Environmental Science & Technology* 54(9): 5419–5428. https://doi.org/10.1021/acs.est.0c00094.

Liang, Y., and Y. Xu. 2014. Emission of phthalates and phthalate alternatives from vinyl flooring and crib mattress covers: The influence of temperature. *Environmental Science & Technology* 48(24): 14228–14237. https://doi.org/10.1021/es504801x.

Licina, D., G. C. Morrison, G. Beko, C. J. Weschler, and W. W. Nazaroff. 2019. Clothing-mediated exposures to chemicals and particles. *Environmental Science & Technology* 53(10): 5559–5575. https://doi.org/10.1021/acs.est.9b00272.

Lim, C. Y., and J. P. Abbatt. 2020. Chemical composition, spatial homogeneity, and growth of indoor surface films. *Environmental Science & Technology* 54(22): 14372–14379. https://doi.org/10.1021/acs.est.0c04163.

Liu, C., Y. Zhang, and C. J. Weschler. 2014. The impact of mass transfer limitations on size distributions of particle associated SVOCs in outdoor and indoor environments. *Science of the Total Environment* 497–498:401–411. https://doi.org/10.1016/j.scitotenv.2014.07.095.

Liu, Q. T., R. Chen, B. E. McCarry, M. L. Diamond, and B. Bahavar. 2003. Characterization of polar organic compounds in the organic film on indoor and outdoor glass windows. *Environmental Science & Technology* 37(11): 2340–2349. https://doi.org/10.1021/es020848i.

Liu, Y., A. G. Bé, V. W. Or, M. R. Alves, V. H. Grassian, and F. M. Geiger. 2020. Challenges and opportunities in molecular-level indoor surface chemistry and physics. *Cell Reports Physical Science* 1:100256. https://doi.org/10.1016/j.xcrp.2020.100256.

Liu, Y., H. M. Chase, and F. M. Geiger. 2019. Partially (resp. fully) reversible adsorption of monoterpenes (resp. alkanes and cycloalkanes) to fused silica. *Journal of Chemical Physics* 150(7): 074701. https://doi.org/10.1063/1.5083585.

Lucattini, L., G. Poma, A. Covaci, J. de Boer, M. H. Lamoree, and P. E. G. Leonards. 2018. A review of semi-volatile organic compounds (SVOCs) in the indoor environment: Occurrence in consumer products, indoor air and dust. *Chemosphere* 201:466–482. https://doi.org/10.1016/j.chemosphere.2018.02.161.

Lunderberg, D. M., K. Kristensen, Y. Tian, C. Arata, P. K. Misztal, Y. Liu, N. Kreisberg, E. F. Katz, P. F. DeCarlo, S. Patel, M. E. Vance, W. W. Nazaroff, and A. H. Goldstein. 2020. Surface emissions modulate indoor SVOC concentrations through volatility-dependent partitioning. *Environmental Science & Technology* 54(11): 6751–6760. https://doi.org/10.1021/acs.est.0c00966.

Manuja, A., J. Ritchie, K. Buch, Y. Wu, C. M. A. Eichler, J. C. Little, and L. C. Marr. 2019. Total surface area in indoor environments. *Environmental Science: Processes & Impacts* 21(8): 1384–1392. https://doi.org/10.1039/c9em00157c.

Meininghaus, R., and E. Uhde. 2002. Diffusion studies of VOC mixtures in a building material. *Indoor Air* 12(4): 215–222. https://doi.org/10.1034/j.1600-0668.2002.01131.x.

Melymuk, L., P. Bohlin-Nizzetto, S. Vojta, M. Kratka, P. Kukucka, O. Audy, P. Pribylova, and J. Klanova. 2016. Distribution of legacy and emerging semivolatile organic compounds in five indoor matrices in a residential environment. *Chemosphere* 153:179–186. https://doi.org/10.1016/j.chemosphere.2016.03.012.

Morrison, G., H. Li, S. Mishra, and M. Buechlein. 2015a. Airborne phthalate partitioning to cotton clothing. *Atmospheric Environment* 115:149–152. https://doi.org/10.1016/j.atmosenv.2015.05.051.

Morrison, G., N. V. Shakila, and K. Parker. 2015b. Accumulation of gas-phase methamphetamine on clothing, toy fabrics, and skin oil. *Indoor Air* 25(4): 405–414. https://doi.org/10.1111/ina.12159.

Morrison, G. C., G. Bekö, C. J. Weschler, T. Schripp, T. Salthammer, J. Hill, A.-M. Andersson, J. Toftum, G. Clausen, and H. Frederiksen. 2017a. Dermal uptake of benzophenone-3 from clothing. *Environmental Science & Technology* 51(19): 11371–11379. https://doi.org/10.1021/acs.est.7b02623.

Morrison, G. C., C. J. Weschler, and G. Beko. 2017b. Dermal uptake of phthalates from clothing: Comparison of model to human participant results. *Indoor Air* 27(3): 642–649. https://doi.org/10.1111/ina.12354.

Morrison, G. C., C. J. Weschler, G. Beko, H. M. Koch, T. Salthammer, T. Schripp, J. Toftum, and G. Clausen. 2016. Role of clothing in both accelerating and impeding dermal absorption of airborne SVOCs. *Journal of Exposure Science & Environmental Epidemiology* 26(1): 113–118. https://doi.org/10.1038/jes.2015.42.

Nazaroff, W. W., and G. R. Cass. 1987. Particle deposition from a natural convection flow onto a vertical isothermal flat plate. *Journal of Aerosol Science* 18(4): 445–455. https://doi.org/10.1016/0021-8502(87)90042-5.

Nazaroff, W. W., and C. J. Weschler. 2020. Indoor acids and bases. *Indoor Air* 30(4): 559–644. https://doi.org/10.1111/ina.12670.

O'Brien, R. E., Y. Li, K. J. Kiland, E. F. Katz, V. W. Or, E. Legaard, E. Q. Walhout, C. Thrasher, V. H. Grassian, P. F. DeCarlo, A. K. Bertram, and M. Shiraiwa. 2021. Emerging investigator series: Chemical and physical properties of organic mixtures on indoor surfaces during HOMEChem. *Environmental Science: Processes & Impacts* 23(4): 559–568. https://doi.org/10.1039/d1em00060h.

Ongwandee, M., and P. Sawanyapanich. 2012. Influence of relative humidity and gaseous ammonia on the nicotine sorption to indoor materials. *Indoor Air* 22(1): 54–63. https://doi.org/10.1111/j.1600-0668.2011.00737.x.

Or, V. W., M. R. Alves, M. Wade, S. Schwab, R. L. Corsi, and V. H. Grassian. 2018. Crystal clear? Microspectroscopic imaging and physicochemical characterization of indoor depositions on window glass. *Environmental Science & Technology Letters* 5(8): 514–519. https://doi.org/10.1021/acs.estlett.8b00355.

Or, V. W., M. Wade, S. Patel, M. R. Alves, D. Kim, S. Schwab, H. Przelomski, R. O'Brien, D. Rim, R. L. Corsi, M. E. Vance, D. K. Farmer, and V. H. Grassian. 2020. Glass surface evolution following gas adsorption and particle deposition from indoor cooking events as probed by microspectroscopic analysis. *Environmental Science: Processes & Impacts* 22:1698–1709. https://doi.org/10.1039/d0em00156b.

Pan, S. H., J. Li, T. Lin, G. Zhang, X. D. Li, and H. Yin. 2012. Polycyclic aromatic hydrocarbons on indoor/outdoor glass window surfaces in Guangzhou and Hong Kong, south China. *Environmental Pollution* 169:190–195. https://doi.org/10.1016/j.envpol.2012.03.015.

Parker, K., and G. Morrison. 2016. Methamphetamine absorption by skin lipids: accumulated mass, partition coefficients, and the influence of fatty acids. Indoor Air 26(4):634–641. https://doi.org/10.1111/ina.12229.

Parnis, J. M., T. Taskovic, A. K. D. Celsie, and D. Mackay. 2020. Indoor dust/air partitioning: Evidence for kinetic delay in equilibration for low-volatility SVOCs. *Environmental Science & Technology* 54(11): 6723–6729. https://doi.org/10.1021/acs.est.0c00632.

Rauert, C., and S. Harrad. 2015. Mass transfer of PBDEs from plastic TV casing to indoor dust via three migration pathways—A test chamber investigation. *Science of the Total Environment* 536:568–574. https://doi.org/10.1016/j.scitotenv.2015.07.050.

Rubasinghege, G., and V. H. Grassian. 2013. Role(s) of adsorbed water in the surface chemistry of environmental interfaces. *Chemical Communications* 49(30): 3071–3094. https://doi.org/10.1039/c3cc38872g.

Saini, A., C. Rauert, M. J. Simpson, S. Harrad, and M. L. Diamond. 2016. Characterizing the sorption of polybrominated diphenyl ethers (PBDEs) to cotton and polyester fabrics under controlled conditions. *Science of the Total Environment* 563–564:99–107. https://doi.org/10.1016/j.scitotenv.2016.04.090.

Salthammer, T., and K. U. Goss. 2019. Predicting the gas/particle distribution of SVOCs in the indoor environment using poly parameter linear free energy relationships. *Environmental Science & Technology* 53(5): 2491–2499. https://doi.org/10.1021/acs.est.8b06585.

Schripp, T., C. Fauck, and T. Salthammer. 2010. Chamber studies on mass-transfer of di(2-ethylhexyl)phthalate (DEHP) and di-n-butylphthalate (DnBP) from emission sources into house dust. *Atmospheric Environment* 44(24): 2840–2845. https://doi.org/10.1016/j.atmosenv.2010.04.054.

Sheu, R., C. Stonner, J. C. Ditto, T. Klupfel, J. Williams, and D. R. Gentner. 2020. Human transport of thirdhand tobacco smoke: A prominent source of hazardous air pollutants into indoor nonsmoking environments. *Science Advances* 6(10): eaay4109. https://doi.org/10.1126/sciadv.aay4109.

Shu, S., and G. C. Morrison. 2011. Surface reaction rate and probability of ozone and alpha-terpineol on glass, polyvinyl chloride, and latex paint surfaces. *Environmental Science & Technology* 45(10): 4285–4292. https://doi.org/10.1021/es200194e.

Sinclair, J. D., L. A. Psota-Kelty, C. J. Weschler, and H. C. Shields. 1990. Deposition of airborne sulfate, nitrate, and chloride salts as it relates to corrosion of electronics. *Journal of the Electrochemical Society* 137:1200–1206. https://doi.org/10.1149/1.2086631.

Sleiman, M., L. A. Gundel, J. F. Pankow, P. Jacob, 3rd, B. C. Singer, and H. Destaillats. 2010. Formation of carcinogens indoors by surface-mediated reactions of nicotine with nitrous acid, leading to potential thirdhand smoke hazards. *Proceedings of the National Academy of Sciences USA* 107(15): 6576–6581. https://doi.org/10.1073/pnas.0912820107.

Stokes, G. Y., A. M. Buchbinder, J. M. Gibbs-Davis, K. A. Scheidt, and F. M. Geiger. 2008. Heterogeneous ozone oxidation reactions of 1-pentene, cyclopentene, cyclohexene, and a menthenol derivative studied by sum frequency generation. *Journal of Physical Chemistry A* 112(46): 11688–11698. https://doi.org/10.1021/jp803277s.

Stokes, G. Y., E. H. Chen, S. R. Walter, and F. M. Geiger. 2009. Two reactivity modes in the heterogeneous cyclohexene ozonolysis under tropospherically relevant ozone-rich and ozone-limited conditions. *Journal of Physical Chemistry A* 113(31): 8985–8993. https://doi.org/10.1021/jp904104s.

Thatcher, T. L., A. C. Lai, R. Moreno-Jackson, R. G. Sextro, and W. W. Nazaroff. 2002. Effects of room furnishings and air speed on particle deposition rates indoors. *Atmospheric Environment* 36(11): 1811–1819. https://doi.org/10.1016/S1352-2310(02)00157-7.

Tichenor, B., Z Guo, J. Dunn, L. Sparks, and M A. Mason. 1991. The interaction of vapour phase organic compounds with indoor sinks. *Indoor Air* 1(1): 23–35. https://doi.org/10.1111/j.1600-0668.1991.03-11.x.

Uhde, E., D. Varol, B. Mull, and T. Salthammer. 2019. Distribution of five SVOCs in a model room: Effect of vacuuming and air cleaning measures. *Environmental Science: Processes & Impacts* 21(8): 1353–1363. https://doi.org/10.1039/c9em00121b.

U.S. Census. 2021. American Housing Survey. https://www.census.gov/programs-surveys/ahs.html.

Van Loy, M. D., W. W. Nazaroff, and J. M. Daisey. 1998. Nicotine as a marker for environmental tobacco smoke: Implications of sorption on indoor surface materials. *Journal of the Air & Waste Management Association* 48(10): 959–968. https://doi.org/10.1080/10473289.1998.10463742.

Venier, M., O. Audy, S. Vojta, J. Becanova, K. Romanak, L. Melymuk, M. Kratka, P. Kukucka, J. Okeme, A. Saini, M. L. Diamond, and J. Klanova. 2016. Brominated flame retardants in the indoor environment—Comparative study of indoor contamination from three countries. *Environment International* 94:150–160. https://doi.org/10.1016/j.envint.2016.04.029.

Wallace, L. A., W. R. Ott, C. J. Weschler, and A. C. K. Lai. 2017. Desorption of SVOCs from heated surfaces in the form of ultrafine particles. *Environmental Science & Technology* 51(3): 1140–1146. https://doi.org/10.1021/acs.est.6b03248.

Wang, C., B. Bottorff, E. Reidy, C. M. F. Rosales, D. B. Collins, A. Novoselac, D. K. Farmer, M. E. Vance, P. S. Stevens, and J. P. D. Abbatt. 2020a. Cooking, bleach cleaning, and air conditioning strongly impact levels of HONO in a house. *Environmental Science & Technology* 54(21): 13488–13497. https://doi.org/10.1021/acs.est.0c05356.

Wang, C., D. B. Collins, C. Arata, A. H. Goldstein, J. M. Mattila, D. K. Farmer, L. Ampollini, P. F. DeCarlo, A. Novoselac, M. E. Vance, W. W. Nazaroff, and J. P. D. Abbatt. 2020b. Surface reservoirs dominate dynamic gas-surface partitioning of many indoor air constituents. *Science Advances* 6(8): eaay8973. https://doi.org/10.1126/sciadv.aay8973.

Wang, C., T. Yuan, S. A. Wood, K. U. Goss, J. Li, Q. Ying, and F. Wania. 2017. Uncertain Henry's law constants compromise equilibrium partitioning calculations of atmospheric oxidation products. *Atmospheric Chemistry and Physics* 17(12): 7529–7540. https://doi.org/10.5194/acp-17-7529-2017.

Watkins, D. J., M. D. McClean, A. J. Fraser, J. Weinberg, H. M. Stapleton, and T. F. Webster. 2013. Associations between PBDEs in office air, dust, and surface wipes. *Environ Int* 59:124–132. https://doi.org/10.1016/j.envint.2013.06.001.

Wei, W., C. Mandin, and O. Ramalho. 2018. Influence of indoor environmental factors on mass transfer parameters and concentrations of semi-volatile organic compounds. *Chemosphere* 195:223–235. https://doi.org/10.1016/j.chemosphere.2017.12.072.

Weschler, C. J. 1980. Characterization of selected organics in size-fractionated indoor aerosols. *Environmental Science & Technology* 14(4): 428–431. https://doi.org/10.1021/es60164a008.

Weschler, C. J., G. Beko, H. M. Koch, T. Salthammer, T. Schripp, J. Toftum, and G. Clausen. 2015. Transdermal uptake of diethyl phthalate and di(n-butyl) phthalate directly from air: Experimental verification. *Environmental Health Perspectives* 123(10): 928–934. https://doi.org/10.1289/ehp.1409151.

Weschler, C. J., and W. W. Nazaroff. 2008. Semivolatile organic compounds in indoor environments. *Atmospheric Environment* 42(40): 9018–9040. https://doi.org/10.1016/j.atmosenv.2008.09.052.

Weschler, C. J., and W. W. Nazaroff. 2010. SVOC partitioning between the gas phase and settled dust indoors. *Atmospheric Environment* 44(30): 3609–3620. https://doi.org/10.1016/j.atmosenv.2010.06.029.

Weschler, C. J., and W. W. Nazaroff. 2012. SVOC exposure indoors: Fresh look at dermal pathways. *Indoor Air* 22(5): 356–377. https://doi.org/10.1111/j.1600-0668.2012.00772.x.

Weschler, C. J., and W. W. Nazaroff. 2017. Growth of organic films on indoor surfaces. *Indoor Air* 27(6): 1101–1112. https://doi.org/10.1111/ina.12396.

Won, D., R. L. Corsi, and M. Rynes. 2001. Sorptive interactions between VOCs and indoor materials. *Indoor Air* 11(4): 246–256. https://doi.org/10.1034/j.1600-0668.2001.110406.x.

Wu, Y., C. M. Eichler, W. Leng, S. S. Cox, L. C. Marr, and J. C. Little. 2017. Adsorption of phthalates on impervious indoor surfaces. *Environmental Science & Technology* 51(5): 2907–2913. https://doi.org/10.1021/acs.est.6b05853.

Zhou, X., J. Lian, Y. Cheng, and X. Wang. 2021. The gas/particle partitioning behavior of phthalate esters in indoor environment: Effects of temperature and humidity. *Environmental Research* 194:110681. https://doi.org/10.1016/j.envres.2020.110681.

4

Chemical Transformations

Chemical transformations can be defined as chemical processes that lead to the loss or removal of certain substances (e.g., reactants) and the generation or formation of new substances (e.g., products). The products that arise from these reactions frequently have very different properties from the reactants in terms of partitioning, toxicity, etc. There are different types of chemical reactions that are relevant indoors, including photolysis, hydrolysis, acid-base reactions, and redox reactions. Some of these processes are irreversible, leading to permanent loss of species, while others are reversible, resulting in temporary loss and eventual regeneration of reactants. These chemical processes are complex and extensive, with numerous species involved as precursors, intermediates, or products.

As outlined in Chapter 3, indoor chemical compounds partition into a variety of compartments that may contain a variety of phases; hence, chemical transformations occur at different locations indoors, including the gas phase, airborne particles, and indoor surfaces, as well as hidden places such as ducts and the heating, ventilation, and air-conditioning (HVAC) system. The partitioning of semivolatile and low-volatility molecules to indoor surfaces can increase their indoor residence times. Surface-adsorbed molecules may diffuse into the bulk of indoor surfaces and materials, where they may undergo chemical transformations. The relative rates of ventilation, gas-phase loss, and loss to surfaces are important to compare when evaluating the fate of an indoor air molecule. Reactions on surfaces can be very important, even if relatively slow, if the species is partitioned strongly to the surface.

This chapter covers the chemical transformations that occur in indoor environments, starting with those in the air and followed by those that occur on surfaces, noting the different classes of important multiphase processes. It discusses the modeling of indoor environments that needs to incorporate our knowledge of chemical reactions and partitioning processes described in Chapter 3. It concludes by listing several priority research areas in indoor chemical transformations.

AIRBORNE CHEMISTRY

In the outdoor atmosphere, chemical transformations are mostly driven by photochemistry. In contrast, indoor settings are generally dark, with much less ultraviolet and overall light (levels a couple of orders of magnitude lower than outdoors), even during daylight hours (Abbatt and Wang, 2020). An important exception is direct sunlight, which can drive chemistry on window glass and other directly illuminated surfaces, as this is where the solar flux, although still diminished in intensity relative to outdoors, is greatest. Note that solar radiation at wavelengths shorter than ~330 nanometers (nm) is completely attenuated by windows, which precludes many photochemical reactions that are important outdoors, including the formation of hydroxyl (OH) radicals by ozone (O_3) photolysis (Young et al., 2019). Furthermore, some indoor photochemistry can also be promoted by specific light sources, such as some bare fluorescent lights (Kowal et al., 2017). The spatial, temporal, and spectral variability need to be taken into account when considering the role of photochemistry in indoor environments (Kowal et al., 2017; Weschler and Carslaw, 2018; Zhou et al., 2020).

The most important indoor oxidant is considered to be O_3, which is largely transported from outdoors: an indoor-to-outdoor ratio of O_3 is commonly between 0.2 and 0.7 as a function of the air exchange rate (Nazaroff and Weschler, 2022). Depending on outdoor O_3 levels, the O_3 mixing ratio indoors is typically around 1 to 30 ppb, which is sufficiently high to trigger gas-phase oxidation of unsaturated volatile organic compounds (VOCs). This represents a major source of OH radicals in dark conditions, leading to typical OH concentrations of 1 to 5×10^5 cm^{-3}. OH concentrations can be elevated as high as 10^6 to 10^7 cm^{-3} for special events, such as cooking and bleach cleaning with substantial release of nitrous acid (HONO) and hypochlorous acid (HOCl), which may go on to be photolyzed in air that is directly illuminated by sunlight (and not by reflected light) to form OH radicals (Young et al., 2019). Due to their very high reactivity, OH radicals can drive indoor chemistry at relatively low concentrations by oxidizing both saturated and unsaturated compounds. Chlorine radicals can be generated via photolysis of a number of inorganic chlorinated species (e.g., HOCl, chlorine [Cl_2], nitryl chloride [$ClNO_2$]) (Wong et al., 2017). Given its high reactivity, Cl-initiated chemistry may lead to the formation of oxygenated semivolatile products as well as secondary organic aerosol (SOA; Mattila et al., 2020a). The concentration of nitrate (NO_3) radicals, which are important outdoors for nighttime chemistry, is likely negligible (<0.01 ppt) in residential indoor settings for most conditions with low O_3 and relatively high NO levels (Arata et al., 2018; Young et al., 2019).

Gas-phase oxidation of VOCs leads to the formation of a myriad of semivolatile compounds, driven by complex reactions of peroxy and alkoxy radicals involving hydrogen (HO_x) and nitrogen oxides (NO_x). Recently, it has been shown that peroxy radicals can undergo isomerization by internal hydrogen shifts, resulting in the generation of highly oxygenated organic molecules (HOMs) (Crounse et al., 2013) (see Figure 4-1). These compounds are extremely low volatility, contributing to new particle formation and growth of SOA particles (Ehn et al., 2014). A recent theoretical study calculated the rate coefficients of the possible unimolecular reactions of the first-generation peroxy radicals formed by limonene ozonolysis, finding that they react unimolecularly with rates that are competitive indoors, especially with low concentrations of hydroperoxy (HO_2) and nitric oxide (NO) (Chen et al., 2021). HOMs generated by O_3-initiated autoxidation of limonene were indeed detected in an art museum; the HOM molar yield of 11 percent and the SOA mass yield of 47 percent were determined, indicating that limonene autoxidation efficiently forms SOA indoors (Pagonis et al., 2019). Inclusion of HOM formation improved the performance of an indoor chemistry model for simulating SOA mass concentrations against measurements (Kruza et al., 2020). Organic hydroperoxides may be labile in the condensed phase, potentially undergoing decomposition to yield reactive oxygen species (ROS) including OH and superoxide (Wei et al., 2021); quantification of ROS in indoor aerosols would be important for evaluation of human exposure (Morrison et al., 2021).

Indoor organic aerosol (OA) may generally be dominated by transport of outdoor OA and primary emissions by cooking, while some specific events can lead to substantial formation of indoor SOA. For example, formation of indoor SOA can be triggered by indoor illumination of bleach

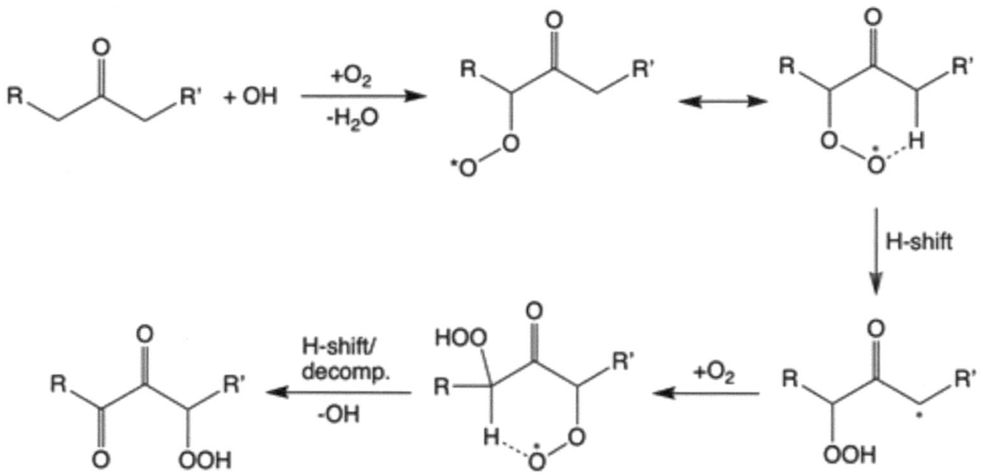

FIGURE 4-1 Example of auto-oxidation pathway for OH-initiated transformation of a carbonyl.
SOURCE: Reprinted (adapted) with permission from Crounse, J. D., L. B. Nielsen, S. Jørgensen, H. G. Kjærgaard, and P. O. Wennberg. 2013. Autoxidation of Organic Compounds in the Atmosphere. *The Journal of Physical Chemistry Letters* 4(20):3513–3520. DOI: 10.1021/jz4019207. Copyright 2013 American Chemical Society.

emissions via chlorine and OH oxidation of terpenes (Wang et al., 2019), and one study showed that ozonolysis of human skin lipids can lead to new particle formation (Yang et al., 2021). A modeling study indicated that oxidative aging can affect indoor OA concentrations when air temperature and OH concentrations are high, and air exchange rates and OA concentrations are low (Cummings and Waring, 2019). Indoor OA may often exist as amorphous semisolids, reflecting low water content under low or moderate relative humidity indoors (Cummings et al., 2020). This may call the assumption of equilibrium SOA partitioning into question, and kinetic limitations of bulk diffusion may need to be properly accounted for. Indoor OA properties, including morphology, mixing state, and phase state, are still largely unexplored and further studies are desired.

While incompletely mixed conditions of indoor air constituents have been recognized as a topic of interest and concern for indoor air quality (e.g., Lambert et al., 1993), indoor air constituents often have been treated as well mixed and homogeneously distributed in ventilated indoor environments. Hence, indoor measurements are mostly conducted at a single location in a room and at a fixed height, and indoor chemistry models often employ a box model assuming homogeneous mixing. A recent study demonstrated, however, that heterogeneous distributions of indoor air pollutants arise indoors, dependent on their temporal and spatial scales as controlled by chemical reactions and deposition rates, as well as indoor air flow and ventilation (Lakey et al., 2021). Short-lived radical species (e.g., OH, Cl, NO_3) exhibit sharp spatial gradients, and their temporal scales are determined mainly by reaction rates, affected only marginally by deposition and ventilation. Moderately long-lived species such as ammonia (NH_3), Cl_2, and O_3 will exhibit spatial gradients within a room as controlled by both chemical processes and indoor air flow conditions. Long-lived species, including carbon dioxide (CO_2) and VOCs, will be mostly well mixed within the indoor space, possibly affecting other rooms and the environment surrounding the building by circulation and transport outdoors. The widely applied concept of deposition velocity, which expresses the species' flux density to the surface divided by its concentration in the uniformly mixed core region (Nazaroff et al., 1993), may need to be revisited for compounds exhibiting spatial gradients. Note that spatial gradients can also arise from localized primary emissions sources even for chemically inert species

(Mahyuddin et al., 2014; Song et al., 2021). A better understanding of spatial distributions of indoor species is critical for accurate assessments of human exposure to indoor oxidants and pollutants. Surface interactions (see Chapter 3) can impact spatial distributions but are still poorly characterized, despite their importance becoming increasingly clear (Ault et al., 2020).

SURFACE CHEMISTRY

This section considers important classes of reactions that occur via surface chemistry. Especially with rapid gas-to-surface loss, the overall rate of the process may be controlled by mass transfer in the gas phase (i.e., the movement of the molecule through the gas-phase concentration boundary layer to the surface interface). However, if the multiphase reaction rate is slow, then the overall rate is determined by the surface chemistry and not by the gas-phase mass transfer.

Oxidation Reactions

Oxidation reactions are driven by a variety of atmospheric oxidants, including ozone, hydroxyl (OH), and nitrate (NO_3) (Gligorovski and Weschler, 2013; Young et al., 2019). This section explores common oxidation reactions occurring on indoor surfaces.

Common Atmospheric Oxidants, Including Ozone

Given a sizable supply arising from outdoor-to-indoor air exchange (Stephens et al., 2012), ozone multiphase chemistry has been studied extensively with numerous recent findings. The new work builds upon extensive previous literature that established that gas-phase ozone is irreversibly lost with variable deposition velocities upon exposure to a wide range of building materials (Grøntoft and Raychaudhuri, 2004). Most indoor ozone is removed via reactions with indoor surfaces as opposed to gas-phase reactions (Nazaroff and Weschler, 2022; Weschler, 2000). Ozone can be lost on some inorganic surfaces, such as components of mineral dust (Hanisch and Crowley, 2003; Mogili et al., 2006) and manganese oxide-based catalysts (Li et al., 2021; Lian et al., 2015), forming molecular oxygen. Ozone reactions with molecules containing carbon-carbon double bonds are notable because they lead to significant chemical transformation. Reactions of this type with electron-rich functional groups are referred to as ozonolysis reactions, for which a general mechanism is provided in Figure 4-2.

As seen in Figure 4-2, a wide range of oxygenated products arises in such reactions:

$$O_3 + \text{alkene} \rightarrow \text{functionalized products}$$
$$\text{(e.g., organic acids, carbonyls, secondary ozonides, peroxides)} \qquad (1)$$

As described below, major advances have arisen in our understanding of the multiphase ozonolysis of unsaturated oils, which are frequently present on the surface of human skin. These molecules also contaminate our clothing, which acts as both a shield to prevent ozone from reaching the underlying skin and a potential source of semivolatile oxidation products close to the body (Lakey et al., 2019; Licina et al., 2019; Morrison et al., 2016). Similar chemistry has also been studied as it pertains to natural products present in cooking oils, essential oils, cannabis, and some cleaning products (Huang et al., 2012; Liu et al., 2017; Shu and Morrison, 2012; Springs et al., 2011; Wylie and Abbatt, 2020; Zhou et al., 2019a). Many studies have recognized that these interactions drive a large reactive flux of ozone to indoor surfaces, with the simultaneous formation of numerous carbonyl-containing products (Abbass et al., 2017; Coleman et al., 2008; Gall et al., 2013; Rai et al., 2014; Wang et al., 2012). While this work focused largely on building materials and furnishings, other research has focused on detailed studies associated with the role of human occupants as the

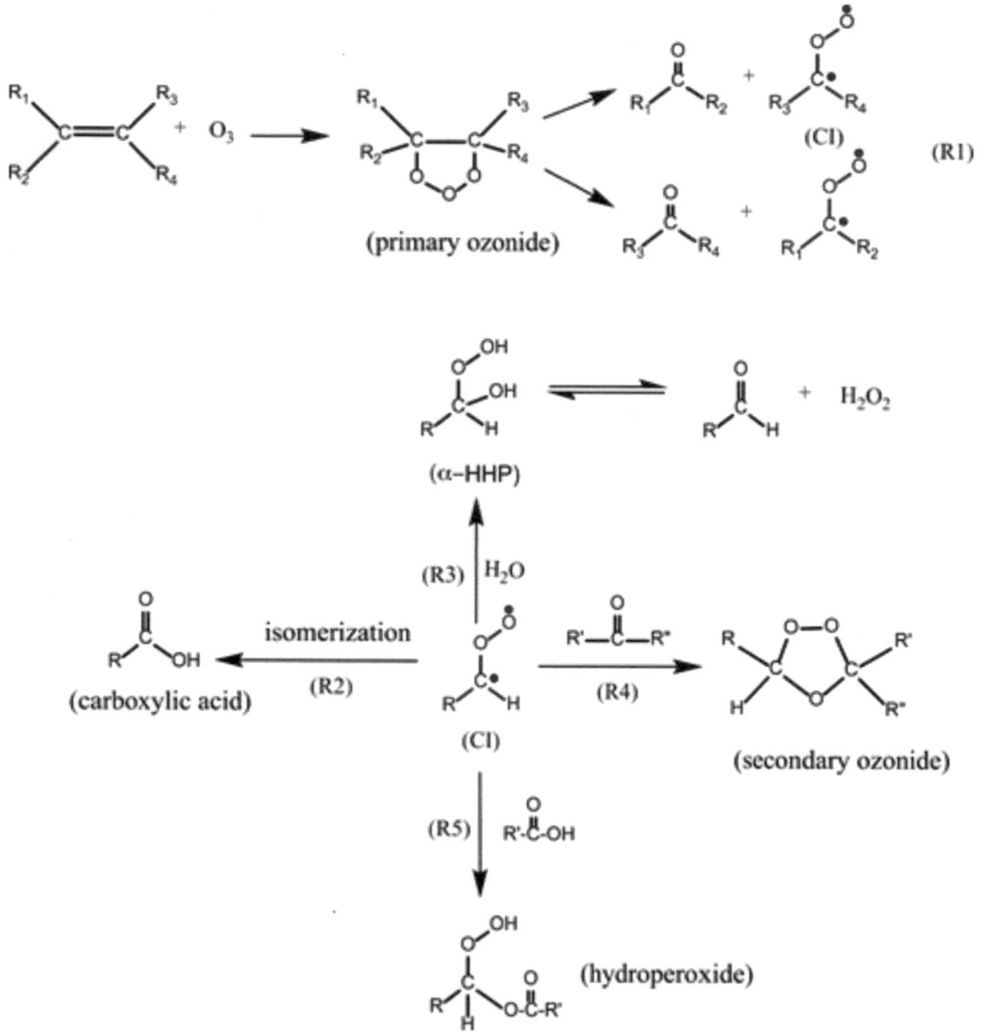

FIGURE 4-2 Ozonolysis mechanism of a model alkene, forming Criegee intermediates indicated in red. SOURCE: Reprinted (adapted) with permission from Zhou, S., M. W. Forbes, and J. P. D. Abbatt. 2016. Kinetics and Products from Heterogeneous Oxidation of Squalene with Ozone. *Environmental Science & Technology* 50(21):11688-11697. DOI: 10.1021/acs.est.6b03270. Copyright 2016 American Chemical Society.

source reaction sites (Coleman et al., 2008; Pandrangi and Morrison, 2008; Tamás et al., 2006; Weschler et al., 2007; Wisthaler and Weschler, 2010; Wisthaler et al., 2005; Zannoni et al., 2021).

Pivotal experiments have involved the exposure of human subjects (or their contaminated clothing) to ozone in controlled settings, clearly demonstrating significant loss of ozone and simultaneous formation of oxygenated VOCs via laboratory, field, and modeling studies (Bekö et al., 2020; Lakey et al., 2017; Morrison et al., 2021; Wisthaler and Weschler, 2010). One study demonstrated across a range of human subjects that ozonolysis occurs not only with human sebum materials but also with exogenous compounds (e.g., lipids from cooking) (Morrison et al., 2021). At a mechanistic level, it is now known that ozone undergoes multiphase chemistry with squalene, a highly unsaturated alkene that is a major component of skin oil (Fu et al., 2013; Petrick and

---

BOX 4-1
Skin Oil Oxidation Chemistry in Indoor Environments

An important connection of these fundamental chemistry studies to genuine indoor environments has been made in a variety of settings (Liu et al., 2016, 2021). For example, in a residential setting, the oxygenated volatile organic compound (VOC) products of squalene ozonolysis were observed not only when human occupants were present in the house but also when they were not (Liu et al., 2021). This is an indication that soiling of indoor surfaces by shedding of skin oil and flakes can lead to lasting chemical impacts indoors. It has also been calculated for a large number of Chinese cities that the ozone VOC oxidation product exposure will be comparable or even higher than ozone exposure itself (Yao et al., 2020). This has important potential implications for health impacts arising from exposure to ozone that have not been fully evaluated.

Dubowski, 2009; Wells et al., 2008; Zhou et al., 2016a), and with unsaturated triglycerides, which are also present in cooking oil as well as skin oil (Zhou et al., 2019c). The chemistry is fast, with significant chemical change in the composition of skin oil occurring on timescales of hours under ambient ozone mixing ratios (Zhou et al., 2016b). Oxygenated VOCs and hydrogen peroxide are formed (Arata et al., 2019; Zhou and Abbatt, 2021), some of which are able to react in the gas phase and form SOA (Avery et al., 2019; Wang and Waring, 2014), leaving behind less volatile, highly oxygenated species on the surface (Zhou et al., 2016a). Importantly, this chemistry has been shown to occur in genuine indoor environments (see Box 4-1).

The detailed mechanism and product distribution for multiphase ozonolysis reactions of molecules containing carbon-carbon double bonds hinges on the behavior of the highly reactive Criegee intermediate. Although long recognized to be important for organic synthesis and in gas-phase atmospheric chemistry, the chemistry of condensed-phase Criegee intermediates under indoor conditions is now being explored. Studies performed as a function of relative humidity have shown that formation yields of VOCs and hydrogen peroxide (H_2O_2) are higher when there is more water present in the gas phase and presumably on surfaces as well (Arata et al., 2019; Zhou and Abbatt, 2021). The form that this surface water may take indoors is described in this chapter's section on Reactions Involving Water. By contrast, when water abundance is low, multiphase loss of Criegee intermediates leads to the formation of secondary ozonides and other complex organic products arising from the Criegee intermediate combining with protic molecules, such as alcohols and carboxylic acids (Heine et al., 2017; Zhao et al., 2018; Zhou et al., 2019b). The lifetimes and toxicity of these highly oxygenated, functionalized, and higher molecular weight products on surfaces are poorly understood, with additional oxidation reactions possible, along with slow self-reactions, hydrolytic, and/or photochemical reactions. The formation of organic surface films described in Chapter 3 may occur not only through gas-to-surface partitioning of semivolatile molecules but also through contributions of such high molecular weight oxidation products and deposition of particles.

An additional class of ozonolysis reactions long explored to understand outdoor chemistry is the reaction of polycyclic aromatic hydrocarbons (PAHs) with ozone. These molecules arise from incomplete combustion activities, such as those that occur during cooking and smoking. They are especially prevalent in poorly ventilated residential settings where open burning and cookstoves are used. PAHs from outdoor wildfires can contribute to indoor concentrations as well (Messier et al., 2019). The larger PAHs are expected to partition strongly to indoor surfaces, making them susceptible to ozonolysis reactions and formation of a suite of products, including redox-active species, such as quinones. A recent study has shown that the surface reactivity of PAHs can be controlled by the viscosity and phase of the surface film, with non-reactive layers of oxidation products impeding the reactivity of buried PAHs (Zhou et al., 2019a). This is an illustration of the importance of mass transfer processes in controlling the rates of multiphase reactive chemistry. The products of these reactions and their toxicity deserve additional attention. For example, ozonolysis of benzo[a]pyrene by indoor air produced a class of highly reactive products with both epoxide and di-alcohol functional groups (Zhou et al., 2017). This is the first demonstration that abiotic, multiphase oxidation processes can also form these carcinogenic compounds, widely known to form biotically in humans (Xue and Warshawsky, 2005).

In addition to reactions with unsaturated organic molecules, ozone reacts with reduced forms of heteroatoms, such as nitrogen, as present in nicotine, forming species such as nicotine oxide and SOA (Destaillats et al., 2006; Petrick et al., 2010; Sleiman et al., 2010; Wang et al., 2018). Indeed, commercial ozonolysis is a process employed in the removal of unwanted third-hand smoking odors from contaminated spaces (see Chapter 5).

By contrast to O_3, gas-phase OH and NO_3 are too short-lived to be transported from outside; rather, they are generated indoors at lower mixing ratios than outdoors (Young et al., 2019). The potential for multiphase indoor oxidation by OH has been explored recently, where potential impacts were examined for the slow (i.e., weeks or longer) loss of low-volatility species, such as phthalates and long-chain carboxylic acids that exist in organic surface films (Alwarda et al., 2018). An area that is in need of more study is the potential for autoxidation of unsaturated oils, initiated by OH oxidation and propagated by Criegee radicals (Zeng et al., 2020). As described above, it is important to assess the rate of air-to-surface mass transfer for reactive species like OH that can be generated within a concentration boundary layer close to a surface (Morrison et al., 2019). The multiphase oxidation chemistry of NO_3 has not been addressed indoors, despite studies showing fast, multiphase reactions with certain classes of molecules, such as those containing carbon-carbon double bonds.

Cleaning Agents, Including Chlorine Bleach

Oxidizing cleaning agents are frequently applied indoors in an aqueous form, either as mists, through surface wipes, or via mopping. They are also sometimes used in dishwashers and washing machines. The molecular oxidants in these cleaning agents include hypochlorite (OCl^-), in commercial chlorine bleach, and H_2O_2, in peroxide-based cleaners. Chlorine dioxide (OClO) has also been used in select environments, usually for microbial remediation.

Indoor surfaces are sufficiently acidic to form HOCl when bleach washing occurs (Mattila et al., 2020b; Wong et al., 2017). As a result, recent work has addressed the multiphase reactions of HOCl. Well-known within the water treatment community, this oxidant reacts with unsaturated organic molecules, as well as with other functional groups such as thiols. Unlike the case with ozone, the carbon-carbon double bond in unsaturated organics is not broken by this chemistry. Rather, the organic molecules become chlorinated with, for example, chlorohydrins forming when HOCl reacts with alkenes such as squalene (Schwartz-Narbonne et al., 2019b), as shown in Reaction (2):

$$\text{(2)}$$

This multiphase chemistry has been found to be rapid with squalene and oleic acids, demonstrating that bleach washing is likely leading to chlorination of skin and cooking oils (Schwartz-Narbonne et al., 2019b). The chlorohydrins are reactive, forming ester linkages with carboxylic acids, and lead to higher molecular weight substances that are likely to contribute to organic film growth. HOCl also reacts with terpenes, such as limonene, forming higher molecular weight substances and potentially SOA in the presence of light (Wang et al., 2019). It is known to be reactive with reduced nitrogen as well, forming chloraldimines, for example (Finewax et al., 2021).

Chlorine bleach can promote additional reactions in the indoor environment. Although chloramines are found in headspace analysis of bleach (Wong et al., 2017), they also form when ammonia dissolves in the cleaning solution, leading to the production of monochloramine (NH_2Cl), chlorimide ($NHCl_2$), and nitrogen trichloride (NCl_3) from sequential reactions with aqueous HOCl (Mattila et al., 2020b). This chemistry is well documented in the water treatment field but has only recently been demonstrated to occur in indoor environments including an indoor aquatic center (Wu et al., 2021). In addition, HOCl can react with surface nitrite, forming $ClNO_2$, which evaporates to the gas phase and lowers HONO levels (Mattila et al., 2020b; Wang et al., 2020). Similarly, Cl_2 is present in the headspace of bleach solutions (Wong et al., 2017), but it can also be formed by HOCl reacting with chloride, either present in indoor surface reservoirs or aerosol particles (Mattila et al., 2020b). Cl_2, $ClNO_2$, and HOCl are generally inert in the gas phase but can be potential sources of radicals in air that is directly exposed to sunlight or some forms of fluorescent lights, as described in the earlier section on Airborne Chemistry.

H_2O_2 is a known antimicrobial agent, indicating its ability to react with organic matter. Gas-phase H_2O_2 is readily lost to surfaces (Zhou et al., 2020), but the specific reactions in which it participates with indoor organic surface molecules are unknown. Specifically, the degree to which Fenton chemistry, which involves the reactions of iron with condensed-phase peroxides, can generate sufficient OH radicals within surface reservoirs to promote additional transformation processes is unknown. Another set of reactions, involving H_2O_2 reacting with carbonyls (especially aldehydes) to form more oxygenated organic products (Zhao et al., 2013), also may potentially occur on surfaces.

Reactions Involving Water

In addition to gas-phase reactions of water vapor with Criegee intermediates arising from ozonolysis reactions, aqueous-phase reactions play an important role in indoor environments. Water can adsorb onto surfaces from the gas phase in amounts determined by temperature and relative humidity. This water adsorption can lead to microscopic thin films of water or water adsorbed in nanometer- to micrometer-sized pores. Water vapor can also absorb into permeable surfaces (Schwartz-Narbonne and Donaldson, 2019). Observable water in the form of macroscopic thick films and bulk water reservoirs can also be present on window glass, in bathrooms, and in other areas within indoor spaces. Several different fields of chemistry are beginning to recognize that water in confined space environments, such as thin films, microdroplets, and inside the pores of porous materials, exhibits unique physicochemical properties compared to bulk phase water (Knight et al., 2020; Wei et al., 2020; Wilson et al., 2020). These differences between the properties of bulk water and water in confined systems (thin films, microdroplets, and pores) also need to be considered in indoor reactions involving water.

Aqueous-phase bulk reactions often involve acid-base chemistry. Acidic conditions are defined by pH values less than 7, whereas basic conditions occur at pH values above 7. These are environments where typically hydroxide ion (OH^-), a base, or hydronium ion (H_3O^+), an acid, play a role in the reaction chemistry. However, there are other important bases in indoor environments, including ammonia, amines, and nicotine, and a number of soluble acidic substances including CO_2 and formic acid (Nazaroff and Weschler, 2020). A major challenge in understanding aqueous-phase reactions in indoor environments is that, besides macroscopic thick films of water and bulk water reservoirs, thin water films and water adsorbed in small pores may be present; the concept of "pH" breaks down in these confined-space environments, as pH, defined as $-\log[H^+]$, is applied to ideal dilute bulk water solutions. In these confined scenarios, activities, a_{H^+}, not concentrations, $[H^+]$, need to be considered. Furthermore, it has been proposed that stable pH gradients can exist on micrometer length scales, making these concepts of equilibrium pH measurements more difficult to apply to indoor environments where porous materials and thin water films are prevalent (Wei et al., 2018).

The above paragraphs outline the challenges in fully understanding chemical reactions of water in indoor environments. Although many of the concepts above have been implied in different review articles, these issues have not been investigated experimentally to any great extent. However, there have been studies of some reactions involving water that are relevant to indoor environments. A few cases are discussed in more detail below.

Reactions with Water and Common Indoor Inorganic Gases

It is well known that trace atmospheric gases can partition into water droplets or films and hydrolyze to yield acids, which will acidify water films (Nazaroff and Weschler, 2020). This includes reactions of carbon dioxide plus water to yield carbonic acid, H_2CO_3, which dissociates to carbonate and bicarbonate in water depending on solution pH. It also includes sulfur dioxide plus water to yield sulfurous acid (H_2SO_3), although sulfur dioxide levels in homes are relatively low compared to outdoors (Spengler et al., 1979). H_2SO_3 then oxidizes to sulfuric acid, leading to sulfate and bisulfate in solution, the exact speciation again depending on solution pH. In the House Observations of Microbial and Environmental Chemistry (HOMEChem) study, addition of acetic acid (in the form of vinegar) sufficiently acidified surfaces so that moderately strong acids such as formic acid, fulminic acid, and nitrous acid were strongly re-partitioned to the gas phase, implying they existed in dissociated forms in polar surface reservoirs (Wang et al., 2020).

Besides partitioning into water phases present indoors, one of the most consequential reactions with water in indoor environments that has been studied for many decades involves the disproportionation reaction of nitrogen dioxide with water (Pitts et al., 1985). Interestingly, the reaction can occur much more readily on surfaces than in the gas phase. Spectroscopic (Finlayson-Pitts et al., 2003; Goodman et al., 1999) and theoretical studies (Finlayson-Pitts, 2009) have shown that the reaction on hydrated silica surfaces leads to two products, nitrous acid and nitric acid, as in Equation (3).

$$2NO_2 + H_2O \rightarrow HONO + HNO_3 \qquad (3)$$

HONO in indoor environments has long been recognized as an important indoor air pollutant (Gligorovski, 2016), for which there is still intense interest. Indoor measurements find that HONO sometimes correlates with indoor NO_x levels and anticorrelates with ozone (e.g., Lee et al., 2002), perhaps because of the reaction of nitrite on surfaces. The correlation with NO_x does not necessarily occur on short timescales (Collins et al., 2018). HONO readily partitions between the surface and gas compartments, and it can undergo gas-phase photodissociation, as described in the earlier section on Airborne Chemistry. Important aspects of HONO multiphase chemistry are emerging. For example, a recent study showed that, similar to many organic compounds present indoors (Wang et al., 2020), large indoor surface reservoirs exist that lead to emissions of gas-phase HONO (Collins et al., 2018). The molecular form in which HONO exists on surfaces is not well known (i.e., whether it is as nitrite or some other species). In addition, gas-phase HONO is known to be reactive with a variety of materials on surfaces, including metal oxides and mineral dust, forming NO_x (El Zein et al., 2013). A prominent example of its multiphase reactivity with surface organics is with nicotine, leading to the production of carcinogenic nitrosamines (Sleiman et al., 2010).

Reactions with Water: Ester Hydrolysis

Because of their properties as plasticizers and flame retardants, esters are used in many consumer products and are present in large quantities indoors (Wensing et al., 2005). For example, Wang et al. (2012) found different esters and their hydrolysis products in indoor dust samples collected in North America and Asia. Furthermore, the formation of alcohols from the hydrolysis of different phthalates has been observed (Castagnoli et al., 2019; Sjoberg and Ramnas, 2007). The hydrolysis reaction results in smaller, lower molar mass compounds, an alcohol and carboxylic acid, as the forward direction of Reaction (4) illustrates.

$$RCOOR' + H_2O \leftrightarrow RCOOH + R'OH \tag{4}$$

The rate of this reaction depends on the availability of water (Bope et al., 2019). Although this process has been identified for a long time, its importance in indoor environments still has not been fully determined.

Overall, the products are more volatile than the parent ester and have the potential to partition into the gas phase. Although esters and their hydrolysis products have been detected in indoor dust samples, few studies have measured the rates of these reactions under conditions found in indoor environments. Notably, this hydrolysis reaction in bulk water is catalyzed by acids and bases, but little is known about the rates of these reactions in thin water films and small pores present in indoor materials. Initial findings of acidity effects have been indicated by fast reaction on wet alkaline concrete surfaces (Uhde and Salthammer, 2007), and with faster emission of formaldehyde from urea formaldehyde glues and resins that are used in building materials (Wolkoff and Kjaergaard, 2007).

Photochemical Reactions

The daytime chemistry of the outdoor environment is driven by photochemical reactions. In polluted environments, the term "photochemical smog" describes the interplay between sunlight (solar radiation) and the presence of hydrocarbons, reactive nitrogen oxides, and oxidants in the atmosphere that lead to particle formation and unhealthy air. Similarly, photochemical reactions can occur indoors to drive chemistry that is not thermally activated (Kowal et al., 2017; Young et al., 2019). Photosensitizers present in various forms, including semiconductor oxide particles such as titanium dioxide (TiO_2), play an important role in photochemical reactions in outdoor chemistry (Chen et al., 2012) as well as indoor environments (Gligorovski, 2016) as discussed in more detail below.

Case Studies of Nitrogen Oxide Photochemistry

One of the most investigated reactions involving light-initiated chemistry is the formation of HONO from other nitrogen-containing species. In particular, photochemical conversion of both nitrogen dioxide and surface-adsorbed nitrates can be important in the formation of HONO indoors. For example, photolysis of surface nitrate leads to nitrogen dioxide (NO_2) and nitrite (NO_2^-), the latter of which is then protonated to form gas-phase HONO. The photochemical loss of nitrate is enhanced on semiconductor oxides such as TiO_2 relative to insulator oxides (Gankanda and Grassian, 2013). The formation of nitrogen oxides from irradiation of nitrates mixed with TiO_2 using all widely available indoor light sources has been suggested as a potential NO_x source indoors (Schwartz-Narbonne et al., 2019a). Additionally, solar light in the range from 330 nm to 400 nm passes through glass windows and can initiate light-mediated conversion of NO_2 to HONO on indoor surfaces, such as gypsum (Gligorovski, 2016; Pandit et al., 2021), and potentially on the skin via TiO_2 in personal care products. Incandescent and fluorescent light sources can drive this chemistry as well (Pandit et al., 2021). Langridge et al. (2009) showed that self-cleaning window glass with a TiO_2 nanoparticle coating can be a strong daytime source of indoor HONO. Several studies have demonstrated photo-induced HONO formation on white wall paints (with 7 percent TiO_2) by both direct solar light and indoor light bulbs near the sources (Gandolfo et al., 2015, 2020; Gemayel et al., 2017). Additionally, photocatalytic paints can degrade NO_2 and formaldehyde, with selective efficiency for other VOCs (Salthammer and Fuhrmann, 2007). Bartolomei et al. (2014) have seen UV-induced HONO formation from NO_2 and common household products, including detergents and lacquers, although the exact photoactive chemicals in these household products were not identified. In one study by Depoorter et al. (2021), specific organic photosensitizers such as furfural were shown to contribute to HONO formation in indoor environments. Therefore, wavelengths less than 400 nm, surface acidity, and the light-absorbing properties of photosensitizers (which can be organic compounds and inorganic materials) are important parameters that can alter indoor HONO concentrations.

Importantly, photosensitizers produce ROS, including OH. It has been suggested that on TiO_2, adsorbed nitrogen dioxide can photodissociate to NO and O (Gligorovski, 2016). This is followed by the reaction of adsorbed forms of NO and OH, present on the surface from reactions of water on irradiated TiO_2, to yield HONO. ROS can also readily react with organic compounds to yield more oxygenated, less volatile organic compounds that will have a higher affinity for indoor surfaces.

MODELING INDOOR CHEMISTRY

Indoor chemistry models are essential for quantifying chemical transformations and partitioning by treating a variety of highly complex chemical and physical processes. Models can be used to

- assess gaps in our fundamental understanding of indoor chemistry processes and evaluate major uncertainties,
- guide measurements and design laboratory experiments through identification of key parameters and estimates of expected concentrations of species,
- predict under what conditions indoor air chemistry processes might cause deleterious impacts to human health and well-being,
- design effective operation of buildings to mitigate such risks, and
- provide a foundation for chemical exposure assessment (see Chapter 6).

More than 30 years ago, Nazaroff and Cass (1986, 1989) conducted a pioneering study to develop a general mathematical model of reactive gas-phase species and particulate matter, accounting for gas-phase chemical reactions and photolysis as well as the effects of ventilation, filtration, deposition onto surfaces, direct emission, and coagulation. They further developed a conceptual model for the rate of deposition of reactive gas-phase pollutants by combining mass transport and surface kinetics under different airflow conditions, including laminar convection flow and homogeneous turbulence (Cano-Ruiz et al., 1993).

Later investigators developed increasingly detailed models of gas chemistry and particle composition (Carslaw, 2007; Sarwar et al., 2003; Wang and Waring, 2014). Some models adopt representations for outdoor atmospheric chemistry including detailed gas-phase chemistry mechanisms and OA model frameworks, further developing them for indoor scenarios by adding indoor-relevant processes (Carslaw et al., 2012; Cummings and Waring, 2019). Models improve aerosol representation by providing a better understanding of different aerosol fractions by season or region, as well as by exploring whether the assumption of equilibrium is relevant for typical buildings with short residence times and large temperature variations (Weschler et al., 2008). The impacts of the wide variety of indoor sources such as cooking and cleaning on indoor aerosol composition warrant further investigation.

Efforts to develop and apply molecular- and process-based models to a variety of indoor chemical and physical processes are increasing. Molecular dynamics simulations are used to investigate interactions of oxidants and organic compounds with indoor surfaces. Numerical and analytical approaches are used to describe mass transport of species through the indoor boundary layer as controlled by diffusion and turbulence (Morrison et al., 2019). The formation and growth of indoor organic surface films have also been modeled with respect to partitioning (Weschler and Nazaroff, 2017), multilayer adsorption (Eichler et al., 2019), and heterogeneous reactions (Lakey et al., 2021).

There are emerging and concerted efforts to coordinate the different types of models into integrated indoor chemistry modeling frameworks (see Figure 4-3). This effort requires expertise in chemistry, engineering, and building science; proficiency in modeling approaches that operate across a wide range of timescales and space scales; and active integration of building- and people-related factors, such as air exchange rate, ventilation strategy, and occupants' behavior and activities. Recent modeling efforts on ozone reactions with human skin lipids represent a good example for integration of different modeling approaches with molecular-to-room scales (von Domaros et al., 2020) as well as model applications to experimental observations showing human impacts on ozone and semivolatile organic compound concentrations (Wisthaler and Weschler, 2010). Molecular dynamics simulations were applied to simulate ozone interactions with squalene, determining key kinetic parameters that can be used directly in a kinetic process model to resolve mass transport and chemical reactions in the gas phase, in clothing, and on skin (Lakey et al., 2019). In clothing, diffusion can be slowed down due to partitioning of species to skin oils and other substances covering the fibers, as simulated by human envelope models (Morrison et al., 2017). For examination of the spatial distributions of ozone and reaction products, computational fluid dynamics (CFD) modeling can account for convection, diffusion, chemical reactions, and source emissions (Won et al., 2020). As it is computationally too expensive to resolve detailed surface interactions within the CFD model, outputs from the kinetic flux model including the ozone uptake coefficient and the product yields are input to constrain the model and alleviate the computational burden. The model results show that primary ozonolysis products are concentrated in the human envelope and the breathing zone, while secondary products are relatively well distributed throughout the room (Won et al., 2020). Such a combined approach can simulate complex indoor chemical processes and evaluate human exposure to secondary pollutants.

To validate and evaluate indoor chemistry models, comparisons with observations and experiments are crucial. Recently, a growing number of targeted laboratory experiments and indoor field

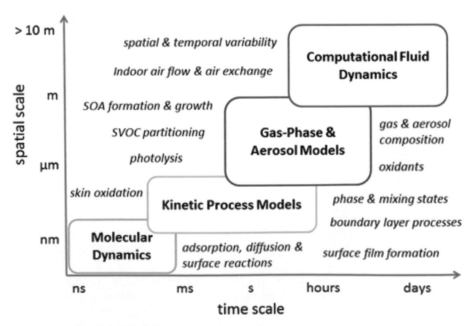

FIGURE 4-3 The wide range of temporal and spatial scales involved in various indoor chemistry modeling tools (in boxes) applied to crosscutting themes (in italics).
SOURCE: Modified from Shiraiwa et al. (2019).
NOTE: SOA = secondary organic aerosol; SVOC = semivolatile organic compounds.

observations have been conducted, such as the HOMEChem study and the Indoor Chemical Human Emissions and Reactivity experiments, providing unique datasets measured by state-of-the-art experimental techniques. Combining models with such experimental data is beneficial to test hypotheses and gain mechanistic and quantitative interpretation of observations, as demonstrated by several recent advances. For example, kinetic modeling revealed that multiphase chemistry in aqueous bleach and aerosol/surface uptake are essential in controlling reactive chlorine and nitrogen species after bleach applications in HOMEChem (Mattila et al., 2020b). Integration of surface spectroscopic measurements with molecular dynamics simulations and kinetic modeling has provided a molecular picture of the interaction of hydrophobic molecules with hydrophilic surfaces (Fang et al., 2019). A combined spectroscopic and atomistic modeling approach has also elucidated the conformational and orientational preferences of squalene at the air/oil interface and their implications for reactions with ozone (von Domaros et al., 2020). Applications of kinetic and thermodynamic modeling revealed that the multiphase reactivity of PAHs is driven by phase separation and diffusion limitations, affecting their fates in indoor environments (Zhou et al., 2019a).

Models also allow for extrapolation of experimental results to indoor conditions, spaces, and scenarios that are inaccessible by measurement. Laboratory experiments are often conducted at high concentrations with short reaction times, which contrasts the relatively low concentrations with longer reaction times in real indoor environments. Once the model is constrained, it can simulate concentrations and species properties under indoor-relevant conditions. While measurements are often conducted at one location, applications of CFD modeling can visualize spatial distributions of indoor species. If experiments and observations are unavailable to validate models, probabilistic modeling can be applied, whereby model input parameters are varied over their most likely values and the model sensitivity to these changes is explored (Cummings et al., 2020). As well as providing an understanding of the model sensitivity to specific parameters, this technique allows ranking of model parameters in terms of importance for prioritizing future measurements and experiments.

CONCLUSIONS

Major findings from the past several years have illustrated the complexity of chemical reactions that occur in indoor environments. In particular, gas-phase oxidation reactions, some occurring via auto-oxidation mechanisms, lead to the formation of a suite of highly oxygenated gas-phase species that may form SOA. In addition, much of reactive indoor chemistry occurs on surfaces via multiphase chemistry. Although long acknowledged to be important, ozonolysis reactions of unsaturated organics have now been demonstrated to form highly oxygenated species, such as secondary ozonides and volatile oxygenates, on surfaces. This chemistry is now known conclusively to occur on humans and their clothing and on other surfaces contaminated by cooking or smoking.

Another important conclusion is that the complexity of such reactions presently precludes a quantitative understanding of these processes under genuine indoor conditions, where substrate composition and environmental parameters (e.g., relative humidity) have been shown to affect the mechanisms and kinetics. Without a better understanding of the identities and amounts of many indoor chemicals, especially in surface reservoirs, an accurate toxicological and epidemiological evaluation of chemical dose and health outcomes is not yet possible in indoor environments. Furthermore, such uncertainties in reactive chemistry when coupled to uncertainties in partitioning make it challenging to determine the relative importance of the major exposure pathways for many indoor chemicals.

Recently, new chemistry has been identified when chemical cleaning agents, such as chlorine bleach, are used on indoor surfaces. The suite of chemical products that arise from such activities is only just starting to be studied. The reactive chemistry that occurs with some other common cleaning agents, such as hydrogen peroxide, has yet to be investigated under indoor conditions.

Photochemistry has yet to be definitely demonstrated to be of importance in genuine indoor settings, except when the air or surfaces are directly illuminated with sunlight. While infrequent in many indoor settings, high levels of oxidants can be generated, and other reactive photochemistry can occur in such situations. It is possible that important, yet slow, photochemistry occurs elsewhere on indoor surfaces that are not exposed to direct sunlight, but this has yet to be confirmed.

Water is an important molecule indoors for facilitating chemical transformations. These can include acid-base reactions, slow hydrolysis of organic compounds such as esters, reactions with Criegee intermediates that form during ozonolysis of unsaturated organics, and the NO_2 disproportionation reaction.

Important progress has been made in the past few years to develop models that integrate our growing knowledge of chemical transformations, partitioning between different indoor reservoirs, mass transfer, and indoor-outdoor air exchange. However, these models remain limited in their predictive capabilities owing to uncertainties in the underlying fundamental chemistry, especially on surfaces.

RESEARCH NEEDS

Given its findings about the current state of the science, the committee has identified priority research areas to help drive future advances in chemical transformations relevant to indoor environments:

- **Expand research into the chemistry associated with human occupancy, behavior, and activities, especially to identify processes that alter exposure to chemicals.** Common human activities, such as cooking, cleaning, smoking, and personal care product use, lead to chemical change that needs to be fully investigated. The complete suite of transformation products that arise when these primary emissions react in the indoor environment is unknown.

- **Investigate transformations of long-lived contaminants.** Many chemical contaminants, such as phthalates, are frequently viewed as being chemically inert, with ventilation a more important removal process for semivolatile species. This may not be the case for some species. There is a need to assess the degree to which potentially toxic contaminants, especially those with low volatility that strongly partition to indoor surfaces, are removed and transformed via chemical reactions. As described in Chapter 2, the indoor environment is the receptor of thousands of new chemicals used in consumer products. It needs to be determined which of these species are chemically unstable or reactive, and which can transform into potentially more toxic chemicals.

- **Apply advanced instrumentation and analytical techniques to study chemistry taking place in a broader range of building types, including their air, contents, and surfaces.** The recent use of highly advanced techniques in analytical science has illustrated how reactive chemistry can drive the chemical complexity of indoor environments. These studies include measurements of the gas phase and suspended particles, with instruments providing detailed chemical information *in situ* and with high time resolution. It is recommended that such detailed studies are used to characterize a wider range of indoor environments. While prior work has frequently characterized indoor constituents in a time-averaged sense, chemical reactions occur in a dynamic manner with their rates dependent on changing environmental conditions and oxidant levels. It is important to examine the dynamical behavior of indoor environments that may give rise to spatial and temporal composition gradients and, consequently, variable exposure. Detailed studies of the surface composition of passively collected samples from indoor environments could also be expanded. Given the dynamic nature of indoor environments, it will be fruitful to apply advanced surface analysis techniques *in situ* to outstanding questions such as the identification of nitrogen-containing compounds on surfaces that lead to the formation of nitrous acid and the assessment of the dynamic mass balance of skin oil materials that are always being deposited and chemically transformed.

- **Broaden our understanding of chemistry taking place on and within the complex surface materials and interfaces present within buildings.** Surface materials are highly complex, with variable chemical composition, morphology, and porosity. It is not known how the structure of such surfaces affects the rates and products of surface reactions. For example, studies in microdroplets and micron-thick water films have shown chemical reactions to be greatly accelerated relative to the bulk, which may be due to the unique environment of the microdroplet, especially the interface. This could be due to partial solvation of reactants, fast diffusion of reactants and products, different chemical speciation at the interface compared to the bulk, orientation of interfacial reactants, and surface potential. It is important to determine how such driving forces apply to porous materials indoors. It is likely that comparable effects arise with organic surface films and materials.

- **Expand, improve, and integrate models across different timescales and spatial scales.** Timescales with direct relevance to human exposure can range from short, with variations in indoor air composition on the scale of seconds to minutes, to very long, with slow release from indoor surface reservoirs occurring over decades. Characterization of reactive behavior at the molecular level, especially on surfaces, needs to be increasingly coupled to models that describe the overall, room-, and building-scale behavior. Multiphase modeling could capture the coupled, complex condensed-phase mass transfer and chemistry that occurs within permeable surface reservoirs. To resolve the spatial and temporal gradients that exist for chemicals indoors, there are new opportunities to couple CFD to detailed chemistry models. Models are also limited due to uncertainties regarding the parameterization of surface interactions, the propagation of light through indoor environments, and the concentrations and identity of a suite of secondary pollutants formed through indoor chemical reactions.

REFERENCES

Abbass, O. A., D. J. Sailor, and E. T. Gall. 2017. Effect of fiber material on ozone removal and carbonyl production from carpets. *Atmospheric Environment* 148:42–48. https://doi.org/10.1016/j.atmosenv.2016.10.034.

Abbatt, J. P. D., and C. Wang. 2020. The atmospheric chemistry of indoor environments. *Environmental Science: Processes & Impacts* 22(1):25–48. https://doi.org/10.1039/c9em00386j.

Alwarda, R., S. Zhou, and J. P. D. Abbatt. 2018. Heterogeneous oxidation of indoor surfaces by gas-phase hydroxyl radicals. *Indoor Air* 28(5):655–664. https://doi.org/10.1111/ina.12476.

Arata, C., N. Heine, N. Wang, P. K. Misztal, J. Williams, W. W. Nazaroff, K. R. Wilson, and A. H. Goldstein. 2019. Heterogeneous ozonolysis of squalene: Gas-phase products depend on water vapor concentration. *Environmental Science & Technology* 53(24):14441–14448. https://doi.org/10.1021/acs.est.9b05957.

Arata, C., K. J. Zarzana, P. K. Misztal, Y. Liu, S. S. Brown, W. W. Nazaroff, and A. H. Goldstein. 2018. Measurement of NO_3 and N_2O_5 in a residential kitchen. *Environmental Science & Technology Letters* 5(10):595–599. https://doi.org/10.1021/acs.estlett.8b00415.

Ault, A. P., V. H. Grassian, N. Carslaw, D. B. Collins, H. Destaillats, D. J. Donaldson, D. K. Farmer, J. L. Jimenez, V. F. McNeill, G. C. Morrison, R. E. O'Brien, M. Shiraiwa, M. E. Vance, J. R. Wells, and W. Xiong. 2020. Indoor surface chemistry: Developing a molecular picture of reactions on indoor interfaces. *Chem* 6(12):3203–3218. https://doi.org/10.1016/j.chempr.2020.08.023.

Avery, A. M., M. S. Waring, and P. F. DeCarlo. 2019. Human occupant contribution to secondary aerosol mass in the indoor environment. *Environmental Science: Processes & Impacts* 21(8):1301–1312. https://doi.org/10.1039/c9em00097f.

Bartolomei, V., M. Sorgel, S. Gligorovski, E. G. Alvarez, A. Gandolfo, R. Strekowski, E. Quivet, A. Held, C. Zetzsch, and H. Wortham. 2014. Formation of indoor nitrous acid (HONO) by light-induced NO_2 heterogeneous reactions with white wall paint. *Environmental Science and Pollution Research International* 21(15):9259–9269. https://doi.org/10.1007/s11356-014-2836-5.

Bekö, G., P. Wargocki, N. Wang, M. Li, C. J. Weschler, G. Morrison, S. Langer, L. Ernle, D. Licina, S. Yang, N. Zannoni, and J. Williams. 2020. The Indoor Chemical Human Emissions and Reactivity (ICHEAR) project: Overview of experimental methodology and preliminary results. *Indoor Air* 30(6):1213–1228. https://doi.org/10.1111/ina.12687.

Bope, A., S. R. Haines, B. Hegarty, C. J. Weschler, J. Peccia, and K. C. Dannemiller. 2019. Degradation of phthalate esters in floor dust at elevated relative humidity. *Environmental Science: Processes & Impacts* 21(8):1268–1279. https://doi.org/10.1039/c9em00050j.

Cano-Ruiz, J. A., D. Kong, R. Balas, and W. W. Nazaroff. 1993. Removal of reactive gases at indoor surfaces: Combining mass transport and surface kinetics. *Atmospheric Environment. Part A. General Topics* 27:2039–2050. https://doi.org/10.1016/0960-1686(93)90276-5.

Carslaw, N. 2007. A new detailed chemical model for indoor air pollution. *Atmospheric Environment* 41(6):1164–1179. https://doi.org/10.1016/j.atmosenv.2006.09.038.

Carslaw, N., T. Mota, M. E. Jenkin, M. H. Barley, and G. McFiggans. 2012. A significant role for nitrate and peroxide groups on indoor secondary organic aerosol. *Environmental Science & Technology* 46(17):9290–9298. https://doi.org/10.1021/es301350x.

Castagnoli, E., P. Backlund, O. Talvitie, T. Tuomi, A. Valtanen, R. Mikkola, H. Hovi, K. Leino, J. Kurnitski, and H. Salonen. 2019. Emissions of DEHP-free PVC flooring. *Indoor Air* 29(6):903–912. https://doi.org/10.1111/ina.12591.

Chen, H., C. E. Nanayakkara, and V. H. Grassian. 2012. Titanium dioxide photocatalysis in atmospheric chemistry. *Chem Rev* 112(11):5919–5948. https://doi.org/10.1021/cr3002092.

Chen, J., K. H. Møller, P. O. Wennberg, and H. G. Kjaergaard. 2021. Unimolecular reactions following indoor and outdoor limonene ozonolysis. *The Journal of Physical Chemistry A* 125(2):669–680. https://doi.org/10.1021/acs.jpca.0c09882.

Coleman, B. K., H. Destaillats, A. T. Hodgson, and W. W. Nazaroff. 2008. Ozone consumption and volatile byproduct formation from surface reactions with aircraft cabin materials and clothing fabrics. *Atmospheric Environment* 42(4):642–654. https://doi.org/10.1016/j.atmosenv.2007.10.001.

Collins, D. B., R. F. Hems, S. Zhou, C. Wang, E. Grignon, M. Alavy, J. A. Siegel, and J. P. D. Abbatt. 2018. Evidence for gas-surface equilibrium control of indoor nitrous acid. *Environmental Science & Technology* 52(21):12419–12427. https://doi.org/10.1021/acs.est.8b04512.

Crounse, J. D., L. B. Nielsen, S. Jørgensen, H. G. Kjærgaard, and P. O. Wennberg. 2013. Autoxidation of organic compounds in the atmosphere. *The Journal of Physical Chemistry Letters* 4(20):3513–3520. https://doi.org/10.1021/jz4019207.

Cummings, B. E., Y. Li, P. F. DeCarlo, M. Shiraiwa, and M. S. Waring. 2020. Indoor aerosol water content and phase state in U.S. residences: Impacts of relative humidity, aerosol mass and composition, and mechanical system operation. *Environmental Science: Processes & Impacts* 22(10):2031–2057. https://doi.org/10.1039/D0EM00122H.

Cummings, B. E., and M. S. Waring. 2019. Predicting the importance of oxidative aging on indoor organic aerosol concentrations using the two-dimensional volatility basis set (2D-VBS). *Indoor Air* 29(4):616–629. https://doi.org/10.1111/ina.12552.

Depoorter, A., C. Kalalian, C. Emmelin, C. Lorentz, and C. George. 2021. Indoor heterogeneous photochemistry of furfural drives emissions of nitrous acid. *Indoor Air* 31(3):682–692. https://doi.org/10.1111/ina.12758.

Destaillats, H., B. C. Singer, S. K. Lee, and L. A. Gundel. 2006. Effect of ozone on nicotine desorption from model surfaces: Evidence for heterogeneous chemistry. *Environmental Science & Technology* 40(6):1799–1805. https://doi.org/10.1021/es050914r.

Ehn, M., J. A. Thornton, E. Kleist, M. Sipilä, H. Junninen, I. Pullinen, M. Springer, F. Rubach, R. Tillmann, B. Lee, F. Lopez-Hilfiker, S. Andres, I.-H. Acir, M. Rissanen, T. Jokinen, S. Schobesberger, J. Kangasluoma, J. Kontkanen, T. Nieminen, T. Kurtén, L. B. Nielsen, S. Jørgensen, H. G. Kjærgaard, M. Canagaratna, M. D. Maso, T. Berndt, T. Petäjä, A. Wahner, V.-M. Kerminen, M. Kulmala, D. R. Worsnop, J. Wildt, and T. F. Mentel. 2014. A large source of low-volatility secondary organic aerosol. *Nature* 506(7489):476–479. https://doi.org/10.1038/nature13032.

Eichler, C. M. A., J. Cao, G. Isaacman-VanWertz, and J. C. Little. 2019. Modeling the formation and growth of organic films on indoor surfaces. *Indoor Air* 29(1):17–29. https://doi.org/10.1111/ina.12518.

El Zein, A., M. N. Romanias, and Y. Bedjanian. 2013. Kinetics and products of heterogeneous reaction of HONO with Fe_2O_3 and Arizona Test Dust. *Environmental Science & Technology* 47(12):6325–6331. https://doi.org/10.1021/es400794c.

Fang, Y., P. S. J. Lakey, S. Riahi, A. T. McDonald, M. Shrestha, D. J. Tobias, M. Shiraiwa, and V. H. Grassian. 2019. A molecular picture of surface interactions of organic compounds on prevalent indoor surfaces: Limonene adsorption on SiO_2. *Chemical Science* 10(10):2906–2914. https://doi.org/10.1039/C8SC05560B.

Finewax, Z., D. Pagonis, M. S. Claflin, A. V. Handschy, W. L. Brown, O. Jenks, B. A. Nault, D. A. Day, B. M. Lerner, J. L. Jimenez, P. J. Ziemann, and J. A. de Gouw. 2021. Quantification and source characterization of volatile organic compounds from exercising and application of chlorine-based cleaning products in a university athletic center. *Indoor Air* 31(5):1323–1339. https://doi.org/10.1111/ina.12781.

Finlayson-Pitts, B. J. 2009. Reactions at surfaces in the atmosphere: Integration of experiments and theory as necessary (but not necessarily sufficient) for predicting the physical chemistry of aerosols. *Physical Chemistry Chemical Physics* 11(36):7760–7779. https://doi.org/10.1039/b906540g.

Finlayson-Pitts, B. J., L. M. Wingen, A. L. Sumner, D. Syomin, and K. A. Ramazan. 2003. The heterogeneous hydrolysis of NO_2 in laboratory systems and in outdoor and indoor atmospheres: An integrated mechanism. *Physical Chemistry Chemical Physics* 5(2):223–242. https://doi.org/10.1039/B208564J.

Fu, D., C. Leng, J. Kelley, G. Zeng, Y. Zhang, and Y. Liu. 2013. ATR-IR study of ozone initiated heterogeneous oxidation of squalene in an indoor environment. *Environmental Science & Technology* 47(18):10611–10618. https://doi.org/10.1021/es4019018.

Gall, E., E. Darling, J. Siegel, G. Morrison, and R. Corsi. 2013. Evaluation of three common green building materials for ozone removal, and primary and secondary emissions of aldehydes. *Atmospheric Environment* 77. https://doi.org/10.1016/j.atmosenv.2013.06.014.

Gandolfo, A., V. Bartolomei, E. Alvarez, T. Sabrine, S. Gligorovski, J. Kleffmann, and H. Wortham. 2015. The effectiveness of indoor photocatalytic paints on NOx and HONO levels. *Applied Catalysis B: Environmental* 166-167:84–90. https://doi.org/10.1016/j.apcatb.2014.11.011.

Gandolfo, A., V. Bartolomei, D. Truffier-Boutry, B. Temime-Roussel, G. Brochard, V. Berge, H. Wortham, and S. Gligorovski. 2020. The impact of photocatalytic paint porosity on indoor NOx and HONO levels. *Physical Chemistry Chemical Physics* 22(2):589–598. https://doi.org/10.1039/c9cp05477d.

Gankanda, A., and V. H. Grassian. 2013. Nitrate photochemistry in NaY zeolite: Product formation and product stability under different environmental conditions. *Journal of Physical Chemistry A* 117(10):2205–2212. https://doi.org/10.1021/jp312247m.

Gemayel, R., B. Temime-Roussel, N. Hayeck, A. Gandolfo, S. Hellebust, S. Gligorovski, and H. Wortham. 2017. Development of an analytical methodology for obtaining quantitative mass concentrations from LAAP-ToF-MS measurements. *Talanta* 174:715–724. https://doi.org/10.1016/j.talanta.2017.06.050.

Gligorovski, S. 2016. Nitrous acid (HONO): An emerging indoor pollutant. *Journal of Photochemistry and Photobiology A: Chemistry* 314:1–5. https://doi.org/10.1016/j.jphotochem.2015.06.008.

Gligorovski, S., and C. J. Weschler. 2013. The oxidative capacity of indoor atmospheres. *Environmental Science & Technology* 47(24):13905–13906. https://doi.org/10.1021/es404928t.

Goodman, A. L., G. M. Underwood, and V. H. Grassian. 1999. Heterogeneous reaction of NO_2: Characterization of gas-phase and adsorbed products from the reaction, $2NO_2(g) + H_2O(a) \rightarrow HONO(g) + HNO_3(a)$ on hydrated silica particles. *The Journal of Physical Chemistry A* 103(36):7217–7223. https://doi.org/10.1021/jp9910688.

Grøntoft, T., and M. R. Raychaudhuri. 2004. Compilation of tables of surface deposition velocities for O_3, NO_2 and SO_2 to a range of indoor surfaces. *Atmospheric Environment* 38(4):533–544. https://doi.org/10.1016/j.atmosenv.2003.10.010.

Hanisch, F., and J. N. Crowley. 2003. Ozone decomposition on Saharan dust: An experimental investigation. *Atmospheric Chemistry and Physics* 3(1):119–130. https://doi.org/10.5194/acp-3-119-2003.

Heine, N., F. A. Houle, and K. R. Wilson. 2017. Connecting the elementary reaction pathways of Criegee intermediates to the chemical erosion of squalene interfaces during ozonolysis. *Environmental Science & Technology* 51(23):13740–13748. https://doi.org/10.1021/acs.est.7b04197.

Huang, H.-L., T.-J. Tsai, N.-Y. Hsu, C.-C. Lee, P.-C. Wu, and H.-J. Su. 2012. Effects of essential oils on the formation of formaldehyde and secondary organic aerosols in an aromatherapy environment. *Building and Environment* 57:120–125. https://doi.org/10.1016/j.buildenv.2012.04.020.

Knight, A. W., P. Ilani-Kashkouli, J. A. Harvey, J. A. Greathouse, T. A. Ho, N. Kabengi, and A. G. Ilgen. 2020. Interfacial reactions of Cu(ii) adsorption and hydrolysis driven by nano-scale confinement. *Environmental Science: Nano* 7(1):68–80. https://doi.org/10.1039/C9EN00855A.

Kowal, S. F., S. R. Allen, and T. F. Kahan. 2017. Wavelength-resolved photon fluxes of indoor light sources: Implications for HO_x production. *Environmental Science & Technology* 51(18):10423–10430. https://doi.org/10.1021/acs.est.7b02015.

Kruza, M., G. McFiggans, M. S. Waring, J. R. Wells, and N. Carslaw. 2020. Indoor secondary organic aerosols: Towards an improved representation of their formation and composition in models. *Atmospheric Environment* 240:117784. https://doi.org/10.1016/j.atmosenv.2020.117784.

Lakey, P. S. J., G. C. Morrison, Y. Won, K. M. Parry, M. von Domaros, D. J. Tobias, D. Rim, and M. Shiraiwa. 2019. The impact of clothing on ozone and squalene ozonolysis products in indoor environments. *Communications Chemistry* 2(1):56. https://doi.org/10.1038/s42004-019-0159-7.

Lakey, P. S. J., A. Wisthaler, T. Berkemeier, T. Mikoviny, U. Poschl, and M. Shiraiwa. 2017. Chemical kinetics of multi-phase reactions between ozone and human skin lipids: Implications for indoor air quality and health effects. *Indoor Air* 27(4):816–828. https://doi.org/10.1111/ina.12360.

Lakey, P. S. J., Y. Won, D. Shaw, F. F. Østerstrøm, J. Mattila, E. Reidy, B. Bottorff, C. Rosales, C. Wang, L. Ampollini, S. Zhou, A. Novoselac, T. F. Kahan, P. F. DeCarlo, J. P. D. Abbatt, P. S. Stevens, D. K. Farmer, N. Carslaw, D. Rim, and M. Shiraiwa. 2021. Spatial and temporal scales of variability for indoor air constituents. *Communications Chemistry* 4(1):110. https://doi.org/10.1038/s42004-021-00548-5.

Lambert, W. E., J. M. Samet, and J. D. Spengler. 1993. Environmental tobacco smoke concentrations in no-smoking and smoking sections of restaurants. *American Journal of Public Health* 83:1339–1341. https://doi.org/10.2105/AJPH.83.9.1339.

Langridge, J., R. Gustafsson, P. Griffiths, R. Cox, R. Lambert, and R. Jones. 2009. Solar driven nitrous acid formation on building material surfaces containing titanium dioxide: A concern for air quality in urban areas? *Atmospheric Environment* 43:5128–5131. https://doi.org/10.1016/j.atmosenv.2009.06.046.

Li, L., R. Gao, and P. Zhang. 2021. Catalytic decomposition of gaseous ozone at room temperature. *Progress in Chemistry* 33(7):1174–1186. https://doi.org/10.7536/PC200716.

Lian, Z., J. Ma, and H. He. 2015. Decomposition of high-level ozone under high humidity over Mn–Fe catalyst: The influence of iron precursors. *Catalysis Communications* 59:156–160. https://doi.org/10.1016/j.catcom.2014.10.005.

Licina, D., G. C. Morrison, G. Beko, C. J. Weschler, and W. W. Nazaroff. 2019. Clothing-mediated exposures to chemicals and particles. *Environmental Science & Technology* 53(10):5559–5575. https://doi.org/10.1021/acs.est.9b00272.

Liu, S., R. Li, R. J. Wild, C. Warneke, J. A. de Gouw, S. S. Brown, S. L. Miller, J. C. Luongo, J. L. Jimenez, and P. J. Ziemann. 2016. Contribution of human-related sources to indoor volatile organic compounds in a university classroom. *Indoor Air* 26(6):925–938. https://doi.org/10.1111/ina.12272.

Liu, T., Z. Li, M. Chan, and C. K. Chan. 2017. Formation of secondary organic aerosols from gas-phase emissions of heated cooking oils. *Atmospheric Chemistry and Physics* 17(12):7333–7344. https://doi.org/10.5194/acp-17-7333-2017.

Liu, Y., P. K. Misztal, C. Arata, C. J. Weschler, W. W. Nazaroff, and A. H. Goldstein. 2021. Observing ozone chemistry in an occupied residence. *Proceedings of the National Academy of Sciences USA* 118(6). https://doi.org/10.1073/pnas.2018140118.

Mahyuddin, N., H. B. Awbi, and M. Alshitawi. 2014. The spatial distribution of carbon dioxide in rooms with particular application to classrooms. *Indoor and Built Environment* 23:433–448.

Mattila, J. M., C. Arata, C. Wang, E. F. Katz, A. Abeleira, Y. Zhou, S. Zhou, A. H. Goldstein, J. P. D. Abbatt, P. F. DeCarlo, and D. K. Farmer. 2020a. Dark chemistry during bleach cleaning enhances oxidation of organics and secondary organic aerosol production indoors. *Environmental Science & Technology Letters* 7(11):795–801. https://doi.org/10.1021/acs.estlett.0c00573.

Mattila, J. M., P. S. J. Lakey, M. Shiraiwa, C. Wang, J. P. D. Abbatt, C. Arata, A. H. Goldstein, L. Ampollini, E. F. Katz, P. F. DeCarlo, S. Zhou, T. F. Kahan, F. J. Cardoso-Saldana, L. H. Ruiz, A. Abeleira, E. K. Boedicker, M. E. Vance, and D. K. Farmer. 2020b. Multiphase chemistry controls inorganic chlorinated and nitrogenated compounds in indoor air during bleach cleaning. *Environmental Science & Technology* 54(3):1730–1739. https://doi.org/10.1021/acs.est.9b05767.

Messier, K. P., L. G. Tidwell, C. C. Ghetu, D. Rohlman, R. P. Scott, L. M. Bramer, H. M. Dixon, K. M. Waters, and K. A. Anderson. 2019. Indoor versus outdoor air quality during wildfires. *Environmental Science & Technology Letters* 6(12):696–701. https://doi.org/10.1021/acs.estlett.9b00599.

Mogili, P. K., P. D. Kleiber, M. A. Young, and V. H. Grassian. 2006. Heterogeneous uptake of ozone on reactive components of mineral dust aerosol: An environmental aerosol reaction chamber study. *The Journal of Physical Chemistry A* 110(51):13799–13807. https://doi.org/10.1021/jp063620g.

Morrison, G., P. S. J. Lakey, J. Abbatt, and M. Shiraiwa. 2019. Indoor boundary layer chemistry modeling. *Indoor Air* 29(6):956–967. https://doi.org/10.1111/ina.12601.

Morrison, G. C., A. Eftekhari, F. Majluf, and J. E. Krechmer. 2021. Yields and variability of ozone reaction products from human skin. *Environmental Science & Technology* 55(1):179–187. https://doi.org/10.1021/acs.est.0c05262.

Morrison, G. C., C. J. Weschler, and G. Bekö. 2017. Dermal uptake of phthalates from clothing: Comparison of model to human participant results. *Indoor Air* 27(3):642–649. https://doi.org/10.1111/ina.12354.

Morrison, G. C., C. J. Weschler, G. Bekö, H. M. Koch, T. Salthammer, T. Schripp, J. Toftum, and G. Clausen. 2016. Role of clothing in both accelerating and impeding dermal absorption of airborne SVOCs. *Journal of Exposure Science & Environmental Epidemiology* 26(1):113–118. https://doi.org/10.1038/jes.2015.42.

Nazaroff, W. W., and G. R. Cass. 1986. Mathematical modeling of chemically reactive pollutants in indoor air. *Environmental Science & Technology* 20(9):924–934. https://doi.org/10.1021/es00151a012.

Nazaroff, W. W., and G. R. Cass. 1989. Mathematical modeling of indoor aerosol dynamics. *Environmental Science & Technology* 23(2):157–166. https://doi.org/10.1021/es00179a003.

Nazaroff, W. W., A. J. Gadgil, and C. J. Weschler. 1993. Critique of the use of deposition velocity in modeling indoor air quality. *Modeling of Indoor Air Quality and Exposure*. ASTM International. https://www.astm.org/stp13101s.html.

Nazaroff, W. W., and C. J. Weschler. 2020. Indoor acids and bases. *Indoor Air* 30(4):559–644. https://doi.org/10.1111/ina.12670.

Nazaroff, W. W., and C. J. Weschler. 2022. Indoor ozone: Concentrations and influencing factors. *Indoor Air* 32:e12942.

Pagonis, D., L. B. Algrim, D. J. Price, D. A. Day, A. V. Handschy, H. Stark, S. L. Miller, J. A. de Gouw, J. L. Jimenez, and P. J. Ziemann. 2019. Autoxidation of limonene emitted in a university art museum. *Environmental Science & Technology Letters* 6(9):520–524. https://doi.org/10.1021/acs.estlett.9b00425.

Pandit, S., S. L. Mora Garcia, and V. H. Grassian. 2021. HONO Production from gypsum surfaces following exposure to NO_2 and HNO_3: Roles of relative humidity and light source. *Environmental Science & Technology* 55(14):9761–9772. https://doi.org/10.1021/acs.est.1c01359.

Pandrangi, L. S., and G. C. Morrison. 2008. Ozone interactions with human hair: Ozone uptake rates and product formation. *Atmospheric Environment* 42:5079–5089. https://doi.org/10.1016/j.atmosenv.2008.02.009.

Petrick, L., H. Destaillats, I. Zouev, S. Sabach, and Y. Dubowski. 2010. Sorption, desorption, and surface oxidative fate of nicotine. *Physical Chemistry Chemical Physics* 12(35):10356–10364. https://doi.org/10.1039/c002643c.

Petrick, L., and Y. Dubowski. 2009. Heterogeneous oxidation of squalene film by ozone under various indoor conditions. *Indoor Air* 19(5):381–391. https://doi.org/10.1111/j.1600-0668.2009.00599.x.

Pitts, J. N., T. J. Wallington, H. W. Biermann, and A. M. Winer. 1985. Identification and measurement of nitrous acid in an indoor environment. *Atmospheric Environment* 19(5):763–767. https://doi.org/10.1016/0004-6981(85)90064-2.

Rai, A. C., B. Guo, C. H. Lin, J. Zhang, J. Pei, and Q. Chen. 2014. Ozone reaction with clothing and its initiated VOC emissions in an environmental chamber. *Indoor Air* 24(1):49–58. https://doi.org/10.1111/ina.12058.

Salthammer, T., and F. Fuhrmann. 2007. Photocatalytic surface reactions on indoor wall paint. *Environmental Science & Technology* 41(18):6573–6578. https://doi.org/10.1021/es070057m.

Sarwar, G., R. Corsi, D. Allen, and C. Weschler. 2003. The significance of secondary organic aerosol formation and growth in buildings: Experimental and computational evidence. *Atmospheric Environment* 37(9):1365–1381. https://doi.org/10.1016/S1352-2310(02)01013-0.

Schwartz-Narbonne, H., and D. J. Donaldson. 2019. Water uptake by indoor surface films. *Scientific Reports* 9(1):11089. https://doi.org/10.1038/s41598-019-47590-x.

Schwartz-Narbonne, H., S. H. Jones, and D. J. Donaldson. 2019a. Indoor lighting releases gas phase nitrogen oxides from indoor painted surfaces. *Environmental Science & Technology Letters* 6(2):92–97. https://doi.org/10.1021/acs.estlett.8b00685.

Schwartz-Narbonne, H., C. Wang, S. Zhou, J. P. D. Abbatt, and J. Faust. 2019b. Heterogeneous chlorination of squalene and oleic acid. *Environmental Science & Technology* 53(3):1217–1224. https://doi.org/10.1021/acs.est.8b04248.

Shiraiwa, M., N. Carslaw, D. J. Tobias, M. S. Waring, D. Rim, G. Morrison, P. S. J. Lakey, M. Kruza, M. von Domaros, B. E. Cummings, and Y. Won. 2019. Modelling consortium for chemistry of indoor environments (MOCCIE): Integrating chemical processes from molecular to room scales. *Environmental Science: Processes & Impacts* 21(8):1240–1254. https://doi.org/10.1039/c9em00123a.

Shu, S., and G. Morrison. 2012. Rate and reaction probability of the surface reaction between ozone and dihydromyrcenol measured in a bench scale reactor and a room-sized chamber. *Atmospheric Environment* 47:421–427. https://doi.org/10.1016/j.atmosenv.2011.10.068.

Sjoberg, A., and O. Ramnas. 2007. An experimental parametric study of VOC from flooring systems exposed to alkaline solutions. *Indoor Air* 17(6):450–457. https://doi.org/10.1111/j.1600-0668.2007.00492.x.

Sleiman, M., L. A. Gundel, J. F. Pankow, P. Jacob, 3rd, B. C. Singer, and H. Destaillats. 2010. Formation of carcinogens indoors by surface-mediated reactions of nicotine with nitrous acid, leading to potential thirdhand smoke hazards. *Proceedings of the National Academy of Sciences USA* 107(15):6576–6581. https://doi.org/10.1073/pnas.0912820107.

Song, Y., Q. Yang, H. Li, and S. Shen. 2021. Simulation of indoor cigarette smoke particles in a ventilated room. *Air Quality Atmosphere & Health* 14:1837–1847.

Spengler, J. D., B. G. Ferris, and D. W. Dockery. 1979. Sulfur dioxide and nitrogen dioxide levels inside and outside homes and the implications on health effects research. *Environmental Science & Technology* 13(10):1276–1280. https://doi.org/10.1021/es60158a013.

Springs, M., J. R. Wells, and G. C. Morrison. 2011. Reaction rates of ozone and terpenes adsorbed to model indoor surfaces. *Indoor Air* 21(4):319–327. https://doi.org/10.1111/j.1600-0668.2010.00707.x.

Stephens, B., E. T. Gall, and J. A. Siegel. 2012. Measuring the penetration of ambient ozone into residential buildings. *Environmental Science & Technology* 46(2):929–936. https://doi.org/10.1021/es2028795.

Tamás, G., C. J. Weschler, Z. Bakó-Biró, D. P. Wyon, and P. Strøm-Tejsen. 2006. Factors affecting ozone removal rates in a simulated aircraft cabin environment. *Atmospheric Environment* 40(32):6122–6133. https://doi.org/10.1016/j.atmosenv.2006.05.034.

Uhde, E., and T. Salthammer. 2007. Impact of reaction products from building materials and furnishings on indoor air quality—A review of recent advances in indoor chemistry. *Atmospheric Environment* 41:3111–3128. https://doi.org/10.1016/j.atmosenv.2006.05.082.

von Domaros, M., P. S. J. Lakey, M. Shiraiwa, and D. J. Tobias. 2020. Multiscale modeling of human skin oil-induced indoor air chemistry: Combining kinetic models and molecular dynamics. *The Journal of Physical Chemistry B* 124(18):3836–3843. https://doi.org/10.1021/acs.jpcb.0c02818.

Wang, C., B. Bottorff, E. Reidy, C. M. F. Rosales, D. B. Collins, A. Novoselac, D. K. Farmer, M. E. Vance, P. S. Stevens, and J. P. D. Abbatt. 2020. Cooking, bleach cleaning, and air conditioning strongly impact levels of HONO in a house. *Environmental Science & Technology* 54(21):13488–13497. https://doi.org/10.1021/acs.est.0c05356.

Wang, C., D. B. Collins, and J. P. D. Abbatt. 2019. Indoor illumination of terpenes and bleach emissions leads to particle formation and growth. *Environmental Science & Technology* 53(20):11792–11800. https://doi.org/10.1021/acs.est.9b04261.

Wang, C., D. B. Collins, R. F. Hems, N. Borduas, M. Antinolo, and J. P. D. Abbatt. 2018. Exploring conditions for ultrafine particle formation from oxidation of cigarette smoke in indoor environments. *Environmental Science & Technology* 52(8):4623–4631. https://doi.org/10.1021/acs.est.7b06608.

Wang, C., and M. S. Waring. 2014. Secondary organic aerosol formation initiated from reactions between ozone and surface-sorbed squalene. *Atmospheric Environment* 84:222–229. https://doi.org/10.1016/j.atmosenv.2013.11.009.

Wang, L., C. Liao, F. Liu, Q. Wu, Y. Guo, H. B. Moon, H. Nakata, and K. Kannan. 2012. Occurrence and human exposure of p-hydroxybenzoic acid esters (parabens), bisphenol A diglycidyl ether (BADGE), and their hydrolysis products in indoor dust from the United States and three East Asian countries. *Environmental Science & Technology* 46(21):11584–11593. https://doi.org/10.1021/es303516u.

Wei, H., E. P. Vejerano, W. Leng, Q. Huang, M. R. Willner, L. C. Marr, and P. J. Vikesland. 2018. Aerosol microdroplets exhibit a stable pH gradient. *Proceedings of the National Academy of Sciences USA* 115(28):7272–7277. https://doi.org/10.1073/pnas.1720488115.

Wei, J., T. Fang, C. Wong, P. S. J. Lakey, S. A. Nizkorodov, and M. Shiraiwa. 2021. Superoxide formation from aqueous reactions of biogenic secondary organic aerosols. *Environmental Science & Technology* 55(1):260–270. https://doi.org/10.1021/acs.est.0c07789.

Wei, Z., Y. Li, R. G. Cooks, and X. Yan. 2020. Accelerated reaction kinetics in microdroplets: Overview and recent developments. *Annual Review of Physical Chemistry* 71:31–51. https://doi.org/10.1146/annurev-physchem-121319-110654.

Wells, J., G. C. Morrison, and B. K. Coleman. 2008. Kinetics and reaction products of ozone and surface-bound squalene. *Journal of ASTM International* 5(7):1–12. https://doi.org/10.1520/JAI101629.

Wensing, M., E. Uhde, and T. Salthammer. 2005. Plastics additives in the indoor environment—flame retardants and plasticizers. *Science of the Total Environment* 339(1–3):19–40. https://doi.org/10.1016/j.scitotenv.2004.10.028.

Weschler, C. J. 2000. Ozone in indoor environments: Concentration and chemistry. *Indoor Air* 10(4):269–288. https://doi.org/10.1034/j.1600-0668.2000.010004269.x.

Weschler, C. J., and N. Carslaw. 2018. Indoor chemistry. *Environmental Science & Technology* 52(5):2419–2428. https://doi.org/10.1021/acs.est.7b06387.

Weschler, C. J., and W. W. Nazaroff. 2017. Growth of organic films on indoor surfaces. *Indoor Air* 27:1101–1112.

Weschler, C. J., T. Salthammer, and H. Fromme. 2008. Partitioning of phthalates among the gas phase, airborne particles and settled dust in indoor environments. *Atmospheric Environment* 42(7):1449–1460. https://doi.org/10.1016/j.atmosenv.2007.11.014.

Weschler, C. J., A. Wisthaler, S. Cowlin, G. Tamas, P. Strom-Tejsen, A. T. Hodgson, H. Destaillats, J. Herrington, J. Zhang, and W. W. Nazaroff. 2007. Ozone-initiated chemistry in an occupied simulated aircraft cabin. *Environmental Science & Technology* 41(17):6177–6184. https://doi.org/10.1021/es0708520.

Wilson, K. R., A. M. Prophet, G. Rovelli, M. D. Willis, R. J. Rapf, and M. I. Jacobs. 2020. A kinetic description of how interfaces accelerate reactions in micro-compartments. *Chemical Science* 11(32):8533–8545. https://doi.org/10.1039/d0sc03189e.

Wisthaler, A., G. Tamas, D. P. Wyon, P. Strom-Tejsen, D. Space, J. Beauchamp, A. Hansel, T. D. Mark, and C. J. Weschler. 2005. Products of ozone-initiated chemistry in a simulated aircraft environment. *Environmental Science & Technology* 39(13):4823–4832. https://doi.org/10.1021/es047992j.

Wisthaler, A., and C. J. Weschler. 2010. Reactions of ozone with human skin lipids: Sources of carbonyls, dicarbonyls, and hydroxycarbonyls in indoor air. *Proceedings of the National Academy of Sciences USA* 107(15):6568–6575. https://doi.org/10.1073/pnas.0904498106.

Wolkoff, P., and S. K. Kjaergaard. 2007. The dichotomy of relative humidity on indoor air quality. *Environment International* 33(6):850–857. https://doi.org/10.1016/j.envint.2007.04.004.

Won, Y., P. S. J. Lakey, G. Morrison, M. Shiraiwa, and D. Rim. 2020. Spatial distributions of ozonolysis products from human surfaces in ventilated rooms. *Indoor Air* 30(6):1229–1240. https://doi.org/10.1111/ina.12700.

Wong, J. P. S., N. Carslaw, R. Zhao, S. Zhou, and J. P. D. Abbatt. 2017. Observations and impacts of bleach washing on indoor chlorine chemistry. *Indoor Air* 27(6):1082–1090. https://doi.org/10.1111/ina.12402.

Wu, T., T. Földes, L. T. Lee, D. N. Wagner, J. Jiang, A. Tasoglou, B. E. Boor, and E. R. Blatchley. 2021. Real-time measurements of gas-phase trichloramine (NCl_3) in an indoor aquatic center. *Environmental Science & Technology* 55:8097–8107. https://doi.org/10.1021/acs.est.0c07413.

Wylie, A. D. L., and J. P. D. Abbatt. 2020. Heterogeneous ozonolysis of tetrahydrocannabinol: Implications for thirdhand cannabis smoke. *Environmental Science & Technology* 54(22):14215–14223. https://doi.org/10.1021/acs.est.0c03728.

Xue, W., and D. Warshawsky. 2005. Metabolic activation of polycyclic and heterocyclic aromatic hydrocarbons and DNA damage: A review. *Toxicology and Applied Pharmacology* 206(1):73–93. https://doi.org/10.1016/j.taap.2004.11.006.

Yao, M., C. J. Weschler, B. Zhao, L. Zhang, and R. Ma. 2020. Breathing-rate adjusted population exposure to ozone and its oxidation products in 333 cities in China. *Environment International* 138:105617. https://doi.org/10.1016/j.envint.2020.105617.

Yang, S., D. Licina, C. J. Weschler, N. Wang, N. Zannoni, M. Li, J. Vanhanen, S. Langer, P. Wargocki, J. Williams, and G. Bekö. 2021. Ozone initiates human-derived emission of nanocluster aerosols. *Environmental Science & Technology* 55(21):14536–14545. https://doi.org/10.1021/acs.est.1c03379.

Young, C. J., S. Zhou, J. A. Siegel, and T. F. Kahan. 2019. Illuminating the dark side of indoor oxidants. *Environmental Science: Processes & Impacts* 21(8):1229–1239. https://doi.org/10.1039/c9em00111e.

Zannoni, N., M. Li, N. Wang, L. Ernle, G. Bekö, P. Wargocki, S. Langer, C. J. Weschler, G. Morrison, and J. Williams. 2021. Effect of ozone, clothing, temperature, and humidity on the total OH reactivity emitted from humans. *Environmental Science & Technology* 55:13614–13624. https://doi.org/10.1021/acs.est.1c01831.

Zeng, M., N. Heine, and K. R. Wilson. 2020. Evidence that Criegee intermediates drive autoxidation in unsaturated lipids. *Proceedings of the National Academy of Sciences* 117(9):4486–4490. https://doi.org/10.1073/pnas.1920765117.

Zhao, R., C. M. Kenseth, Y. Huang, N. F. Dalleska, X. M. Kuang, J. Chen, S. E. Paulson, and J. H. Seinfeld. 2018. Rapid aqueous-phase hydrolysis of ester hydroperoxides arising from Criegee intermediates and organic acids. *Journal of Physical Chemistry A* 122(23):5190–5201. https://doi.org/10.1021/acs.jpca.8b02195.

Zhao, R., A. K. Y. Lee, R. Soong, A. J. Simpson, and J. P. D. Abbatt. 2013. Formation of aqueous-phase α-hydroxyhydroperoxides (α-HHP): Potential atmospheric impacts. *Atmospheric Chemistry and Physics* 13(12):5857–5872. https://doi.org/10.5194/acp-13-5857-2013.

Zhou, Z., and J. P. D. Abbatt. 2021. Formation of gas-phase hydrogen peroxide via multiphase ozonolysis of unsaturated lipids. *Environmental Science & Technology Letters* 8(2):114–120. https://doi.org/10.1021/acs.estlett.0c00757.

Zhou, S., M. W. Forbes, and J. P. Abbatt. 2016a. Kinetics and products from heterogeneous oxidation of squalene with ozone. *Environmental Science & Technology* 50(21):11688–11697. https://doi.org/10.1021/acs.est.6b03270.

Zhou, S., M. W. Forbes, Y. Katrib, and J. P. D. Abbatt. 2016b. Rapid oxidation of skin oil by ozone. *Environmental Science & Technology Letters* 3(4):170–174. https://doi.org/10.1021/acs.estlett.6b00086.

Zhou, S., B. C. H. Hwang, P. S. J. Lakey, A. Zuend, J. P. D. Abbatt, and M. Shiraiwa. 2019a. Multiphase reactivity of polycyclic aromatic hydrocarbons is driven by phase separation and diffusion limitations. *Proceedings of the National Academy of Sciences USA* 116(24):11658–11663. https://doi.org/10.1073/pnas.1902517116.

Zhou, S., S. Joudan, M. W. Forbes, Z. Zhou, and J. P. D. Abbatt. 2019b. Reaction of condensed-phase Criegee intermediates with carboxylic acids and perfluoroalkyl carboxylic acids. *Environmental Science & Technology Letters* 6(4):243–250. https://doi.org/10.1021/acs.estlett.9b00165.

Zhou, S., Z. Liu, Z. Wang, C. J. Young, T. C. VandenBoer, B. B. Guo, J. Zhang, N. Carslaw, and T. F. Kahan. 2020. Hydrogen peroxide emission and fate indoors during non-bleach cleaning: A chamber and modeling study. *Environmental Science & Technology* 54(24):15643–15651. https://doi.org/10.1021/acs.est.0c04702.

Zhou, S., L. W. Y. Yeung, M. W. Forbes, S. Mabury, and J. P. D. Abbatt. 2017. Epoxide formation from heterogeneous oxidation of benzo[a]pyrene with gas-phase ozone and indoor air. *Environmental Science: Processes & Impacts* 19(10):1292–1299. https://doi.org/10.1039/c7em00181a.

Zhou, Z., S. Zhou, and J. P. D. Abbatt. 2019c. Kinetics and condensed-phase products in multiphase ozonolysis of an unsaturated triglyceride. *Environmental Science & Technology* 53(21):12467–12475. https://doi.org/10.1021/acs.est.9b04460.

5

Management of Chemicals in Indoor Environments

Effective management of chemicals in the indoor environment is critical to human health. This chapter considers management approaches to the control of pollutants in indoor air as well as the indoor chemistry associated with these management approaches. The chapter provides an overview of the hierarchy of controls as a framework for considering risk-reduction strategies. With this framework in mind, the committee first considers management approaches that result in minimal changes in indoor chemistry, followed by management approaches that rely on chemical transformations. This chapter also discusses environmental factors, human behavior, and other management considerations. In addition, the current state of design standards addressing management of chemicals is noted where applicable. Finally, the committee summarizes key knowledge gaps and recommendations for future research. The chapter is not intended to be an exhaustive review of methods for modifying indoor chemistry; instead, the committee's goal is to highlight general approaches that may be considered, especially recent findings related to removal approaches. This chapter not only focuses on gas and particle removal but also provides insights into the impact of surface cleaning and pathogen inactivation on indoor air quality.

TYPES OF CONTROL

Approaches to the management of chemicals in the indoor environment are generally consistent with the industrial hygiene hierarchy of controls (Table 5-1) for chemical exposure (Schulte et al., 2013). This hierarchy of controls (Figure 5-1) includes elimination of the hazard, substitution of the hazard, engineering controls, administrative controls, and personal protective equipment (PPE)—each control approach has various advantages and disadvantages. Each control method can have different levels of success, with complete elimination being considered the most effective control. It should be noted that not every type of control may be feasible in a given situation, and the hierarchy of effectiveness is subject to variation. Additionally, some controls may be classified in more than one way. For example, elimination of indoor smoking in public buildings can be viewed both as source control and an administrative control depending on one's perspective.

TABLE 5-1 Hierarchy of Controls in the Context of Indoor Chemistry

Control Methodology	Examples	Comments
Elimination of the hazard	Elimination of lead paint use in indoor environments. Removal of flame retardants from furniture or clothing manufacturing.	Often easiest to implement in the design or development stage of a process. Afterward, changes in equipment and procedures may be required to eliminate or substitute for a hazard.
Substitution	Reformulating cosmetic nail products to replace toxic or irritating ingredients.	
Engineering controls	Local exhaust including the use of cooking stove hoods. General dilution ventilation systems (e.g., HVAC and HEPA systems). Air cleaning using particle filters. Simultaneous heating and exhausting a building to increase chemical emission rates prior to building occupancy ("bake-out"). Installation of vapor barriers to reduce intrusion of VOCs or radon into a building.	Exhaust hoods are designed to remove the hazard at the source. HVAC systems can dilute or filter air contaminants to reduce hazard.
Administrative controls	Educating consumers about chemical sources and their effects and operation of the ventilation system. Placing occupancy limits or other management rules in occupational settings.	Administrative controls and PPE may be needed when hazards are poorly controlled.
PPE	Use of respirators and other PPE.	

NOTE: HEPA = high-efficiency particulate air; HVAC = heating, ventilation, and air conditioning; PPE = personal protective equipment; VOC = volatile organic compound.

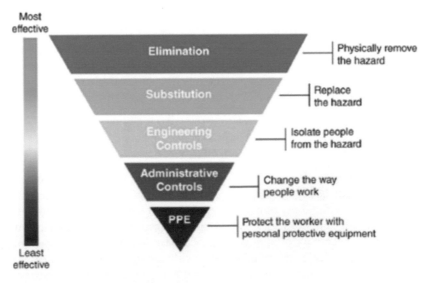

FIGURE 5-1 Hierarchy of controls. Less effective controls rely on individual compliance. For example, evidence indicates that mask wearing (a form of PPE) reduces the transmission of COVID-19 (Howard et al., 2021). However, low rates of compliance can reduce the effectiveness of this control strategy.
NOTE: PPE = personal protective equipment. SOURCE: NIOSH (2018).

Elimination of indoor contaminants could involve the removal of a chemical from a consumer product or another material, eliminating high emission rate sources (e.g., ozone generator). Elimination is also a viable approach to manage those chemicals (e.g., air fresheners or fragrances) found indoors that are introduced by human activity. Several specialized environments rely on eliminating chemicals of concern from construction materials. For example, the National Aeronautics and Space Administration has developed test procedures for the evaluation of off-gassing of chemicals from materials that could be used onboard the International Space Station or the Orion Multipurpose Crew Vehicle (NASA, 2011). Likewise, the U.S. Navy evaluates off-gassing from materials used in the construction of submarine decks (NAVSEA System Communications, 2016). Only dock materials that pass the testing requirements are permitted for subsequent use on a submarine. Both of these environments have limited access to replacement air and are associated with continuous exposure of occupants, making elimination of potentially hazardous materials from these specialized environments of paramount importance. The next most effective control method is substitution, which is the replacement of a chemical of concern found in a consumer product, furnishing, construction material, or other product with a less hazardous chemical. Because indoor chemistry is complex and often incompletely understood, caution is needed with this approach—replacement chemicals may contribute to unforeseen chemistry or later be found to be equally or more hazardous than the original chemical. Frameworks to guide the selection of alternative chemicals are available and need to be applied to decision making to reduce the likelihood of a regrettable substitution (NRC, 2014).

Engineering controls, such as mechanical ventilation or general exhaust systems that bring in outdoor air, filtration of gas-phase and particulate contaminants, and other types of air cleaners, further reduce the risk when elimination or substitution of a chemical hazard is not practical or incompletely mitigates the risk. Ventilation, exhaust, and filtration are commonly combined in heating, ventilation, and air-conditioning (HVAC) systems that also heat, cool, humidify, and dehumidify indoor spaces. Local exhausts, such as range hoods, control emissions from cooking at the source. For most offices and other non-manufacturing building work environments, these approaches have historically been the primary method to control indoor air quality (Woods, 1991). To reduce the entry of volatile organic compounds (VOCs) or radon through cracks or other defects in foundation slabs or basement walls into an overlying building (i.e., vapor intrusion), engineering controls include installation of fans, blowers, or physical and chemical barriers (EPA, 2015; Khan et al., 2019; Verginelli et al., 2017). Engineering controls seek to isolate and remove potentially hazardous material from the indoor environment.

Administrative controls include a range of regulations, policies, and procedures that limit exposure but require compliance on the part of the person. Reducing chemical exposure by altering human activity patterns can be an effective administrative control. For example, this could be accomplished by changing the application times for building maintenance chemicals and services (e.g., pesticide applications and air handler unit cleaning) to times that buildings are unoccupied. Other administrative controls include product safety labeling and training of end users to reduce chemical misuse. Another administrative control is the use of warning alarms (e.g., carbon monoxide or radon detectors) indoors.

PPE is considered the least effective of all control measures for chemical exposure in the workplace because it relies most heavily on individual compliance. Effective use of PPE requires individuals to be trained on its proper use and limitations. Well-fit respirators and other PPE offer direct protection for people who work with specific chemicals of concern. PPE is often used in occupational settings (Keer et al., 2018), with specialized masks for reducing exposure to paint fumes, volatile solvents, cleaning products, and other chemicals in a non-occupational setting. Masks and respirators can be effective for reducing exposure to some gases and particulate matter (PM), and the pollutants and pathogens associated with that PM.

However, no classification system is perfect and that this system is not always applicable. For example, while masks and respirators are typically considered PPE in terms of chemical exposure, they are more effectively considered source control in the case of exhaled airborne pathogens and thus are an engineering control.

MANAGEMENT THROUGH CAPTURE AND REMOVAL

Air contaminants can be removed, decomposed, or otherwise inactivated by ventilation, filtration, sorption, physical cleaning, and passive surface removal. These approaches have different levels of efficiency for removing gas- and particle-phase contaminants and are most effective for controlling risk when used in combination for comprehensive improvement of indoor air quality (i.e., the "swiss cheese model"). This section briefly describes each approach and summarizes its potential impacts on indoor chemistry.

Ventilation

For the purposes of this chapter, ventilation is defined as "intentional introduction of air from the outdoors into a building," primarily for the purpose of removing indoor air contaminants (ASHRAE, 2021). In some cases, ventilation air may be brought into a building without conditioning, but often ventilation air is heated, cooled, dehumidified, or humidified to indoor conditions, which is a major energy demand on building HVAC systems. This energy load can cause building managers or homeowners to reduce usage and lower ventilation rates.

Ventilation is the primary control for indoor air quality in most buildings, and, in principle, it can remove any type of indoor air contaminant that can be entrained in air. Removal by ventilation of volatile pollutants from continuous sources results in dilution and a temporary reduction in their indoor concentrations (Hult et al., 2015) but cannot reduce concentrations to zero. Once ventilation ceases, many pollutants may return to their pre-ventilation values (Wang et al., 2020). This is generally the case for VOCs emitted from building materials. Thus, while ventilation can increase dilution-driven emission rates of semivolatile compounds, the efficacy of this process for surface cleaning is unlikely to be substantial. However, pre-occupancy ventilation, sometimes combined with use of HVAC systems to raise the indoor temperature, could help to remove contaminants emitted during the curing of various building materials, resulting in lower indoor concentrations at occupancy (Kang et al., 2010; Kim et al., 2008). Particle filtration (see the next section) is incorporated into ventilation systems to remove both particles from the outdoor air being brought in and particles from indoor sources. In addition to being a source of particle-phase contaminants, ventilation air also brings in ozone and other ambient air pollutants from outdoors, potentially degrading indoor air quality and promoting undesirable chemical transformations (see Chapter 4).

Ventilation rates can affect indoor chemistry in multiple ways. Higher air change rates increase the rate at which contaminants in outdoor air, such as ozone and PM, enter a building, but they also shorten the average residence time of air indoors, which reduces the time available for reactions to occur in air. Thus, the consequence of increased ventilation rates may be both lower concentrations of indoor contaminants and reaction products and higher concentrations of outdoor contaminants (Weschler and Shields, 2000). On the other hand, surface chemistry is less affected. Moreover, the type of ventilation system may impact chemical reactions between outdoor and indoor air. Some systems do not recirculate indoor air (i.e., they supply 100 percent outdoor air to conditioned spaces and remove an approximately equal amount of air). Other types of systems bring in outdoor air and mix it with a generally much larger volume flow rate of recirculated indoor air, which is then conditioned to heat or cool. This may have an effect on indoor chemistry because ozone-containing outdoor air mixes with and can react with recirculated indoor air in the HVAC system before the mixture is supplied to occupied spaces.

The effectiveness of ventilation is greatly influenced by how outdoor air is brought into a building and distributed within spaces. Some buildings have "natural ventilation," in which air enters through openings in the building enclosure that may be engineered for that purpose or through operable windows (Izadyar et al., 2020). Natural ventilation via operable windows is more common in single-family residences and older buildings in the United States. Natural ventilation flow rate can be highly variable and depends on the placement and size of openings as well as the magnitude of driving forces due to wind and temperature differences (Li and Delsante, 2001). Larger, naturally ventilated buildings may have openings that are actively controlled (Saber et al., 2021). A primary reason for using natural ventilation is to reduce the energy used for cooling while maintaining occupant comfort and indoor air quality. Because natural ventilation could result in direct introduction of outdoor air into a building without any filtration or air treatment, its use in polluted areas raises serious concerns about impacts on indoor air quality (Chen et al., 2019).

Most new nonresidential buildings, and an increasing number of residential buildings, have "mechanical ventilation," which utilizes fans to draw outdoor air into the building. In some cases, 100 percent outdoor air is delivered to spaces via ductwork; in others, it is mixed with recirculating air before being distributed. The most common approach for room air distribution from mechanical systems is to mix it thoroughly with indoor air to dilute contaminants to a relatively uniform concentration. Stratified ventilation pushes contaminated air upward and out of the occupied zone to be removed at the top of the space. In theory, this more effectively removes contaminants for the same ventilation rate as mixing ventilation (Arghand et al., 2015; ASHRAE, 2021; Fatollahzadeh et al., 2015). Personal ventilation supplies ventilation air directly to occupants, achieving further increases in efficiency. Our understanding of the influence of these ventilation approaches on exposure that results from indoor chemistry is still developing.

Ventilation rates are set by building codes that rely on consensus standards, such as ASHRAE Standard 62.1, to achieve "acceptable" indoor air quality. The current definition of "acceptability" could be described as striving for safety with respect to air contaminants with adverse health effects (but excluding allergens, pathogens, and other biocontaminants) and occupant satisfaction in terms of odor control (80 percent "not dissatisfied" in ASHRAE Standard 62.1). Based on extensive experimental studies of response to body odor, an outdoor air flow rate of roughly 7.5 L/s (15 ft^3/min) per sedentary adult is required to achieve a rate of 20 percent of persons dissatisfied with air quality (Fanger and Berg-Munch, 1983). Although ASHRAE standards are the basis of codes primarily in the United States, ventilation standards used in other countries are similar in most respects, so ASHRAE standards are used as examples of the state of practice throughout this chapter (see Box 5-1). Historically, ventilation rates set in building regulations have at times been both much higher, when infection prevention was a consideration, and much lower, when influenced by the desire to reduce the energy consumption of buildings. As a result, minimum ventilation rates have varied by a factor of six since the 1830s (Janssen, 1999). Consideration of indoor chemistry has rarely entered into, or influenced, the standard-setting process. Very low ventilation rates adopted in the United States circa 1980 (as low as 2.5 L/s or 5 ft^3/min per person) were in a range correlated with high incidence of sick building syndrome symptoms and contemporaneous with the emergence of sick building syndrome as a significant indoor air quality problem (Fisk et al., 2009). Although many factors, both social and environmental, may contribute to sick building syndrome, inadequate ventilation rate (a surrogate for high exposures to contaminants) has been suggested by multiple studies. Compliance with ventilation codes is mainly prescriptive, meaning that acceptable indoor air quality is assumed if requirements for ventilation rates are met; however, performance-based approaches that require documentation of control of specific contaminants also exist and could predominate in the future. Design of mechanical ventilation and HVAC systems to meet applicable codes is necessary but not sufficient for ensuring healthy indoor environments: by setting the thermostat, turning HVAC systems on or off, and opening or closing windows, building occupants and facility managers can exert strong influence over how both mechanical and natural ventilation are utilized in the spaces they occupy (see Box 5-2).

BOX 5-1
The Role of Standards and Codes

Building codes are the regulations adopted by authorities that have jurisdiction over building construction, typically at the state or local level in the United States. Adopted building codes may be model building codes such as those published by the International Code Council, which are the basis for many state codes. Much of the content of codes originates in standards developed under a consensus process such as that of the American National Standards Institute (ANSI) by various standards developing organizations (SDOs). To the extent that standards specify equipment performance, they may cite certification programs that often depend on methods of test that are also developed by SDOs. A consensus process such as that of ANSI requires that all affected parties participate in the development of standards, that there is opportunity for public comment, and that comments are addressed. Key indoor air quality standards in the United States are developed by ASHRAE (formerly the American Society of Heating, Refrigerating and Air-Conditioning Engineers), an international nonprofit technical society that develops standards and educational programs and materials, and supports its own research program. ASHRAE ventilation standards, either by reference or by excerpt, are the basis for ventilation rates, air filtration efficiency, and air cleaner requirements in U.S. codes. The path from science to practice is, therefore, one that can take many years—from clear scientific findings, to commercial feasibility, to inclusion in standards, to adoption in codes. Significant time and effort are needed to put what may seem like clear scientific evidence into practical use in buildings.

The role of ventilation in mitigating risk of respiratory infection transmission has been widely discussed during the COVID-19 pandemic. Ventilation lowers the airborne concentration of infectious aerosols. Risk is related to the quantity of pathogen-containing air exhaled by an infected person that is inhaled by a susceptible one. Rudnick and Milton (2003) derived a relationship between secondary cases of an airborne disease and the rebreathed fraction of air (air inhaled that was previously exhaled by another person) and showed that it could be related to indoor carbon dioxide concentration because occupants are indoor carbon dioxide sources. This approach has also been applied to assessment of COVID-19 infection risk (Peng and Jimenez, 2021). A typical approach is to determine the critical rebreathed fraction resulting in a basic reproductive number, R_0, of 1. The basic reproductive number is the number of new infections resulting from a given

---

BOX 5-2
Human Behavior

Building occupants respond to and influence their environment in many ways, sometimes making choices that have more influence than other mitigation efforts. Efforts to control indoor climate by operating windows and doors has a substantial effect on outdoor-to-indoor air exchange (Becker et al., 2014; Bekö et al., 2004; Howard-Reed et al., 2002; Iwashita and Akasaka, 1997; Offermann, 2009). In most parts of the world, it is common for homes to have at least one door or window open sometime during the day during comfortable weather (Andersen et al., 2013; Calì et al., 2016; Johnson and Long, 2005; Lai et al., 2018; Morrison et al., 2022; Price and Sherman, 2006; Rijal et al., 2008, 2018; Tsang and Klepeis, 1996; Yao and Zhao, 2017). Even in very cold or hot weather, some homes keep windows open (Morrison et al., 2022). Open windows dilute indoor-sourced pollutants but also increase the indoor concentration of outdoor pollutants, including particulate matter and ozone. Occupant-determined thermostat settings for heating and cooling can influence not only temperature (and therefore partitioning, chemistry, and humidity; see Chapters 2-4) but also the cycling of heating and air-conditioning systems that filter the air (Stephens et al., 2011).

case and is a key parameter in the spread of an epidemic. Theoretically, if R_0 is greater than 1, an epidemic will spread at a rate that increases with R_0; if it is less than 1, it will die out. From the critical rebreathed air fraction, the necessary ventilation rate can be determined, as well as the resulting indoor carbon dioxide concentration that can be used in ventilation system control.

Outdoor air is a source of contaminants to the indoor environment, creating caveats for ventilation. Outdoor air has to meet certain minimum criteria to be considered suitable for use. For example, ASHRAE Standard 62.1 provides criteria for nonresidential buildings during regionally elevated ozone events. However, these standards do not apply to residential buildings and could be challenging to implement. Similarly, ASHRAE standards require enhanced particle filtration of outdoor air in nonresidential buildings when regional PM_{10} or $PM_{2.5}$ exceed national maximum standards, but such practices are neither required nor logistically straightforward in residential buildings. Adding these filtration or sorption controls is straightforward for mechanical ventilation systems but may be difficult in naturally ventilated buildings, particularly those that rely on windows to admit outdoor air. Furthermore, the extent to which nonresidential buildings adhere to these standards requires consideration when evaluating exposure. Extreme pollution events introduce pollution episodically, but residential environments typically rely on occupants to implement mitigation approaches. For example, wildfire smoke occurs sporadically and can have substantial impacts on indoor air quality and human exposure to pollutants. Building resilience to such events is important. However, modifying buildings to reduce outdoor air pollution is an engineering control, while reduction of outdoor air pollution, including both background and episodic events, constitutes removal of the hazard; improving outdoor air quality is thus consistent with the hierarchy of controls.

Filtration

Filters used in indoor air quality applications remove a wide range of particles from the air from indoor and outdoor sources (described in detail in Chapter 2). The range of sizes of these particles can cover at least four orders of magnitude, from less than 0.01 to more than 100 micrometers (μm). Not only are larger particles easier to filter but they also settle out of the air (i.e., deposit) more rapidly than smaller particles. Particles of greatest concern are those with adverse health effects because of size or chemical composition. Exposure to PM is associated with premature death, cardiac arrhythmias and heart attacks, and respiratory effects including asthma and bronchitis (Anderson et al., 2012). Translocation of inhaled ultrafine particles to the brain and other tissues may also contribute to systemic disease (Schraufnagel, 2020). Mechanical filtration is most commonly used to remove smaller particles from air in indoor air quality applications; however, electrostatic capture is also used, either independently (electrostatic precipitation) or as an enhancement to mechanical filtration.

Mechanical filters remove particles by capturing them on filter media as air passes through. Most filter media are mats of randomly oriented fibers of glass, synthetic, metal, and other materials. Open cell foam is also used as a filter media. Depending upon particle size and mass, particle capture results from a number of mechanisms. Large particles may deviate from the flow path of air moving around a fiber and be captured by impaction. Particles may also be captured by impingement if the flow path they are following brings them within less than one particle radius from the filter surface. Both impingement and impaction efficiency (i.e., fraction of particles removed on a single pass) decrease as particle size decreases. For smaller particles, diffusion becomes the predominant mechanism of capture and increases in efficiency as particle size decreases. Ultimately, a typical mechanical filter has its highest efficiency for the largest and smallest particles with a lower efficiency in an intermediate size range, generally 0.1–1 μm.

The performance of mechanical filters varies depending on several factors, including material, density, thickness, and flow rate, and is measured most commonly using ASHRAE Standard 52.2, which tests filters with particles of different sizes and rates them based on the minimum single-pass efficiency within each of three size ranges: 0.3–1, 1–3, and 3–10 μm (ASHRAE, 2017). Minimum Efficiency Reporting Value (MERV) varies from 1 to 16, with a higher MERV indicating better performance. The minimum requirement for nonresidential, non-health care buildings in ASHRAE Standard 62.1 is MERV 8, which has no minimum requirement for 0.3–1 μm particles, 20 percent for 1–3 μm, and 70 percent for 3–10 μm. Because of their low efficiency in the $PM_{2.5}$ range, MERV 6 and 8 filters do little to remove the fine particles associated with health effects. Their primary purpose is to remove larger particles that can foul HVAC equipment. Studies of the potential impact of more efficient filtration on mortality have found that use of higher-efficiency filters could yield large annual economic benefits from morbidity and mortality reductions and reduce incidence of airborne respiratory diseases, such as seasonal influenza (Azimi and Stephens, 2013).

Different standards are used to rate higher-efficiency filters, generally with a single particle size that is near the most penetrating size. High-efficiency particulate air (HEPA) filters are used in small air cleaners, clean rooms, laboratories, and health care spaces designed to isolate or protect patients. While there is some variation across different standards that define multiple levels of HEPA and ultra-high efficiency performance, the definition of a HEPA filter is one that removes at least 99.97 percent of 0.3 μm particles (roughly the most penetrating particle size) (White, 2009). A widely used standard for high performance filters is published by the Institute of Environmental Sciences and Technology (IEST Contamination Control Division, 2016).

The performance of some mechanical filters is enhanced by placing an electrostatic charge on filter fibers during the manufacturing process (electret filters). This enhances the performance of the filter by attracting charged particles in the air stream. The benefit of charging mechanical filters is that a desired performance level can be achieved with a thinner filter that has lower resistance to flow. The use of electret filters is controversial, however, as they may rapidly lose their charge and perform at much lower levels (Lee and Kim, 2020), while other studies indicate that stable performance can be achieved through selection of appropriate filter materials (Cai et al., 2020).

Electrostatic precipitators (ESPs) rely solely on electrostatic forces to remove particles from the air. While mechanical filtration is most common, ESPs are used in both commercial and residential buildings. Typically, a high voltage is applied between wires or pins and collector plates. Corona discharge from the wires ionizes the air and charges particles in the air stream, which move transversely to the air stream and deposit on the collector plates. Because the corona discharge in ESPs may produce ozone (Poppendieck et al., 2014), they need to be tested if intended for indoor use. ESP performance may be adversely affected by fouling. Experiments in which siloxanes (found in a number of personal care products) are present in air created silicon oxide deposits on positive corona discharge wires that could reduce particle collection efficiency (Davidson and McKinney, 1998). Performance of ordinary mechanical filters can also be enhanced by ionizing air upstream of the filter to create charged particles.

Because they collect particles rather than destroy them, mechanical filters need to be replaced periodically when they reach their maximum loading. An additional motivation for regular maintenance is the potential for filters to become sources of indoor pollutants. Particles captured on filter media continue to be exposed to recirculating indoor air and, consequently, can themselves become chemical pollution sources, contributing to odors (Bekö et al., 2004, 2007; Hyttinen et al., 2001; Pasanen et al., 1994; Pejtersen, 1996; Schleibinger and Rüden, 1999) and other adverse effects (Bekö et al., 2004, 2007; Lin and Chen, 2014; Sidheswaran et al., 2013; Siegel, 2016). Additionally, because filter materials or the particles they collect may constitute a food source for bacteria and fungi, it is possible for growth to occur on filters when sufficient moisture is present, creating further potential for filters to be a secondary source of indoor air pollution (Forthomme et al., 2014; Perrier et al., 2008). This is a practical concern of sufficient importance, and use of antimicrobial materials to control growth has been considered (Foarde and Hanley, 2001; Verdenelli et al., 2003).

Antimicrobial coatings also have been shown to reduce airborne microbial levels (Watson et al., 2022) but may or may not contribute to the chemical burden of indoor air.

Sorption

Sorbents are materials with high surface area that have a high capacity for adsorbing (or sometimes chemically reacting with) gas-phase contaminants. As an optional component of HVAC systems, sorbent filters may be included to help control odors and reduce VOC levels. Both physical and chemical sorption are used, with activated carbon being the primary physical adsorbent. Permanganate-impregnated alumina is a commercialized chemically sorbent medium (Han et al., 2017). Activated carbon may be impregnated with other materials to increase its effectiveness. The capacity of sorbents to hold various common air contaminants varies with the material; for example, activated carbon has a much higher capacity to adsorb toluene than permanganate-impregnated alumina but a significantly lower ability to adsorb formaldehyde (Spengler et al., 2001). All sorbents require maintenance as their capacity eventually becomes depleted and they need to be replaced. Quantified breakthrough times for formaldehyde at typical indoor concentrations with no competition from other chemicals through activated carbon filters is typically between 50 and 2,000 hours, depending on loading rates and activated carbon type (Ligotski et al., 2019; Zhu et al., 2019). Temperature and relative humidity can also impact sorbent capacity, with a 35 percent increase in relative humidity decreasing breakthrough times for perchloroethylene on activated carbon by a factor of two (ASTM D5160).

Some materials categorized as sorbents can be more accurately described as chemically transforming materials. For example, permanganate is an oxidant that can reduce concentrations of some VOCs; however, its potential to generate unwanted oxidized products requires more research. In addition to the impact of the sorbent medium, a number of other factors can affect sorbent performance, especially the concentration of chemicals in the air, composition of contaminants, and humidity. In general, low concentration, presence of competing species, and higher humidity all reduce sorbent effectiveness (Spengler et al., 2001; Underhill, 2001). In most HVAC systems, ventilation provides the only active means of removing gas-phase chemical contaminants from occupied spaces, with the possible exception of ambient ozone. Activated carbon can be used to remove ozone and is recommended by ASHRAE Standard 62.1 in circumstances where outdoor ozone levels exceed specified thresholds (Aldred et al., 2016a,b). By reducing indoor ozone concentrations, application of activated carbon filtration can therefore reduce ozone-initiated chemistry indoors (see Chapter 4).

Physical Cleaning of Surfaces

As described in Chapters 2, 3, and 4, surfaces play an important role in the chemical state of the indoor environment, leading to a variety of exposure mechanisms. Not only does touching surfaces or ingesting dust lead to chemical exposure but the gas-phase levels of many indoor air pollutants are controlled by partitioning interactions with a much larger quantity of those chemicals on the surfaces (see Chapter 3). Thus, a crucial point is that physical cleaning is important for lowering not only dermal and ingestion exposure but also inhalation exposure.

A number of processes occur with physical washing, with many relying on dissolving contaminants into the cleaning agent. For example, if the cleaning agent is water or water-based, water-soluble contaminants are easily removed. These include species such as small acids and bases, as well as highly oxygenated organics. If the cleaning agent is acidic (e.g., vinegar) or basic (e.g., ammonia-based), then it will be especially effective at removing basic and acidic molecules, respectively. An example is from cooking where basic amines can be formed. It is expected that these species will be removed using an acidic cleaner. However, water-based cleaners are not effective at removing nonpolar, hydrocarbon-like materials, such as cooking oils. As a result, physical cleaning often employs surfactant-containing materials that are able to solubilize such materials.

While the examples above rely on solubility to remove surface pollutants, other classes of cleaning agents have chemically active agents that are typically designed to oxidize contaminants. Common examples are chlorine-based bleach and hydrogen peroxide solutions. Both of these cleaners contain strong oxidizing agents that are particularly effective against microorganisms but also drive complex chemical reactions on the surface, as described in Chapter 4. Importantly, while the primary contaminant may be effectively destroyed by cleaning with such agents, chemical transformation products are left on the surface and may partition into the gas phase (Collins and Farmer, 2021). In many cases, it is not known whether these products are more or less toxic than the original target. In addition, the unused cleaning agent remains on the surface and may partition to the gas phase. For example, Wong et al. (2017) demonstrated that repeated washing of a floor with chlorine bleach solution led to progressively more and more hypochlorous acid (HOCl) gas added to the air in the room. This occurred because the floor was becoming cleaner and cleaner, such that HOCl did not react on the floor as much, with more available for inhalation exposure in the room instead.

Surfaces Engineered to Improve Indoor Air Quality

As noted in previous chapters, chemicals can deposit, adsorb, absorb, and react with surfaces of building materials, furnishings, and occupants. These phenomena can, in theory, be leveraged, and the materials can be engineered to improve indoor air quality. Removal of pollutants at indoor surfaces is an attractive option because of passive transport of pollutants to the large available surface area. The effective clean air delivery rate for small molecules in an indoor environment is typically ~2.5 m^3/h per square meter of engineered surface based on reported deposition velocities of ozone to passive removal materials (Darling et al., 2016). If all inner walls of a typical building removed contaminants at this rate, this would be equivalent to five or more air changes per hour of fresh air—but only for the specific contaminants removed at that surface.

In general, materials that only absorb or adsorb chemicals will eventually saturate and either require replacement or regeneration to continue to be effective. Products of this sort that have been promoted include odor removing paint and wallboard impregnated with activated carbon for VOC control. Upon saturation, dynamic changes in environmental conditions may effectively alter equilibrium conditions in a way that acts to periodically drive molecules off these materials, turning them into sources that increase exposure. However, it is also possible that day-night cycles of temperature or humidity could be used intentionally to regenerate the materials by desorbing them during non-occupied periods. Feasibility of this approach for carbon dioxide (CO_2)-sorbing coatings has been demonstrated (Rajan et al., 2017) and moisture control (buffering) in buildings through use of hygroscopic materials is the subject of many studies (Zhang et al., 2017).

To overcome saturation, removal at surfaces by chemical transformation has been proposed. Ozone reacts readily on many surfaces, and many available materials already remove ozone, some with minimal formation of byproducts (Cros et al., 2012; Gall et al., 2011; Kunkel et al., 2010; Lamble et al., 2011). To remove VOCs and nitrogen oxides more effectively, photocatalytic paints have been designed and tested for indoor use. Because these paints generally rely on available indoor lighting, photolytic energy and flux is much lower than in photocatalytic oxidation (PCO) units that use intense UV light. Therefore, these paints have limited effectiveness, only partially oxidize molecules (if at all) (Salthammer and Fuhrmann, 2007), can result in the net formation and release of formaldehyde and other VOCs (Gandolfo et al., 2018), and have been shown to effectively convert nitrogen oxides to nitrous acid (Gandolfo et al., 2015). Periodic renewal of engineered surfaces will be necessary since any systems that rely on surface chemistry will degrade over time as the active sites become soiled or "poisoned" with permanent deposits of reaction products. Despite the good intentions of designers, controls such as the use of plants or passive "green walls" are of limited effectiveness in improving air quality (Cummings and Waring, 2020; Irga et al., 2018).

Overall, there are possibilities for passive air cleaning through the use of indoor surfaces, but often the details are more complicated than just surfaces being effective adsorbents for indefinite periods of time. Saturation, re-emission, and reactive chemistry all play roles that require consideration before passive surfaces are implemented to remove indoor air pollutants.

MANAGEMENT THROUGH CHEMICAL TRANSFORMATIONS

Chemically modifying air pollutants to transform them into benign species or increase their removal rates is an increasingly used approach to improve indoor air quality. Devices that use chemical transformations can be additive (e.g., addition of an oxidant or other reactive chemical species to air), photolytic (e.g., application of UV light), or contained (e.g., photocatalysis systems that operate in a confined device, converting polluted air to "cleaner" air). The efficacy of these systems in removing air pollutants and their potential to create unintended byproducts require careful testing and investigation (Collins and Farmer, 2021; Siegel, 2016; Ye et al., 2021; Zhang et al., 2011). This section outlines each approach below.

Chemical Additions

The addition of chemical compounds to indoor air is a commonly proposed approach to air cleaning. These additions—whether through gas-phase or misting additions—raise concerns of secondary chemistry, which have been detailed by Collins and Farmer (2021).

Ozone is a biradical that reacts rapidly with alkenes and select inorganic species, notably including nitric oxide (NO) radicals. The concept behind ozone addition devices is to break down organic molecules, including those causing odors, through these oxidation reactions. However, there are few field studies of how the high levels of ozone from generators impact indoor chemistry. For example, Tang et al. (2021) showed that ozone could be effective at removing compounds associated with third-hand smoke but released other potential gases and induced ultrafine particle formation; as a result, that study proposed calculating minimum times before re-entry for occupants following ozone addition. Due to the chemistry described in Chapter 4, ozone addition to the built environment raises several concerns:

- The subsequent functionalization and fragmentation of organic molecules produces an array of oxygenated products, some of which are more toxic than the parent molecule. For example, ozonolysis of limonene, a common indoor VOC, produces formaldehyde (Weschler, 2006), while ozonolysis of skin oils and building materials produces an array of aldehydes and ketones (Wang and Morrison, 2010; Wisthaler and Weschler, 2010).
- Ozonolysis of VOCs is well established to produce condensable material that forms secondary organic aerosol (Hallquist et al., 2009), which has known health effects (Chowdhury et al., 2018).
- Ozone reacts with elastomers including natural rubbers, causing degradation and other unintended consequences, such as cracking of insulation on wiring.
- Ozone itself is a known air toxic, causing inflammation and cardiorespiratory effects.

Because of these concerns, the California Air Resources Board (CARB) recommends against the use of ozone generators. In 2010, CARB adopted a regulation requiring all indoor air cleaners sold in California to produce less than 50 ppb of ozone (California Code of Regulations, 2010).

Hydroxyl (OH) radicals are typically more broadly reactive with organic species than ozone, oxidizing not only alkenes by OH addition but also hydrocarbons by H abstraction to form water and an alkoxy radical (R) that quickly forms a peroxy radical (RO_2), which can undergo a series of reactions resulting in complex products, including peroxides, carbonyls, and carboxylic acids. Hydroxyl radical oxidation is not yet widely used in indoor environments to intentionally degrade VOCs but raises many of the same concerns as ozone, including the potential for production of secondary organic aerosol and other unintended byproducts due to its rapid oxidation chemistry (Friedman and Farmer, 2018; Lee et al., 2006). A recent study demonstrated that operating hydroxyl generator air-cleaning devices in an office environment increased PM and substantially enhanced oxidized organic compounds in the gas phase and secondary organic aerosol (Joo et al., 2021). However, chemically comprehensive studies of the emissions of these devices are lacking, and the extent to which ozone or other oxidants contributed to observed oxidation chemistry remains unexplored, particularly considering the short lifetime (order of seconds) of hydroxyl radicals in the indoor environment. Another investigation by Ye et al. (2021) studied multiple air cleaners that purported to remove VOCs by sorption and/or oxidative degradation and found that a range of byproducts was produced, including formaldehyde. The health effects of exposure to emissions and subsequent chemistry from these hydroxyl radical generators are not understood.

Odor-masking products and disinfectants are often added to indoor environments through fogging, spraying, or other vapor or droplet dispersal. The addition of scented products typically introduces reactive VOCs (e.g., monoterpenes) that can undergo chemical transformations and create unintended byproducts. The chemical mechanisms of oxidation reactions of individual VOCs are becoming better understood, but the interaction between these molecules and the chemical complexity present in indoor surfaces and air is not. A few commercial products act to trap odiferous compounds in cyclodextrin to reduce obnoxious smells (Hammer et al., 2013).

The addition of vaporized disinfectants, such as hydrogen peroxide, hypochlorous acid, and chlorine dioxide, has been used to decontaminate buildings and materials. Vaporized hydrogen peroxide can be effective in deactivating bacterial contamination (Johnston et al., 2005; Kahnert et al., 2005; Rudnick et al., 2009) but may be photolyzed to produce hydroxyl and peroxy radicals (Zhou et al., 2020), which can undergo further reactions indoors. The deposition of vaporized hydrogen peroxide on building materials induces the emission of various VOCs, although the extent to which this release is the result of reactive chemistry or simple displacement reactions remains unknown (Poppendieck et al., 2021). Hypochlorous acid (HOCl) is the conjugate acid of hypochlorite (OCl^-), the key ingredient in bleach and a well-established disinfectant. Fogging indoor environments with hypochlorous acid has emerged as a disinfectant strategy during the COVID-19 pandemic, but this approach has not yet been established in independent literature to effectively deactivate microbes. Box 5-3 briefly discusses other challenges of evaluating chemically transformative air-cleaning devices that have emerged during the pandemic. The addition of hypochlorous acid into indoor environments will initiate a series of reactions that can produce secondary aerosol (Mattila et al., 2020; Wang et al., 2019) and unintended byproducts, including cyanogen chloride, molecular chlorine, and organic isocyanates (Mattila et al., 2020; Wong et al., 2017). Research is beginning to elucidate the chemical reactions related to hypochlorous acid in the indoor environment, but the underlying chemical mechanisms and subsequent health consequences of these byproducts are largely unknown. HOCl reacts with skin oils (Schwartz-Narbonne et al., 2019) and may react similarly with lung or other tissue, also causing direct health impacts—and producing additional secondary products. Vaporized triethylene glycol (TEG) has long been known to destroy airborne pathogens (Lester Jr et al., 1952; Rosebury et al., 1947) and is far less reactive than the other surface disinfectants described here. The potential for TEG to react with other molecules in the indoor environment has not been substantively explored in the scientific literature, although the high boiling point and viscous nature of TEG suggests that it may accumulate on surfaces and impact surface-air partitioning.

Fumigation with gas-phase chlorine dioxide (ClO_2) has also been used as a building-wide disinfectant for bacteria and fungi, including in response to anthrax attacks or mold remediation

---

BOX 5-3
Air Cleaning for COVID-19

With the growing public understanding that SARS-CoV-2 can be transmitted via exhaled aerosol, the COVID-19 pandemic has created an increasing demand for air cleaning that removes particles or deactivates airborne pathogens. Established technologies for airborne pathogens include enhanced building ventilation and filtration, portable air cleaners with high-efficiency particulate air (HEPA) filters, and germicidal ultraviolet (UV) systems. However, each of these approaches require careful implementation. For example, clean air delivery rates (CADR) for portable air cleaners with filters have to match room size, while germicidal UV systems require careful selection and installation to avoid human exposure or unintended production of ozone. However, many other air-cleaning devices have rapidly entered the market with far less evidence of effectiveness and safety. Some devices use chemically additive approaches as described in this section. The effectiveness of these additive air-cleaning devices in deactivating airborne pathogens is particularly difficult to quantify as these pathogens include viruses or bacteria encapsulated in aerosols made of respiratory fluid. The lack of established testing protocols means that many devices designed to remove airborne particles are tested using surfaces (i.e., petri dishes) that lack the potentially protective aerosol encapsulation. Challenges in evaluating the myriad technologies available include

- a lack of peer-reviewed literature investigating the efficacy of these devices and their claims for particle removal in real-world settings;
- a lack of consistent testing methodology or regulatory oversight to determine potential health risks from hazardous emissions or byproduct formation in real-world settings;
- inconsistent terminology to describe technologies, their efficacy, and the extent of testing;
- a lack of certification programs and regulatory oversight to ensure that only effective and safe products are permitted in the marketplace; and
- a lack of application-relevant metrics (similar to CADR) that clearly communicate performance.

(Hsu et al., 2015; Hubbard et al., 2009). However, ClO_2 is both photolabile and reactive, thus raising concerns over byproduct formation. ClO_2 can also be sorbed to, and react with, material surfaces, raising concerns over material degradation (Derkits et al., 2010).

In summary, the literature suggests that chemical additions and modifications to indoor environments are sometimes effective tools in the arsenal for decontamination, but they can enable secondary chemistry: the potential formation of unintended byproducts with unknown implications for exposure and health. The physical and chemical mechanisms responsible for byproduct formation—and the health effects of those byproducts in real-world indoor environments—warrants further study.

Ultraviolet Light

Photolysis, the decomposition of molecules due to interaction with light, can also be used to control chemical air contaminants and inactivate or kill microorganisms. Indeed, indoor air quality applications of UV light date to at least the 1930s, when it was first used for air disinfection in operating rooms (Hart, 1936) and schools (Wells et al., 1942), and it was in use even earlier for drinking water disinfection (von Recklinghausen, 1914). Because it has been a subject of intensive study for more than a century, much is known about the effectiveness of UV. In practice, photolysis is employed for indoor air quality control via ultraviolet germicidal irradiation (UVGI) systems. UVGI is an accepted adjunct to ventilation and filtration for control of tuberculosis as noted in the Centers for Disease Control and Prevention's guidelines (Jensen et al., 2005; NIOSH, 2009).

Most UVGI systems utilize 254 nanometer (nm) UV-C, the predominant wavelength produced by low pressure mercury vapor lamps. While less hazardous than light in the UV-B range, which can penetrate more deeply into the skin and increase skin cancer risk, direct exposure to UV-C can cause painful, although transient, eye and skin irritation and has a lower (but nonzero) level of carcinogenicity (Sliney and Stuck, 2021). Emerging light-emitting diode (LED) and excimer lamp technology offers the prospect of a wider range of available wavelengths in the future that will lead to safer and more effective UVGI systems (Buonanno et al., 2020; Ma et al., 2021). UV-C air disinfection using 254 nm sources is applied in a number of ways in indoor environments (Kowalski, 2010):

- Upper room systems, in which fixtures are placed in an occupied space to create a disinfection zone above the occupied zone to protect occupants from exposure. Such systems continuously expose air and surfaces in the space. The effectiveness of an upper room system depends on air circulation between the occupied and disinfection zones in a space. This can be driven by thermal plumes from people and equipment and by ventilation systems.
- Airstream disinfection, in which lamps are placed in air distribution ducts or in air-handling units. They may be positioned to simultaneously prevent microbial growth on the wetted surfaces of cooling coils, which not only control air temperature but also dehumidify by condensing moisture on their surfaces.
- Standalone air cleaners (i.e., enclosed air cleaners located in occupied spaces that typically include fans). Some air cleaners that rely primarily on high-efficiency mechanical filters or other technologies to disinfect air include UV lamps for the purpose of disinfecting filter surfaces.

Research into the indoor chemistry associated with the use of UVGI systems is sparse and primarily limited to ozone production. However, very little is known about production of other byproducts by UVGI systems as they are typically applied. UV light is known to photolyze some compounds, but little real-world testing has been done to identify the extent to which this affects indoor air. Degradation of materials that are exposed to UV-C in HVAC systems has been studied but not the impact of these processes on indoor air quality (Kauffman and Wolf, 2012). One study has investigated secondary organic aerosol formation resulting from exposure of toluene to 254 nm UV-C (Choi et al., 2019). Particle formation was observed in a test chamber at high toluene concentration (55 to 85 mg/m^3) and UV doses (19.5 mW/cm^2). However, tests in a bathroom with an upper room fixture yielded no measurable particle generation, possibly owing to both VOC concentrations and UV doses being orders of magnitude smaller than in the test chamber.

Photocatalysis

PCO is an approach that relies on the UV activation of catalysts, such as titanium dioxide (TiO$_2$), to convert molecular oxygen to hydroxyl and superoxide radicals, which then oxidize organic and inorganic gases (Chen et al., 2012; Hay et al., 2015; Huang et al., 2016). In general, the molecule has to adsorb to the catalyst, where it becomes sequentially oxidized. If the molecule and/or resulting oxidized intermediates are retained on the catalyst for a sufficiently long time, the final products are CO$_2$ and water (Tompkins et al., 2005). However, application of PCO is challenged in indoor applications by the formation and release of partially oxidized byproducts, including formaldehyde (Destaillats et al., 2012; Farhanian and Haghighat, 2014; Haghighatmamaghani et al., 2019; Hay et al., 2015; Sleiman et al., 2009). Transformation of nitrogen oxides to nitrous acid has also been demonstrated (Gligorovski, 2016). Removal efficiency decreases and byproduct formation increases as humidity and VOC concentrations increase (Sleiman et al., 2009; Yu et al., 2006). Some light-activated systems rely on visible rather than UV light; these may be even more susceptible to incomplete oxidation and byproduct formation. Fouling of the catalyst from an array of compounds present in indoor air, including benzaldehyde, benzoic acid, and volatile siloxanes, can decrease the lifespan of photocatalytic devices, decrease removal efficiency, and increase byproduct formation, further reducing their effectiveness (Cao et al., 2000; Hay et al., 2015). While PCO efficiency and byproduct formation have been investigated extensively in controlled laboratory environments, more studies of effectiveness, longevity, and byproducts in real-world indoor environments are needed to understand the potential for subsequent chemical transformations. In particular, more accurate evaluations of these devices would take place in occupied buildings, because occupants are a major source of siloxanes that can decrease efficiency and increase byproduct formation (Tang et al., 2015).

Ionizers

Ions are positively or negatively charged species. At ground level, total ion concentrations in the atmosphere are thought to be on the order of hundreds of ions per cubic centimeter (Beig and Brasseur, 2000). In the gas phase, these species are reactive in the atmosphere, with tropospheric lifetimes measured in seconds. This reactivity means that ions can react with trace gases and particles, potentially transforming pollutants into either new or more easily removed forms. For example, charging particles may enhance particle agglomeration processes, leading to larger particles that deposit more rapidly or are filtered more easily than the original size distribution. However, few peer-reviewed studies have investigated whether these ionizers work in real-world environments (Pushpawela et al., 2017), with questions raised over their ability to enhance particle removal (Tang et al., 2021). In addition to questions of efficacy, ion generators face several challenges: negative health effects from exposure to high ion levels (Liu et al., 2021), the potential for ozone formation following ionization of molecular oxygen, the potential for byproduct formation from ion-molecule reactions, and the potential for nucleation of new particles through reactions of ions and molecules (Zeng et al., 2021).

OTHER CONSIDERATIONS FOR MANAGEMENT OF CHEMICALS

Environmental Factors

Ambient and indoor environmental factors can impact the effectiveness and choice of control technologies. While mechanical or natural ventilation are utilized in virtually all buildings, the extent to which they are relied upon is strongly dependent on local climate, which impacts the amount of energy consumption necessary to bring outdoor air to indoor conditions. Particularly in extreme climates, there is a strong economic disincentive to use increased outdoor air supply as a means of enhanced control. Air-to-air energy recovery between outdoor air intake and exhaust streams is required in some systems in most climate zones (ASHRAE, 2019) to mitigate the energy cost of ventilation, but it remains a major contributor to the total energy use of a building. Additionally, as noted earlier in this chapter, outdoor air can be a significant source of contaminants. High ambient humidity is also a concern in some climates because some types of HVAC equipment have limited ability to dehumidify, which can lead to indoor moisture problems affecting chemistry (see Chapter 4) and mold growth. Likewise, temperature and humidity can affect the performance of sorbents. They also affect natural sorption/desorption processes leading to variable emission rates in indoor spaces (Haghighat and De Bellis, 1998; Markowicz and Larsson, 2015).

Occupational Risk and Specialized Environments

Specific built environments have distinct air-cleaning needs that may provide insight into more general settings. While industrial settings are not within the scope of this report, and typically follow the hierarchy of controls summarized earlier, air quality in aircraft has been considered in some detail (NRC, 2002; Spengler and Wilson, 2003; Zhang et al., 2011). Aircraft cabins without ozone catalysts are often subject to high ozone levels, with peak ranges of 30–275 ppb at cruising altitude (Bhangar and Nazaroff, 2013; Bhangar et al., 2008; Bekö et al., 2015; Spengler et al., 2004; Spicer et al., 2004; Weisel et al., 2013). Newer airplanes with properly serviced ozone catalyst systems can have peak ozone concentrations below 50 ppb (Bhangar et al., 2008; Weisel et al., 2013). Ozone concentrations in airplane cabins can be high enough to raise concern over exposure as both a primary air pollutant and initiator of secondary chemistry (Coleman et al., 2008; Gao et al., 2015; Weisel et al., 2013). Aircraft typically moderate air quality with enhanced ventilation and filtration, although chemically additive approaches have been considered. For example, Wisthaler et al. (2007) investigated various PCO and sorption air purifiers in aircraft cabins and found that while many were effective at reducing total VOC loading, photocatalytic devices incompletely oxidized ethanol into unacceptable levels of acetaldehyde and formaldehyde. However, the high ozone levels present in some aircraft can react with soiled filters and sorptive surfaces (Spengler et al., 2004; Spicer et al., 2004; also see Chapter 4 for more on ozone chemistry). Other unique environments that require specialized approaches to air cleaning include submarines and the International Space Station.

Economics and Sustainability

While the primary criteria for selection of any indoor environmental control technology are efficacy and safety, other considerations also weigh heavily in such decisions. The most important of these factors is economics, which includes the cost of equipment, operating costs, and energy use. Costs have always been a major concern for owners of indoor air quality control systems, but, increasingly, emission of greenhouse gases in the production of energy used to operate building systems is also a factor in decisions. What is considered an acceptable cost depends on the purpose of the facility. For example, high expenditures may be justified for systems protecting vulnerable occupants, such as health care facilities. Because buildings are required by building codes to meet minimum outdoor air supply requirements, the cost of a ventilation system is unavoidable. However, this can range from a mechanical ventilation system that delivers outdoor air as required whenever a building is occupied to operable windows that provide a much lower level of reliability because they depend on occupant behavior. Chemical filtration and air cleaning, on the other hand, is not mandatory for most buildings, and some types of equipment add significant initial costs and ongoing maintenance costs to replace consumable components (e.g., sorbent filter media). On the other hand, because outdoor air brought in to control contaminant levels needs to be conditioned to indoor temperature and humidity levels, increasing ventilation rates above minimum requirements for the purpose of control could, depending on climate, result in greater incremental energy use and incur higher operating costs than filtration and cleaning of recirculated air. For the same reason, air cleaners may be viewed as a more sustainable alternative to ventilation for control of indoor contaminants. Despite the attractiveness of filtration and air cleaning, ventilation remains the predominant means of control because of the difficulty of properly specifying air-cleaning systems to achieve overall indoor air quality goals. In order to apply air-cleaning technologies effectively, a definition of indoor air quality in terms of measurable contaminants is needed. An addendum to ASHRAE Standard 62.1 that proposes threshold values for 14 compounds and $PM_{2.5}$ was approved in 2021. Given the possible presence of a much larger number of compounds in indoor environments, however, there is a clear need for advances in both analytical instrumentation and our understanding of indoor chemistry to move such efforts forward. Applied research on energy-optimal use of combinations of control technologies is also needed.

CONCLUSIONS

The management of chemical contaminants in indoor environments includes removal (through ventilation, filtration, sorption, physical cleaning, and passive surface removal) and chemical transformations (including photolysis, ionizers, chemical additions, and photocatalysis). No single management approach can remove all contaminants that are present indoors; therefore, source elimination is always the preferred method of control. However, combinations of management approaches can also be effective at reducing exposure, as can situation-specific choices, such as increasing ventilation to reduce air contaminant exposure. There are different chemical consequences of every management approach. Approaches that include oxidation are particularly prone to generating products of concern. While this chapter highlights the scientific community's knowledge of underlying physical and chemical principles of air cleaning, several knowledge gaps remain, including the fundamental chemistry of many air-cleaning technologies. With the exception of ventilation, particle filtration, and sorption, few air-cleaning approaches are tested in real-world environments, which contain a far more complicated mixture of compounds than most laboratory studies. As outlined in previous chapters, chemical reactions in indoor environments can follow complex mechanisms and result in numerous different products. This makes predicting chemical reactions and the efficacy of air-cleaning devices challenging—and highlights the need for better testing standards for air-cleaning efficacy and chemistry that account for this complexity. There is insufficient chemical research to truly understand and prioritize chemical byproducts in terms of toxicology and health effects, and to identify safe and effective levels of chemical additives for air-cleaning technologies.

RESEARCH NEEDS

In particular, the committee highlights the need for research incorporating real-world testing of management approaches that are anticipated to induce chemical transformations, specifically the following:

- **Testing approaches need to be developed that consider both efficacy and byproduct formation in a representative range of real-world environments** (e.g., ultrafine particles, $PM_{2.5}$, oxygenated VOCs including formaldehyde). Different chemicals induce different types of chemistry, so any testing approach has to be flexible enough to account for likely products, and the complexity of indoor chemistry means that non-targeted analysis approaches could be useful. These tests and measurements can help inform a quantitative assessment of thresholds for health effects for relevant compounds.
- **Developers of air-cleaning technologies need to recognize that many gas-phase molecules in indoor air partition with indoor surface reservoirs.** Cleaning the air may not substantially remove the contamination if a large amount of that molecule remains on the surfaces and re-partitions to the gas phase after air cleaning stops. Surface cleaning must thus be coincident with air cleaning, although comprehensive surface removal is rarely feasible for compounds adsorbed to, for example, paint surfaces.
- **Controlled field experiments are necessary to better understand the fundamental chemistry of emerging air-cleaning technologies, as well as mold and smoke remediation schemes.**

Finally, given the recent public interest in indoor air quality, device manufacturers, researchers, and public health professionals need to communicate clearly to consumers about the efficacy and chemical consequences of different air-cleaning approaches. The lack of testing and regulation has led to rampant unsubstantiated claims about efficacy and health benefits of devices. The potential health risks and benefits resulting from their use warrant further investigation and potential certification or regulatory oversight. Based on the current state of knowledge, the committee cautions against approaches that induce secondary chemistry in occupied settings, unless the benefits demonstrably outweigh the risks of exposure to chemical reactants and byproducts.

REFERENCES

Aldred, J. R., E. Darling, G. Morrison, J. Siegel, and R. Corsi. 2016a. Benefit-cost analysis of commercially available activated carbon filters for indoor ozone removal in single-family homes. *Indoor Air* 26:501–512.

Aldred, J. R., E. Darling, G. Morrison, J. Siegel, and R. L. Corsi. 2016b. Analysis of the cost effectiveness of combined particle and activated carbon filters for indoor ozone removal in buildings. *Science and Technology for the Built Environment* 22:227–236.

Andersen, R., V. Fabi, J. Toftum, S. P. Corgnati, and B. W. Olesen. 2013. Window opening behaviour modelled from measurements in Danish dwellings. *Building and Environment* 69:101–113. https://doi.org/10.1016/j.buildenv.2013.07.005.

Anderson, J. O., J. G. Thundiyil, and A. Stolbach. 2012. Clearing the air: A review of the effects of particulate matter air pollution on human health. *Journal of Medical Toxicology* 8:166–175. https://doi.org/10.1007/s13181-011-0203-1.

Arghand, T., T. Karimipanah, H. B. Awbi, M. Cehlin, U. Larsson, and E. Linden. 2015. An experimental investigation of the flow and comfort parameters for under-floor, confluent jets and mixing ventilation systems in an open-plan office. *Building and Environment* 92:48–60. https://doi.org/10.1016/j.buildenv.2015.04.019.

ASHRAE. 2017. *Standard 52.2—Method of Testing General Ventilation Air-Cleaning Devices for Removal Efficiency by Particle Size*.

ASHRAE. 2019. *Standard 62.1—Ventilation for Acceptable Indoor Air Quality*.

ASHRAE. 2021. *ASHRAE Handbook—Fundamentals*. Atlanta, GA: ASHRAE.

ASTM International. 2019. *ASTM D5160-95 Standard Guide for Gas-Phase Adsorption Testing of Activated Carbon*.

Azimi, P., and B. Stephens. 2013. HVAC filtration for controlling infectious airborne disease transmission in indoor environments: Predicting risk reductions and operational costs. *Building and Environment* 70:150–160. https://doi.org/10.1016/j.buildenv.2013.08.025.

Becker, R., G. Haquin, and K. Kovler. 2014. Air change rates and radon accumulation in rooms with various levels of window and door closure. *Journal of Building Physics* 38(3):234–261. https://doi.org/10.1177/1744259113506071.

Beig, G., and G. P. Brasseur. 2000. Model of tropospheric ion composition: A first attempt. *Journal of Geophysical Research: Atmospheres* 105(D18):22671–22684. https://doi.org/10.1029/2000JD900119.

Bekö, G., J. G. Allen, C. J. Weschler, J. Vallarino, and J. D. Spengler. 2015. Impact of cabin ozone concentrations on passenger reported symptoms in commercial aircraft. *PLOS ONE* 10(5):e0128454. https://doi.org/10.1371/journal.pone.0128454.

Bekö, G., G. Clausen, and C. J. Weschler. 2007. Further studies of oxidation processes on filter surfaces: Evidence for oxidation products and the influence of time in service. *Atmospheric Environment* 41(25):5202–5212. https://doi.org/10.1016/j.atmosenv.2006.07.063.

Bekö, G., O. Halás, G. Clausen, and C. J. Weschler. 2004. Ventilation filters as sources of air pollution–Processes occurring on surfaces of used filters. Presented at Indoor Climate of Buildings '04, Strbske Pleso, November 21–24, 2004.

Bhangar, S., S. C. Cowlin, B. C. Singer, R. G. Sextro, and W. W. Nazaroff. 2008. Ozone levels in passenger cabins of commercial aircraft on North American and transoceanic routes. *Environmental Science & Technology* 42(11):3938–3943. https://doi.org/10.1021/es702967k.

Bhangar, S., and W. W. Nazaroff. 2013. Atmospheric ozone levels encountered by commercial aircraft on transatlantic routes. *Environmental Research Letters* 8(1):014006. https://doi.org/10.1088/1748-9326/8/1/014006.

Buonanno, M., D. Welch, I. Shuryak, and D. J. Brenner. 2020. Far-UVC light (222 nm) efficiently and safely inactivates airborne human coronaviruses. *Scientific Reports* 10:10285. https://doi.org/10.1038/s41598-020-67211-2.

Cai, R.-R., S.-Z. Li, L.-Z. Zhang, and Y. Lei. 2020. Fabrication and performance of a stable micro/nano composite electret filter for effective $PM_{2.5}$ capture. *Science of the Total Environment* 725:138297. https://doi.org/10.1016/j.scitotenv.2020.138297.

Calì, D., R. K. Andersen, D. Müller, and B. W. Olesen. 2016. Analysis of occupants' behavior related to the use of windows in German households. *Building and Environment* 103:54–69. https://doi.org/10.1016/j.buildenv.2016.03.024.

California Code of Regulations. 2010. Regulation for Limiting Ozone Emissions from Indoor Air Cleaning Devices. California Code of Regulations Title 17. Public Health Division 3. Air Resources Chapter 1. Air Resources Board Subchapter 8.7. Indoor Air Cleaning Devices Article 1. Indoor Air Cleaning Devices. https://ww2.arb.ca.gov/sites/default/files/2020-03/air-cleaner-regulation.pdf.

Cao, L., Z. Gao, S. L. Suib, T. N. Obee, S. O. Hay, and J. D. Freihaut. 2000. Photocatalytic oxidation of toluene on nanoscale TiO_2 catalysts: Studies of deactivation and regeneration. *Journal of Catalysis* 196(2): 253–261. https://doi.org/10.1006/jcat.2000.3050.

Chen, H., C. E. Nanayakkara, and V. H. Grassian. 2012. Titanium dioxide photocatalysis in atmospheric chemistry. *Chemical Reviews* 112(11): 5919–5948. https://doi.org/10.1021/cr3002092.

Chen, J., G. S. Brager, G. Augenbroe, and X. Song. 2019. Impact of outdoor air quality on the natural ventilation usage of commercial buildings in the US. *Applied Energy* 235:673–684. https://doi.org/10.1016/j.apenergy.2018.11.020.

Choi, E., Z. Tan, and W. A. Anderson. 2019. Formation of secondary organic aerosols by germicidal ultraviolet light. *Environments* 6(2):17. https://doi.org/10.3390/environments6020017.

Chowdhury, P. H., Q. He, T. Lasitza Male, W. H. Brune, Y. Rudich, and M. Pardo. 2018. Exposure of lung epithelial cells to photochemically aged secondary organic aerosol shows increased toxic effects. *Environmental Science & Technology Letters* 5(7):424–430. https://doi.org/10.1021/acs.estlett.8b00256.

Coleman, B. K., H. Destaillats, A. T. Hodgson, and W. W. Nazaroff. 2008. Ozone consumption and volatile byproduct formation from surface reactions with aircraft cabin materials and clothing fabrics. *Atmospheric Environment* 42(4):642–654. https://doi.org/10.1016/j.atmosenv.2007.10.001.

Collins, D. B., and D. K. Farmer. 2021. Unintended consequences of air cleaning chemistry *Environmental Science & Technology* 55(18):12172–12179. https://doi.org/10.1021/acs.est.1c02582.

Cros, C., G. Morrison, J. Siegel, and R. Corsi. 2012. Long-term performance of passive materials for removal of ozone from indoor air. *Indoor Air* 22(1):43–53. https://doi.org/10.1111/j.1600-0668.2011.00734.x.

Cummings, B. E., and M. S. Waring. 2020. Potted plants do not improve indoor air quality: A review and analysis of reported VOC removal efficiencies. *Journal of Exposure Science & Environmental Epidemiology* 30(2):253–261. https://doi.org/10.1038/s41370-019-0175-9.

Darling, E., G. C. Morrison, and R. L. Corsi. 2016. Passive removal materials for indoor ozone control. *Building and Environment* 106:33–44. https://doi.org/10.1016/j.buildenv.2016.06.018.

Davidson, J. H., and P. J. McKinney. 1998. Chemical vapor deposition in the corona discharge of electrostatic air cleaners. *Aerosol Science and Technology* 29(2):102–110.

Derkits, G. E., M. L. Mandich, W. D. Reents, J. P. Franey, C. Xu, D. Fleming, R. Kopf, and S. Ryan. 2010. Reliability of electronic equipment exposed to chlorine dioxide used for biological decontamination. 2010 IEEE International Reliability Physics Symposium, Anaheim, CA, May 2-6-2010. https://doi.org/10.1109/IRPS.2010.5488715.

Destaillats, H., M. Sleiman, D. P. Sullivan, C. Jacquiod, J. Sablayrolles, and L. Molins. 2012. Key parameters influencing the performance of photocatalytic oxidation (PCO) air purification under realistic indoor conditions. *Applied Catalysis B: Environmental* 128:159–170. https://doi.org/10.1016/j.apcatb.2012.03.014.

EPA (U.S. Environmental Protection Agency). 2015. OSWER technical guide for assessing and mitigating the vapor intrusion pathway from subsurface vapor sources to indoor air. *EPA OSWER Publication* 9200:2–154.

Fanger, P. O., and B. Berg-Munch. 1983. Ventilation and body odor. *Proceedings of an Engineering Foundation Conference on Management of Atmospheres in Tightly Enclosed Spaces*, Santa Barbara, CA, October 17–21, 1983. https://www.aivc.org/resource/ventilation-and-body-odor.

Farhanian, D., and F. Haghighat. 2014. Photocatalytic oxidation air cleaner: Identification and quantification of by-products. *Building and Environment* 72:34–43. https://doi.org/10.1016/j.buildenv.2013.10.014.

Fathollahzadeh, M. H., G. Heidarinejad, and H. Pasdarshahri. 2015. Prediction of thermal comfort, IAQ, and energy consumption in a dense occupancy environment with the under floor air distribution system. *Building and Environment* 90:96–104. https://doi.org/10.1016/j.buildenv.2015.03.019.

Fisk, W. J., A. G. Mirer, and M. J. Mendell. 2009. Quantitative relationship of sick building syndrome symptoms with ventilation rates. *Indoor Air* 19(2):159–165. https://doi.org/10.1111/j.1600-0668.2008.00575.x.

Foarde, K. K., and J. T. Hanley. 2001. Determine the efficacy of antimicrobial treatments of fibrous air filters. *ASHRAE Transactions* 107:156.

Forthomme, A., A. Joubert, Y. Andrès, X. Simon, P. Duquenne, D. Bemer, and L. Le Coq. 2014. Microbial aerosol filtration: Growth and release of a bacteria–fungi consortium collected by fibrous filters in different operating conditions. *Journal of Aerosol Science* 72:32–46. https://doi.org/10.1016/j.jaerosci.2014.02.004.

Friedman, B., and D. K. Farmer. 2018. SOA and gas phase organic acid yields from the sequential photooxidation of seven monoterpenes. *Atmospheric Environment* 187:335–345. https://doi.org/10.1016/j.atmosenv.2018.06.003.

Gall, E. T., R. L. Corsi, and J. A. Siegel. 2011. Barriers and opportunities for passive removal of indoor ozone. *Atmospheric Environment* 45(19):3338–3341. https://doi.org/10.1016/j.atmosenv.2011.03.032.

Gandolfo, A., V. Bartolomei, E. G. Alvarez, S. Tlili, S. Gligorovski, J. Kleffmann, and H. Wortham. 2015. The effectiveness of indoor photocatalytic paints on NOx and HONO levels. *Applied Catalysis B: Environmental* 166:84–90. https://doi.org/10.1016/j.apcatb.2014.11.011.

Gandolfo, A., S. Marque, B. Temime-Roussel, R. Gemayel, H. Wortham, D. Truffier-Boutry, V. Bartolomei, and S. Gligorovski. 2018. Unexpectedly high levels of organic compounds released by indoor photocatalytic paints. *Environmental Science & Technology* 52(19):11328–11337. https://doi.org/10.1021/acs.est.8b03865.

Gao, K., J. Xie, and X. Yang. 2015. Estimation of the contribution of human skin and ozone reaction to volatile organic compounds (VOC) concentration in aircraft cabins. *Building and Environment* 94:12–20. https://doi.org/10.1016/j.buildenv.2015.07.022.

Gligorovski, S. 2016. Nitrous acid (HONO): An emerging indoor pollutant. *Journal of Photochemistry and Photobiology A: Chemistry* 314:1–5. https://doi.org/10.1016/j.jphotochem.2015.06.008.

Haghighat, F., and L. De Bellis. 1998. Material emission rates: Literature review, and the impact of indoor air temperature and relative humidity. *Building and Environment* 33(5):261–277. https://doi.org/10.1016/S0360-1323(97)00060-7.

Haghighatmamaghani, A., F. Haghighat, and C.-S. Lee. 2019. Performance of various commercial TiO_2 in photocatalytic degradation of a mixture of indoor air pollutants: Effect of photocatalyst and operating parameters. *Science and Technology for the Built Environment* 25(5):600–614. https://doi.org/10.1080/23744731.2018.1556051.

Hallquist, M., J. C. Wenger, U. Baltensperger, Y. Rudich, D. Simpson, M. Claeys, J. Dommen, N. M. Donahue, C. George, A. H. Goldstein, J. F. Hamilton, H. Herrmann, T. Hoffmann, Y. Iinuma, M. Jang, M. E. Jenkin, J. L. Jimenez, A. Kiendler-Scharr, W. Maenhaut, G. McFiggans, T. F. Mentel, A. Monod, A. S. H. Prévôt, J. H. Seinfeld, J. D. Surratt, R. Szmigielski, and J. Wildt. 2009. The formation, properties and impact of secondary organic aerosol: Current and emerging issues. *Atmospheric Chemistry and Physics* 9(14):5155–5236. https://doi.org/10.5194/acp-9-5155-2009.

Hammer, T. R., N. Berner-Dannenmann, and D. Hoefer. 2013. Quantitative and sensory evaluation of malodour retention of fibre types by use of artificial skin, sweat and radiolabelled isovaleric acid. *Flavour and Fragrance Journal* 28:238–244.

Han, K. H., J. S. Zhang, and B. Guo. 2017. Toward effective design and adoption of catalyst-based filter for indoor hazards: Formaldehyde abatement under realistic conditions. *Journal of Hazardous Materials* 331:161–170. https://doi.org/10.1016/j.jhazmat.2017.02.021.

Hart, D. 1936. Sterilization of the air in the operating room by special bactericidal radiant energy: Results of its use in extrapleural thoracoplasties. *Journal of Thoracic Surgery* 6(1):45–81.

Hay, S. O., T. Obee, Z. Luo, T. Jiang, Y. Meng, J. He, S. C. Murphy, and S. Suib. 2015. The viability of photocatalysis for air purification. *Molecules* 20(1):1319–1356. https://doi.org/10.3390/molecules20011319.

Howard, J., A. Huang, Z. Li, Z. Tufekci, V. Zdimal, H.-M. van der Westhuizen, A. von Delft, A. Price, L. Fridman, and L.-H. Tang. 2021. An evidence review of face masks against COVID-19. *Proceedings of the National Academy of Sciences* 118(4). https://doi.org/10.1073/pnas.2014564118.

Howard-Reed, C., L. A. Wallace, and W. R. Ott. 2002. The effect of opening windows on air change rates in two homes. *Journal of the Air & Waste Management Association* 52(2):147–159. https://doi.org/10.1080/10473289.2002.10470775.

Hsu, C.-S., M.-C. Lu, and D.-J. Huang. 2015. Disinfection of indoor air microorganisms in stack room of university library using gaseous chlorine dioxide. *Environmental Monitoring and Assessment* 187(2):1–11. https://doi.org/10.1007/s10661-014-4235-2.

Huang, Y., S. S. H. Ho, Y. Lu, R. Niu, L. Xu, J. Cao, and S. Lee. 2016. Removal of indoor volatile organic compounds via photocatalytic oxidation: A short review and prospect. *Molecules* 21(1):56. https://doi.org/10.3390/molecules21010056.

Hubbard, H., D. Poppendieck, and R. L. Corsi. 2009. Chlorine dioxide reactions with indoor materials during building disinfection: Surface uptake. *Environmental Science & Technology* 43(5):1329–1335. https://doi.org/10.1021/es801930c.

Hult, E. L., H. Willem, P. N. Price, T. Hotchi, M. L. Russell, and B. C. Singer. 2015. Formaldehyde and acetaldehyde exposure mitigation in US residences: In-home measurements of ventilation control and source control. *Indoor Air* 25(5):523–535. https://doi.org/10.1111/ina.12160.

Hyttinen, M., P. Pasanen, and P. Kalliokoski. 2001. Adsorption and desorption of selected VOCs in dust collected on air filters. *Atmospheric Environment* 35(33):5709–5716. https://doi.org/10.1016/S1352-2310(01)00376-4.

IEST (Institute of Environmental Sciences and Technology) Contamination Control Division. 2016. *IEST-RP-CC001: HEPA and ULPA Filters*.

Irga, P. J., T. J. Pettit, and F. R. Torpy. 2018. The phytoremediation of indoor air pollution: A review on the technology development from the potted plant through to functional green wall biofilters. *Reviews in Environmental Science and Bio/Technology* 17(2):395–415. https://doi.org/10.1007/s11157-018-9465-2.

Iwashita, G., and H. Akasaka. 1997. The effects of human behavior on natural ventilation rate and indoor air environment in summer—A field study in southern Japan. *Energy and Buildings* 25(3):195–205. https://doi.org/10.1016/S0378-7788(96)00994-2.

Izadyar, N., W. Miller, B. Rismanchi, and V. Garcia-Hansen. 2020. Impacts of façade openings' geometry on natural ventilation and occupants' perception: A review. *Building and Environment* 170:106613. https://doi.org/10.1016/j.buildenv.2019.106613.

Janssen, J. E. 1999. The history of ventilation and temperature control. *ASHRAE Journal* 41:48–72.

Jensen, P. A., L. A. Lambert, M. F. Iademarco, and R. Ridzon. 2005. Guidelines for preventing the transmission of Mycobacterium tuberculosis in health-care settings. *Morbidity and Mortality Weekly Report: Recommendations and Reports* 54(RR-17):1–141.

Johnson, T., and T. Long. 2005. Determining the frequency of open windows in residences: A pilot study in Durham, North Carolina during varying temperature conditions. *Journal of Exposure Science & Environmental Epidemiology* 15(4):329–349. https://doi.org/10.1038/sj.jea.7500409.

Johnston, M. D., S. Lawson, and J. A. Otter. 2005. Evaluation of hydrogen peroxide vapour as a method for the decontamination of surfaces contaminated with Clostridium botulinum spores. *Journal of Microbiological Methods* 60(3):403–411. https://doi.org/10.1016/j.mimet.2004.10.021.

Joo, T., J. C. Rivera-Rios, D. Alvarado-Velez, S. Westgate, and N. L. Ng. 2021. Formation of oxidized gases and secondary organic aerosol from a commercial oxidant-generating electronic air cleaner. *Environmental Science & Technology Letters* 8(8):691–698. https://doi.org/10.1021/acs.estlett.1c00416.

Kahnert, A., P. Seiler, M. Stein, B. Aze, G. McDonnell, and S. H. E. Kaufmann. 2005. Decontamination with vaporized hydrogen peroxide is effective against Mycobacterium tuberculosis. *Letters in Applied Microbiology* 40(6):448–452. https://doi.org/10.1111/j.1472-765X.2005.01683.x.

Kang, D. H., D. H. Choi, S. M. Lee, M. S. Yeo, and K. W. Kim. 2010. Effect of bake-out on reducing VOC emissions and concentrations in a residential housing unit with a radiant floor heating system. *Building and Environment* 45(8):1816–1825. https://doi.org/10.1016/j.buildenv.2010.02.010.

Kauffman, R. E., and J. D. Wolf. 2012. Study of the degradation of typical HVAC materials, filters and components irradiated by UVC energy—Part II; polymers *ASHRAF Transactions* 110(2).

Keer, S., D. McLean, B. Glass, and J. Douwes. 2018. Effects of personal protective equipment use and good workplace hygiene on symptoms of neurotoxicity in solvent-exposed vehicle spray painters. *Annals of Work Exposures and Health* 62(3): 307–320. https://doi.org/10.1093/annweh/wxx100.

Khan, S. M., J. Gomes, and D. R. Krewski. 2019. Radon interventions around the globe: A systematic review. *Heliyon* 5(5):e01737. https://doi.org/10.1016/j.heliyon.2019.e01737.

Kim, S.-S., D.-H. Kang, D.-H. Choi, M.-S. Yeo, and K.-W. Kim. 2008. Comparison of strategies to improve indoor air quality at the pre-occupancy stage in new apartment buildings. *Building and Environment* 43(3):320–328. https://doi.org/10.1016/j.buildenv.2006.03.026.

Kowalski, W. 2010. *Ultraviolet Germicidal Irradiation Handbook: UVGI for Air and Surface Disinfection.* University Park, USA. Springer Science & Business Media.

Kunkel, D. A., E. T. Gall, J. A. Siegel, A. Novoselac, G. C. Morrison, and R. L. Corsi. 2010. Passive reduction of human exposure to indoor ozone. *Building and Environment* 45(2):445–452. https://doi.org/10.1016/j.buildenv.2009.06.024.

Lai, D., S. Jia, Y. Qi, and J. Liu. 2018. Window-opening behavior in Chinese residential buildings across different climate zones. *Building and Environment* 142:234–243. https://doi.org/10.1016/j.buildenv.2018.06.030.

Lamble, S., R. Corsi, and G. Morrison. 2011. Ozone deposition velocities, reaction probabilities and product yields for green building materials. *Atmospheric Environment* 45(38):6965–6972. https://doi.org/10.1016/j.atmosenv.2011.09.025.

Lee, A., A. H. Goldstein, J. H. Kroll, N. L. Ng, V. Varutbangkul, R. C. Flagan, and J. H. Seinfeld. 2006. Gas-phase products and secondary aerosol yields from the photooxidation of 16 different terpenes. *Journal of Geophysical Research: Atmospheres* 111(D17). https://doi.org/10.1029/2006JD007050.

Lee, J., and J. Kim. 2020. Material properties influencing the charge decay of electret filters and their impact on filtration performance. *Polymers* 12(3). https://doi.org/10.3390/polym12030721.

Lester Jr, W., E. Dunklin, and O. Robertson. 1952. Bactericidal effects of propylene and triethylene glycol vapors on airborne Escherichia coli. *Science* 115(2988):379–382. https://doi.org/10.1126/science.115.2988.379.

Li, Y., and A. Delsante. 2001. Natural ventilation induced by combined wind and thermal forces. *Building and Environment* 36(1):59–71. https://doi.org/10.1016/S0360-1323(99)00070-0.

Ligotski, R., U. Sager, U. Schneiderwind, C. Asbach, and F. Schmidt. 2019. Prediction of VOC adsorption performance for estimation of service life of activated carbon based filter media for indoor air purification. *Building and Environment* 149:146–156. https://doi.org/10.1016/j.buildenv.2018.12.001.

Lin, C.-C., and H.-Y. Chen. 2014. Impact of HVAC filter on indoor air quality in terms of ozone removal and carbonyls generation. *Atmospheric Environment* 89:29–34. https://doi.org/10.1016/j.atmosenv.2014.02.020.

Liu, W., J. Huang, Y. Lin, C. Cai, Y. Zhao, Y. Teng, J. Mo, L. Xue, L. Liu, W. Xu, X. Guo, Y. Zhang, and J. Zhang. 2021. Negative ions offset cardiorespiratory benefits of $PM_{2.5}$ reduction from residential use of negative ion air purifiers. *Indoor Air* 31(1):220–228. https://doi.org/10.1111/ina.12728.

Ma, B., P. M. Gundy, C. P. Gerba, M. D. Sobsey, and K. G. Linden. 2021. UV inactivation of SARS-CoV-2 across the UVC spectrum: KrCl* excimer, mercury-vapor, and light-emitting-diode (LED) sources. *Applied and Environmental Microbiology* 87(22):e01532–21. https://doi.org/10.1128/AEM.01532-21.

Markowicz, P., and L. Larsson. 2015. Influence of relative humidity on VOC concentrations in indoor air. *Environmental Science and Pollution Research* 22(8):5772–5779. https://doi.org/10.1007/s11356-014-3678-x.

Mattila, J. M., C. Arata, C. Wang, E. F. Katz, A. Abeleira, Y. Zhou, S. Zhou, A. H. Goldstein, J. P. D. Abbatt, P. F. DeCarlo, and D. K. Farmer. 2020. Dark chemistry during bleach cleaning enhances oxidation of organics and secondary organic aerosol production indoors. *Environmental Science & Technology Letters* 7(11):795–801. https://doi.org/10.1021/acs.estlett.0c00573.

Morrison, G., J. Cagle, and G. Date. 2022. A national survey of window-opening behavior in United States homes. *Indoor Air* 32(1):e12932. https://doi.org/10.1111/ina.12932.

NASA (National Aeronautics and Space Administration). 2011. *Flammability, Offgassing, and Compatibility Requirements and Test Procedures.* https://standards.nasa.gov/standard/nasa/nasa-std-6001.

NAVSEA (Naval Sea Systems Command) System Communications. 2016. *Approved Deck Coverings.* Washington, DC: SUBJ/NAVSEA.

NIOSH (National Institute for Occupational Safety and Health). 2009. *Environmental Control for Tuberculosis: Basic Upper-room Ultraviolet Germicidal Irradiation Guidelines for Healthcare Settings*. https://www.cdc.gov/niosh/docs/2009-105/.

NIOSH. 2018. Deck covering materials, interior, cosmetic polymeric. In *Performance Specification*. http://everyspec.com/MIL-PRF/MIL-PRF-010000-29999/MIL-PRF-24613A_7535/.

NRC (National Research Council). 2002. *The Airliner Cabin Environment and the Health of Passengers and Crew*. Washington, DC: The National Academies Press. https://doi.org/10.17226/10238.

NRC. 2014. *A Framework to Guide Selection of Chemical Alternatives*. Washington, DC: The National Academies Press.

Offermann, F. J. 2009. Ventilation and indoor air quality in new homes. *PIER Collaborative Report*. https://ww2.arb.ca.gov/sites/default/files/classic/research/apr/past/04-310.pdf.

Pasanen, P. O., J. Teijonsalo, O. Seppänen, J. Ruuskanen, and P. Kalliokoski. 1994. Increase in perceived odor emissions with loading of ventilation filters. *Indoor Air* 4(2):106–113. https://doi.org/10.1111/j.1600-0668.1994.t01-2-00005.x.

Pejtersen, J. 1996. Sensory pollution and microbial contamination of ventilation filters. *Indoor Air* 6(4):239–248. https://doi.org/10.1111/j.1600-0668.1996.00003.x.

Peng, Z., and J. L. Jimenez. 2021. Exhaled CO_2 as a COVID-19 infection risk proxy for different indoor environments and activities. *Environmental Science & Technology Letters* 8(5):392–397. https://doi.org/10.1021/acs.estlett.1c00183.

Perrier, J. C. B., L. Le Coq, Y. Andres, and P. Le Cloirec. 2008. SFGP 2007-Microbial growth onto filter media used in air treatment devices. *International Journal of Chemical Reactor Engineering* 6(1). https://doi.org/10.2202/1542-6580.1675.

Poppendieck, D., H. Hubbard, and R. L. Corsi. 2021. Hydrogen peroxide vapor as an indoor disinfectant: Removal to indoor materials and associated emissions of organic compounds. *Environmental Science & Technology Letters* 8(4):320–325. https://doi.org/10.1021/acs.estlett.0c00948.

Poppendieck, D. G., D. Rim, and A. K. Persily. 2014. Ultrafine particle removal and ozone generation by in-duct electrostatic precipitators. *Environmental Science & Technology* 48(3):2067–2074.

Price, P. N., and M. H. Sherman. 2006. *Ventilation Behavior and Household Characteristics in New California Houses*. Berkeley, CA: Ernest Orlando Lawrence Berkeley National Laboratory.

Pushpawela, B., R. Jayaratne, A. Nguy, and L. Morawska. 2017. Efficiency of ionizers in removing airborne particles in indoor environments. *Journal of Electrostatics* 90:79–84. https://doi.org/10.1016/j.elstat.2017.10.002.

Rajan, P. E., A. Krishnamurthy, G. Morrison, and F. Rezaei. 2017. Advanced buffer materials for indoor air CO_2 control in commercial buildings. *Indoor Air* 27(6):1213–1223. https://doi.org/10.1111/ina.12386.

Rijal, H. B., M. A. Humphreys, and J. F. Nicol. 2018. Development of a window opening algorithm based on adaptive thermal comfort to predict occupant behavior in Japanese dwellings. *Japan Architectural Review* 1(3):310–321. https://doi.org/10.1002/2475-8876.12043.

Rijal, H. B., P. Tuohy, M. A. Humphreys, J. F. Nicol, A. Samuel, I. A. Raja, and J. Clarke. 2008. Development of adaptive algorithms for the operation of windows, fans, and doors to predict thermal comfort and energy use in Pakistani buildings. *American Society of Heating, Refrigerating and Air-Conditioning Engineers (ASHRAE) Transactions* 114(2):555–573.

Rosebury, T., G. Meiklejohn, L. C. Kingsland, and M. H. Boldt. 1947. Disinfection of clouds of meningopneumonitis and psittacosis viruses with triethylene glycol vapor. *The Journal of Experimental Medicine* 85(1):65–76. https://doi.org/10.1084/jem.85.1.65.

Rudnick, S., and D. Milton. 2003. Risk of indoor airborne infection transmission estimated from carbon dioxide concentration. *Indoor Air* 13(3): 237–245. https://doi.org/10.1034/j.1600-0668.2003.00189.x.

Rudnick, S. N., J. J. McDevitt, M. W. First, and J. D. Spengler. 2009. Inactivating influenza viruses on surfaces using hydrogen peroxide or triethylene glycol at low vapor concentrations. *American Journal of Infection Control* 37(10):813–819. https://doi.org/10.1016/j.ajic.2009.06.007.

Saber, E. M., I. Chaer, A. Gillich, and B. G. Ekpeti. 2021. Review of intelligent control systems for natural ventilation as passive cooling strategy for UK buildings and similar climatic conditions. *Energies* 14(15):4388. https://doi.org/10.3390/en14154388.

Salthammer, T., and F. Fuhrmann. 2007. Photocatalytic surface reactions on indoor wall paint. *Environmental Science & Technology* 41(18):6573–6578. https://doi.org/10.1021/es070057m.

Schleibinger, H., and H. Rüden. 1999. Air filters from HVAC systems as possible source of volatile organic compounds (VOC)–laboratory and field assays. *Atmospheric Environment* 33(28):4571–4577. https://doi.org/10.1016/S1352-2310(99)00274-5.

Schraufnagel, D. E. 2020. The health effects of ultrafine particles. *Experimental & Molecular Medicine* 52(3):311–317. https://doi.org/10.1038/s12276-020-0403-3.

Schulte, P. A., L. T. McKernan, D. S. Heidel, A. H. Okun, G. S. Dotson, T. J. Lentz, C. L. Geraci, P. E. Heckel, and C. M. Branche. 2013. Occupational safety and health, green chemistry, and sustainability: A review of areas of convergence. *Environmental Health* 12(1):1–9. https://doi.org/10.1186/1476-069X-12-31.

Schwartz-Narbonne, H., C. Wang, S. Zhou, J. P. D. Abbatt, and J. Faust. 2019. Heterogeneous chlorination of squalene and oleic acid. *Environmental Science & Technology* 53(3):1217–1224. https://doi.org/10.1021/acs.est.8b04248.

Sidheswaran, M., W. Chen, A. Chang, R. Miller, S. Cohn, D. Sullivan, W. J. Fisk, K. Kumagai, and H. Destaillats. 2013. Formaldehyde emissions from ventilation filters under different relative humidity conditions. *Environmental Science & Technology* 47(10):5336–5343. https://doi.org/10.1021/es400290p.

Siegel, J. 2016. Primary and secondary consequences of indoor air cleaners. *Indoor Air* 26(1):88–96. https://doi.org/10.1111/ina.12194.

Sleiman, M., P. Conchon, C. Ferronato, and J.-M. Chovelon. 2009. Photocatalytic oxidation of toluene at indoor air levels (ppbv): Towards a better assessment of conversion, reaction intermediates and mineralization. *Applied Catalysis B: Environmental* 86(3–4):159–165. https://doi.org/10.1016/j.apcatb.2008.08.003.

Sliney, D. H., and B. E. Stuck. 2021. A need to revise human exposure limits for ultraviolet UV-C radiation. *Photochemistry and Photobiology* 97(3):485–492. https://doi.org/10.1111/php.13402.

Spengler, J., S. Ludwig, and R. Weker. 2004. Ozone exposures during trans-continental and trans-Pacific flights. *Indoor Air* 14:67–73. https://doi.org/10.1111/j.1600-0668.2004.00275.x.

Spengler, J. D., J. M. Samet, and J. F. McCarthy. 2001. *Indoor Air Quality Handbook*. New York. McGraw-Hill Education. https://www.accessengineeringlibrary.com/content/book/9780074455494.

Spengler, J. D., and D. G. Wilson. 2003. Air quality in aircraft. *Proceedings of the Institution of Mechanical Engineers, Part E: Journal of Process Mechanical Engineering* 217(4):323–335. https://doi.org/10.1243/095440803322611688.

Spicer, C. W., M. J. Murphy, M. W. Holdren, J. D. Myers, I. C. MacGregor, C. Holloman, R. R. James, K. Tucker, and R. Zaborski. 2004. *Relate Air Quality and Other Factors to Comfort and Health Symptoms Reported by Passengers and Crew on Commercial Transport Aircraft (Part I) (ASHRAE Project 1262-TRP)*. Atlanta: American Society for Heating, Refrigerating and Air-Conditioning Engineers. https://www.techstreet.com/ashrae/standards/rp-1262-relate-air-quality-and-other-factors-to-comfort-and-health-symptoms-reported-by-passengers-and-crew-on-commercial-transport-aircraft-part-i?product_id=1717909.

Stephens, B., J. A. Siegel, and A. Novoselac. 2011. Operational characteristics of residential and light-commercial air-conditioning systems in a hot and humid climate zone. *Building and Environment* 46(10):1972–1983. https://doi.org/10.1016/j.buildenv.2011.04.005.

Tang, X., N. R. González, M. L. Russell, R. L. Maddalena, L. A. Gundel, and H. Destaillats. 2021. Chemical changes in thirdhand smoke associated with remediation using an ozone generator. *Environmental Research* 198:110462. https://doi.org/10.1016/j.envres.2020.110462.

Tang, X., P. K. Misztal, W. W. Nazaroff, and A. H. Goldstein. 2015. Siloxanes are the most abundant volatile organic compound emitted from engineering students in a classroom. *Environmental Science & Technology Letters* 2(11):303–307. https://doi.org/10.1021/acs.estlett.5b00256.

Tompkins, D. T., W. A. Zeltner, B. J. Lawnicki, and M. A. Anderson. 2005. Evaluation of photocatalysis for gas-phase air cleaning-Part 1: Process, technical, and sizing considerations. *ASHRAE Transactions* 111:60.

Tsang, A. M., and N. E. Klepeis. 1996. *Descriptive Statistics Tables from a Detailed Analysis of the National Human Activity Pattern Survey (NHAPS) Data*. Las Vegas, NV: Lockheed Martin Environmental Systems and Technologies.

Underhill, D. 2001. Removal of gases and vapors. In *Indoor Air Quality Handbook* 1st ed., edited by J. D. Spengler, J. M. Samet, and J. F. McCarthy. New York: McGRAW-HILL.

Verdenelli, M., C. Cecchini, C. Orpianesi, G. Dadea, and A. Cresci. 2003. Efficacy of antimicrobial filter treatments on microbial colonization of air panel filters. *Journal of Applied Microbiology* 94(1):9–15. https://doi.org/10.1046/j.1365-2672.2003.01820.x.

Verginelli, I., O. Capobianco, N. Hartog, and R. Baciocchi. 2017. Analytical model for the design of in situ horizontal permeable reactive barriers (HPRBs) for the mitigation of chlorinated solvent vapors in the unsaturated zone. *Journal of Contaminant Hydrology* 197:50–61. https://doi.org/10.1016/j.jconhyd.2016.12.010.

Von Recklinghausen, M. 1914. The ultra-violet rays and their application for the sterilization of water. *Journal of the Franklin Institute* 178(6):681–704.

Wang, C., D. B. Collins, and J. P. Abbatt. 2019. Indoor illumination of terpenes and bleach emissions leads to particle formation and growth. *Environmental Science & Technology* 53(20):11792–11800. https://doi.org/10.1021/acs.est.9b04261.

Wang, H., and G. Morrison. 2010. Ozone-surface reactions in five homes: Surface reaction probabilities, aldehyde yields, and trends. *Indoor Air* 20(3):224–234. https://doi.org/10.1111/j.1600-0668.2010.00648.x.

Wang, Z., S. F. Kowal, N. Carslaw, and T. F. Kahan. 2020. Photolysis-driven indoor air chemistry following cleaning of hospital wards. *Indoor Air* 30(6):1241–1255. https://doi.org/10.1111/ina.12702.

Watson, R., M. Oldfield, J. A. Bryant, L. Riordan, H. J. Hill, J. A. Watts, M. R. Alexander, M. J. Cox, Z. Stamataki, D. J. Scurr, and F. de Cogan. 2022. Efficacy of antimicrobial and anti-viral coated air filters to prevent the spread of airborne pathogens. *Scientific Reports* 12:2803. https://doi.org/10.1038/s41598-022-06579-9.

Weisel, C., C. J. Weschler, K. Mohan, J. Vallarino, and J. D. Spengler. 2013. Ozone and ozone byproducts in the cabins of commercial aircraft. *Environmental Science & Technology* 47(9):4711–4717. https://doi.org/10.1021/es3046795.

Wells, W.F., M. W. Wells, and T. S. Wilder. 1942. The environmental control of epidemic contagion. I. An epidemiologic study of radiant disinfection of air in day schools *American Journal of Epidemiology* 35(1):97–121.

Weschler, C. J. 2006. Ozone's impact on public health: Contributions from indoor exposures to ozone and products of ozone-initiated chemistry. *Environmental Health Perspectives* 114(10):1489–1496. https://doi.org/10.1289/ehp.9256.

Weschler, C. J., and H. C. Shields. 2000. The influence of ventilation on reactions among indoor pollutants: Modeling and experimental observations. *Indoor Air* 10(2): 92–100.

White, E. 2009. HEPA and ULPA filters. *Journal of Validation Technology* 15(3):48–55. https://www.proquest.com/openview/955ada5d217bd22cf9a31441253b38ba/1?pq-origsite=gscholar&cbl=29232.

Wisthaler, A., P. Strom-Tejsen, L. Fang, T. J. Arnaud, A. Hansel, T. D. Mark, and D. P. Wyon. 2007. PTR-MS assessment of photocatalytic and sorption-based purification of recirculated cabin air during simulated 7-h flights with high passenger density. *Environmental Science & Technology* 41(1):229–234. https://doi.org/10.1021/es060424e.

Wisthaler, A., and C. J. Weschler. 2010. Reactions of ozone with human skin lipids: Sources of carbonyls, dicarbonyls, and hydroxycarbonyls in indoor air. *Proceedings of the National Academy of Sciences* 107(15):6568–6575. https://doi.org/10.1073/pnas.0904498106.

Wong, J. P. S., N. Carslaw, R. Zhao, S. Zhou, and J. P. D. Abbatt. 2017. Observations and impacts of bleach washing on indoor chlorine chemistry. *Indoor Air* 27(6):1082–1090. https://doi.org/10.1111/ina.12402.

Woods, J. E. 1991. An engineering approach to controlling indoor air quality. *Environmental Health Perspectives* 95:15–21. https://doi.org/10.1289/ehp.919515.

Yao, M., and B. Zhao. 2017. Window opening behavior of occupants in residential buildings in Beijing. *Building and Environment* 124:441–449. https://doi.org/10.1016/j.buildenv.2017.08.035.

Ye, Q., J. E. Krechmer, J. D. Shutter, V. P. Barber, Y. Li, E. Helstrom, L. J. Franco, J. L. Cox, A. I. H. Hrdina, M. B. Goss, N. Tahsini, M. Canagaratna, F. N. Keutsch, and J. H. Kroll. 2021. Real-time laboratory measurements of VOC emissions, removal rates, and byproduct formation from consumer-grade oxidation-based air cleaners. *Environmental Science & Technology Letters* 8(12):1020–1025. https://doi.org/10.1021/acs.estlett.1c00773.

Yu, K.-P., G. W. Lee, W.-M. Huang, C. Wu, and S. Yang. 2006. The correlation between photocatalytic oxidation performance and chemical/physical properties of indoor volatile organic compounds. *Atmospheric Environment* 40(2):375–385. https://doi.org/10.1016/j.atmosenv.2005.09.045.

Zeng, Y., P. Manwatkar, A. Laguerre, M. Beke, I. Kang, A. S. Ali, D. K. Farmer, E. T. Gall, M. Heidarinejad, and B. Stephens. 2021. Evaluating a commercially available in-duct bipolar ionization device for pollutant removal and potential byproduct formation. *Building and Environment* 195:107750. https://doi.org/10.1016/j.buildenv.2021.107750.

Zhang, M., M. Qin, C. Rode, and Z. Chen. 2017. Moisture buffering phenomenon and its impact on building energy consumption. *Applied Thermal Engineering* 124:337–345. https://doi.org/10.1016/j.applthermaleng.2017.05.173.

Zhang, Y., J. Mo, Y. Li, J. Sundell, P. Wargocki, J. Zhang, J. C. Little, R. Corsi, Q. Deng, and M. H. Leung. 2011. Can commonly-used fan-driven air cleaning technologies improve indoor air quality? A literature review. *Atmospheric Environment* 45(26):4329–4343. https://doi.org/10.1016/j.atmosenv.2011.05.041.

Zhou, S., Z. Liu, Z. Wang, C. J. Young, T. C. VandenBoer, B. B. Guo, J. Zhang, N. Carslaw, and T. F. Kahan. 2020. Hydrogen peroxide emission and fate indoors during non-bleach cleaning: a chamber and modeling study. *Environmental Science & Technology* 54(24):15643–15651. https://doi.org/10.1021/acs.est.0c04702.

Zhu, X., M. Lv, and X. Yang. 2019. A test-based method for estimating the service life of adsorptive portable air cleaners in removing indoor formaldehyde. *Building and Environment* 154:89–96. https://doi.org/10.1016/j.buildenv.2019.03.018.

6

Indoor Chemistry and Exposure

This chapter begins by introducing exposure routes and defining some of the factors that influence exposure in indoor environments, drawing in part from the 1991 National Research Council (NRC) report *Human Exposure Assessment for Airborne Pollutants* (NRC, 1991). Some exposure variables are linked to environmental health disparities, and these are discussed in the context of indoor chemistry. Subsequent sections cover the intersection of indoor chemistry and exposure modeling and measurement science for exposure. The chapter concludes with a list of priority research needs identified by the committee.

EXPOSURE ROUTES

Exposure to chemicals indoors can occur by three routes: inhalation, ingestion, and dermal uptake (Figure 6-1) (Feld-Cook and Weisel, 2021). Inhalation includes the uptake of gas- or particle-phase chemicals into the respiratory tract. Particle-phase chemicals deposit into the nasopharynx, bronchi, or deep lung (alveolar) region, with a general trend of smaller particles penetrating deeper into the respiratory tract (Heyder et al., 1986). The most penetrating particle size is 0.1–0.5 micrometers (μm), but even small ultrafine particles (UFPs) are lost by diffusion to the upper respiratory tract (Hofmann, 2011). Some particles are small enough to cross the lung epithelial tissue and pass into the circulatory system (Nakane, 2012), and particles have also been shown to enter the brain directly (Block and Calderón-Garcidueñas, 2009). Dermal uptake involves the deposition of gas-phase chemicals on skin or direct contact with surfaces that contain chemicals. Deposition or contact is followed by migration into the epidermis where chemicals may be metabolized, undergo active transport, and be taken up into the dermis. Chemicals may cross capillary membranes and potentially partition into the bloodstream and be absorbed. Ingestion is mediated by hand-to-mouth behavior and is most common for particles (dust) (Ott et al., 2006). Additionally, some fraction of inhaled aerosols (larger particles) that are moved up the bronchi by mucociliary clearance to be swallowed can be ingested and enter the gastrointestinal tract (Wanner et al., 1996).

Exposure levels and the relative contribution of exposure routes to total exposure burden are influenced by factors such as age, human behaviors, environment and surroundings, and the

FIGURE 6-1 Common exposure routes for chemicals in the indoor environment.

physical and chemical properties of the agents to which people are exposed. For example, the inhalation rate (volume of air inhaled per day) varies with an individual's age and body mass index and can increase significantly during physical activity (EPA, 2011). In addition, estimates of dust ingestion are larger for children than adults (EPA, 2017), while dermal uptake of volatile organic compounds (VOCs) increases with age (Morrison et al., 2016). Dermal uptake involves partitioning of gas-phase organic chemicals into the skin; thus, rates are influenced by physicochemical properties, including molecular weights and gas-phase partitioning coefficients (Weschler and Nazaroff, 2014). Transdermal uptake for some semivolatile organic compounds (SVOCs) can rival that of inhalation intake (Weschler and Nazaroff, 2012).

A number of conceptual frameworks exist to describe how environmental stressors ultimately affect human health. The environmental health paradigm (also known as the source to receptor model) presents a biological response pathway (Figure 6-2). The pathway from a source of exposure to its effect(s) is complex. Exposure to environmental stressors occurs at the interface between sources and receptors. Typically, receptors are people who may subsequently experience physical and/or mental health outcomes, resulting from or being influenced by the exposure. Multiple approaches exist for assessing chemical exposure in individuals. Indirect assessments can include surveys of behavior and time spent near chemical sources. Indoor concentrations may be measured by stationary devices or grab samples (e.g., dust wipes) as proxies for personal exposure.

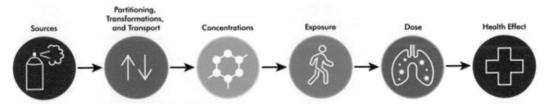

FIGURE 6-2 Exposure pathway schematic, as conceived within the environmental health paradigm, charting the fate and transport of an agent from source to ultimate impacts on health.

Stationary indoor measurements made in more than one microenvironment can be combined with time-weighted activity information to yield estimated personal exposure concentrations; further translation to a dose requires additional information on intake (i.e., inhalation, ingestion, or dermal absorption) rates. The spatial and temporal variability in concentrations and human movement can best be captured by direct personal monitoring, which also minimizes uncertainty or exposure measurement error inherent in using more indirect approaches (e.g., questionnaires, microenvironmental models, and outdoor measurements). Personal exposure monitoring can reduce exposure misclassification; however, because such monitoring can be burdensome and resource-intensive, surrogate and proxy measures of exposure are commonly used but may still be referred to as measures of exposure. Finally, biomonitoring provides a more direct measurement of internal dose and includes measurements of chemicals and chemical metabolites in biospecimens (e.g., blood, urine, saliva, hair, cord blood, breast milk, and nails). Different biological matrices may reflect variable time integration of exposure (e.g., a few days for urine versus a few months for hair). However, biomonitoring data and measures of internal dose cannot provide insight on how exposure occurred and what the source, or sources, may be. The National Health and Nutrition Examination Survey (NHANES) is an example of a biomonitoring dataset that measures chemicals in blood and urine, some of which are present indoors.

While understanding exposure to chemicals in the indoor environment is in a nascent stage, several factors are known to modify human exposure. Important sources of particles and gas-phase chemicals, as discussed in detail in Chapter 2, include combustion, resuspension of particles caused by walking and cleaning, applications of cleaning agents and personal care products, and off-gassing of volatile and semivolatile chemicals from building material surfaces and myriad consumer products. These sources may contribute to indoor chemistry in both nonresidential and residential environments. In residential settings, several additional common activities contribute substantially to indoor chemistry, including cooking, space-heating, and behaviors associated with natural, mechanical, and unintentional ventilation of the occupied space. The location of a building, the integrity of the building envelope, and the presence of building air treatment systems are important factors that influence indoor chemical transformations and reactions, as well as indoor exposures to chemicals that can intrude from outdoors.

EXPOSURE DEFINITIONS, SETTINGS, AND TIMING

Exposure to a given chemical reflects the integration of concentrations of that chemical that an individual comes into contact with over the duration of time spent in contact with the agent across various settings and microenvironments.

To translate an exposure concentration (typically expressed in units of mass per volume or mixing ratio) into an intake dose, the exposure concentration is multiplied by an exposure duration (expressed in units of time) and by an intake rate that depends on individual characteristics (e.g., age, height, body mass index), activities (e.g., sedentary, active), and route (inhalation, ingestion, and dermal absorption as the major pathways of relevance in indoor environments) (EPA, 2011).

Settings, tasks, activities, behaviors, and features and conditions of indoor environments change over time, resulting in indoor environmental exposures that vary both within individuals over time and space and between individuals, even in the same indoor environment. Understanding variability in exposures is important for informing design of measurement and monitoring campaigns; assigning estimates of exposure to participants in a health study; identifying determinants of exposure; evaluating adherence with exposure guidelines; and, ultimately, evaluating the distributions of impacts of interventions and control measures intended to prevent or mitigate harmful exposures.

Given the challenging nature of measuring exposure continuously (or even the concentrations that individuals are exposed to over time), the microenvironmental model posits that dose can be approximated by capturing time-averaged concentrations, time-activity patterns, and intake rates within key or major microenvironments where individuals spend time (Branco et al., 2014). These microenvironments often include indoor locations, like the home; occupational or school settings; transit settings; and outdoor microenvironments. Therefore, according to the microenvironmental model, an individual's total dose of a given chemical is the sum of doses across all known microenvironments. These concepts provide a framework for characterizing and understanding variability in exposures. Specifically, this framework can provide insight on settings, activities, or characteristics that might determine or lead to elevated exposures, especially for sensitive or susceptible individuals and subpopulations across concentrations, time, and intake rates.

Exposure Settings

Exposure settings that fall within the scope of this report include nonindustrial indoor settings, including but not limited to the following: homes and places of permanent and temporary residence, office settings, schools, hospitals, prisons, public venues (indoors), spaces for community gatherings, retail environments, restaurants, and transport environments. Reviews and individual studies have illustrated the variability in the identities and concentrations of harmful compounds in a range of indoor settings, including residential environments (e.g., Diaz Lozano Patino and Siegel, 2018; Logue et al., 2011), light commercial buildings (e.g., Mandin et al., 2017; Ng et al., 2012; Nirlo et al., 2014; Zaatari et al., 2014), early childcare centers and schools (e.g., Bradman et al., 2014, 2017; Erlandson et al., 2019; Gaspar et al., 2014, 2018; Givehchi et al., 2019; Hoang et al., 2017), and hospitals and public utilities (e.g., Chamseddine, 2019; Śmiełowska et al., 2017), among others. The referenced sources serve as examples but are not intended to provide a comprehensive or systematic review. It is important to acknowledge that indoor gas and aerosol phase concentrations are more common in the literature, and equivalent measurements, studies, and reviews of surface and dust composition are less common. Estimating exposures to chemicals in dust via ingestion and/or dermal absorption has been a particular challenge for scientific and practitioner communities alike.

Some settings have potential for more intense indoor exposures; for instance, indoor settings or microenvironments with high concentrations of some chemicals might include residences near major roadways or other strong outdoor point sources, and service-oriented work settings where workers and patrons alike may experience high exposures (e.g., restaurants and beauty and nail salons). Homes where solid fuels are used for cooking and heating pose significant risks to health through exposures to pollutants emitted during incomplete (and sometimes unvented) combustion, including organic carbon, elemental carbon, UFPs, inorganic ions, carbohydrates, and VOCs and SVOCs, in addition to better understood pollutants, such as fine particulate matter ($PM_{2.5}$), carbon monoxide (CO), and nitrogen oxides (NO, NO_2). These settings are more prevalent outside the United States, yet some U.S. households continue to use solid fuels, mostly for space heating, either as a primary or supplemental energy source depending on factors such as access, convenience, cost, and household preferences. A large body of work documents emission rates and factors associated with solid fuel combustion, indoor concentrations of byproducts of incomplete combustion, and exposures and health effects associated with these exposures (Champion, 2017; Noonan et al., 2015; Rogalsky et al., 2014; Semmens et al., 2015). For example, daily average (24-h) concentrations of commonly measured indoor pollutants (e.g., $PM_{2.5}$, CO, and NO_2) in these settings have been measured ranging into the hundreds of micrograms per cubic meter (for $PM_{2.5}$), the tens of parts per million (ppm; for CO), and the tens to hundreds of parts per billion (for NO_2). In the United States, just less than half of households rely on natural gas as a primary heating fuel, while approximately one-third use natural gas as their primary cooking fuel. When in use, natural gas appliances in homes can emit oxides of nitrogen—namely, nitrogen oxide (NO) and nitrogen dioxide (NO_2). Low level exposure to NO_2 has been associated with increased bronchial reactivity in some people with asthma, as well as decreased lung function among individuals with chronic obstructive pulmonary disease and increased risk of respiratory infections, especially in young children. Sustained exposure to moderate to high levels of NO_2 can also contribute to the development of bronchitis. Formaldehyde is another air pollutant for which exposure tends to be intensified indoors relative to outdoors. For instance, living in newer, more modern, energy-efficient homes is associated with higher indoor concentrations of formaldehyde (Huang et al., 2017b; Langer et al., 2016), terpenes, and other VOCs (Derbez et al., 2018; Langer et al., 2015).

In total, indoor environments are chemically diverse. Some settings, like homes, have been the focus of many studies in the scientific literature, and they are indeed important because of how much time people spend in them. Yet, adverse chemical exposures can occur in a much wider range of indoor settings despite, in some cases, the more limited time people spend in those settings.

Exposure Timing, Duration, and Time-Activity Patterns

On average, people in the United States tend to spend the majority (69 percent) of their indoor time in their homes, and overall time spent indoors can exceed 90 percent when transit environments (e.g., cars, buses, trains) are taken into consideration (Klepeis et al., 2001). A range of methods can be applied to develop understanding of these time-activity budgets, which can vary with age and stage of life as well as with other social, cultural, economic, and demographic factors. For example, the Bureau of Labor Statistics and the U.S. Environmental Protection Agency (EPA) gather data through nationally representative survey instruments to characterize distributions of where and how people spend their time, sub-divided among multiple age and socioeconomic and sociodemographic groups. Time-activity budgets might also consider specific subgroups of the population that spend a lot of time in settings with the potential for indoor exposures to reach harmful levels, even some of the time. Examples include children in daycare facilities, newborns in neonatal intensive care units, expectant and new mothers, elderly people in nursing homes, individuals with lower mobility, and adults with asthma in their residence and work locations. Data on time-activity patterns are also routinely gathered in exposure and health-based epidemiological studies. Nationally sourced data provide broadly representative and potentially generalizable information; however, datasets such as these can mute variability that may exist at individual- and sub-group levels. Recently, GPS and location and motion sensors embedded in smartphones and wearables have been used to derive space and time-resolved time-activity data for exposure and health studies. Crowdsourced commercial services and apps are also providing access to large amounts of time-activity data (e.g., Google Timelines); however, data privacy considerations and generalizability/representation across and within regions, populations, and over time remain an issue. Some populations remain difficult to reach. As such, information on time-activity patterns is still limited among some subpopulations in the United States, including refugees, migrant or unhoused populations, populations for whom English is not a primary language, and those who are not connected to smartphones or the internet. The reasons why some people are difficult to access or gather information from are likely associated with other socioeconomic and demographic factors that could also be associated with greater likelihood of experiencing elevated or adverse exposures, including in indoor environments, and/or greater susceptibility to the effects of those chemical exposures.

Exposure Factors, Behaviors, and Intake Rates

Greater insight on the chemical composition of indoor air would be achieved if the scientific community had better data on some indoor activities, including window-opening, cooking, cleaning, and using personal care and leisure products. Furthermore, understanding the periodicity of activities such as these and others is important for determining the acute, chronic, and/or episodic nature of indoor chemical exposures, as well as anticipating how those exposures might change under disruptive circumstances (e.g., shifts to remote work, as have occurred throughout the

COVID-19 pandemic). Improved measurements of intake rates could improve exposure estimates and modeling. Intake rates vary with individual and demographic characteristics, life stage, and comorbidities and health conditions that jointly and independently influence exposure and susceptibility to harm resulting from exposure to indoor chemicals. For example, children's dust ingestion rates are elevated compared to adults' due to crawling on floors and hand-to-mouth behavior. Similarly, inhalation rates of pregnant women increase significantly during the first trimester and stay elevated throughout pregnancy (LoMauro and Aliverti, 2015).

Susceptibility factors also come into play when thinking about exposure and health implications. Given similar levels of chemical exposure, different subpopulations can be more susceptible to adverse effects. As an example, children have less developed immune systems and receive higher doses of chemicals per body weight compared to adults and thus might be more susceptible to adverse effects of chemical contaminants, even if they are exposed to comparable concentrations in the indoor environment. Developing fetuses are also susceptible to chemical exposures, and timing of exposure relative to conception, gestation, and developmental windows of various organ systems can be critical to determine risk of adverse health effects as described in the Developmental Origins of Health and Disease paradigm (Haugen et al., 2015). Chemical exposures can interfere with proper placentation, modify epigenetic programming and gene expression, and act in direct and indirect ways to adversely affect fetal health in-utero, at birth, or later in childhood and across the life course (Almieda et al., 2019; Harley et al., 2017; Haugen et al., 2015; Wigle et al., 2008). In addition, exposure to environmental contaminants may occur simultaneously and repeatedly with exposures to multiple other chronic social stressors, and the interaction of chemical and social exposures may increase biological susceptibility or vulnerability to adverse health outcomes. This has been termed a "double jeopardy," where individuals living in disadvantaged neighborhoods, historically marginalized populations, racial and ethnic minorities, communities of color, and low-income groups can be disproportionately exposed to multiple environmental contaminants and can have higher susceptibility or vulnerability to their adverse effects (Morello-Frosch et al., 2011; Morello-Frosch and Lopez, 2006; Morello-Frosch and Shenassa, 2006). The terms "vulnerability" and "susceptibility" have overlapping meanings and could be used interchangeably. However, this report uses susceptibility to refer more to inherent physiological factors that may predispose an individual or sub-population to an elevated adverse response to an exposure. Vulnerability, on the other hand, refers to external factors (e.g., social, economic, and demographic) that may interact with chemical exposures to influence their effect (Bell et al., 2013). Several studies have documented these environmental health disparities or social inequalities in terms of exposure to chemicals in the environment. Biological mechanisms like allostatic load have been postulated to explain this increased susceptibility or vulnerability on a physiological level (Beckie, 2012; Carlson and Chamberlain, 2005; Szanton et al., 2005). An overview of some of the most commonly reported settings and determinants of environmental health disparities in relation to exposures experienced in the indoor environment is provided in the next section.

ENVIRONMENTAL HEALTH DISPARITIES AND EXPOSURE VARIABLES

Exposure to indoor air pollutants varies across individual households, yet certain exposures affect subsets of the population differently. These differential exposures derive from variables that influence indoor air chemistry. As an example, the location, build quality, age, and condition of housing are variables recognized in the literature as factors that can contribute to disparate exposures. The literature also establishes that certain indoor air exposures are observed disproportionately among communities of color and low-income households, such as particulate matter (PM) and lead (Baxter et al., 2007; Hauptman et al., 2021). The literature on indoor air quality and disparities is still a growing field, however, and not as robust as that on ambient air quality and disparities.

A scoping review of the indoor air pollution literature in high-income countries found significant gaps in the current understanding of inequalities related to indoor exposures and socioeconomic status (Ferguson et al., 2020).

In the United States, indoor chemical exposures coupled with frequent or elevated social stressors are believed to contribute to downstream disparities in health outcomes. The U.S. Office of Disease Prevention and Health Promotion (2008) defines a health disparity as "a particular type of health difference that is closely linked with social, economic, and/or environmental disadvantage." Thus, understanding sources of exposure variability is critical to reducing disparities and achieving health equity. This section focuses on disparities in exposure and exposure determinants in the United States.

The literature on indoor air pollution and variable exposure across different socioeconomic, racial, and ethnic groups is somewhat limited. This is in spite of the environmental justice and heathy housing research which consistently finds that low-income, Black, Hispanic, and Indigenous communities are more likely to live in substandard housing (Jacobs, 2011; Seltenrich, 2012).

To understand exposure disparities thus requires consideration of differences in indoor air pollutant variables, which include nearby sources of outdoor pollutants, construction materials and practices, energy efficiency practices, energy use and home heating, occupancy rates, occupant practices and behaviors, indoor environmental maintenance patterns, and climate change. Indoor exposures to chemicals have been linked to the location, age, and condition of the structures in which people live, work, play, and congregate; and to the poor quality of our residences and work settings, as well as a lack of standardized maintenance and sanitation. These have had a well-documented impact on health that dates back to the 19th century (Adamkiewicz et al., 2011).

This report cannot thoroughly discuss emerging indoor chemistry issues without addressing climate: climate change may influence the condition and quality of indoor environments, the chemistry that arises indoors, and the resulting exposures and exposure variability in numerous ways. A 2016 National Climate Assessment Report recommended consideration of how vulnerable populations experience disproportionate risks in response to climate change (U.S. Global Change Research Program, 2016). Thus, the committee sought to provide examples from emerging science on the role climate change may play in widening or narrowing differences in indoor environmental exposures and related health outcomes. Comprehensive coverage of all documented exposure determinants and their links to environmental health disparities would be too broad to achieve within the scope of this report. However, drawing from the available literature, this section highlights categories of exposure determinants that are especially relevant for health disparities related to indoor environments.

Indoor Pollutants of Outdoor Origin

Indoor air chemistry is influenced by proximity to outdoor polluting sources and ambient air pollutant concentrations, which vary based on land-use types, zoning, topography, and geography. Redlining, a real estate practice dating to the 1930s, produced a pattern in which communities of color are more likely to live in neighborhoods with a high density of polluting sources and land uses and heavily polluted airsheds. This historical pattern creates pervasive differences in exposure to outdoor air pollution by race and ethnicity. Although exposure disparities exist for a wide range of source types, research has shown significant disparities in exposure from four sources: industry, light-duty gasoline vehicles, construction, and diesel PM (Gee and Payne-Sturges, 2004; Miranda et al., 2011; Tessum et al., 2021). In the United States, "non-Hispanic blacks are consistently over-represented in communities with the poorest air quality" with respect to $PM_{2.5}$ and ozone (Miranda et al., 2011). Communities of color are also more likely to occupy housing that is adjacent to major roadways, where mobile sources emit hydrocarbons, carbon monoxide, nitrogen oxides, and fine or UFPs from diesel (HEI, 2010; Perez et al., 2013; Tessum et al., 2021). Mobile sources also emit chemicals classified by EPA as Hazardous Air Pollutants, such as 1,3-butadiene, acetaldehyde, benzene, and formaldehyde, which can infiltrate indoor spaces. Many American Indian and Alaska Natives live on reservations near anthropogenic sources of air pollutants, including oil and gas extraction, mining, and other major industrial emitters (Kramer et al., 2020). While infiltration rates vary spatially and temporally, in substandard housing infiltration rates of chemicals and PM of outdoor origin may be higher due to greater leakiness (e.g., poorly maintained structures, cracks and openings in floors and walls); a lack of mechanical ventilation, or, alternatively, increased reliance on mechanical ventilation; and less air filtration and cleaning compared to more modern or properly maintained residences (Underhill et al., 2018). Shrestha et al. (2019a) found that homes with higher annual average infiltration rates were associated with higher mold growth, higher levels of dust, and unacceptable odor levels. Window-opening was positively associated with lower-income housing, apartment homes, and rental properties, which may increase exposure to air pollutants of outdoor origin while decreasing exposure to indoor-sourced pollutants (Morrison et al., 2022). Few studies are available that document substandard ventilation and filtration in low-income housing, but higher indoor concentrations of ambient pollutants have been observed in low-income and public housing and could be indicative of broader conditions (Colton et al., 2014; Tessum et al., 2021; Zota et al., 2005). As an example, an assessment of residential air leakage in the United States found that older, smaller homes were leakier than newer, larger homes (Chan et al., 2005). Because age and size of housing are linked to housing prices, the authors note that older, smaller homes are more likely to be occupied by low-income families.

Construction Practices and Materials

Construction practices also affect building systems for air handling. In turn, air handling system design, installation, operation, and maintenance all influence the potential for indoor exposure to harmful chemicals, either of indoor or outdoor origin (see Chapter 5 for more details). Low-income, public, and multifamily housing units are more likely to be constructed with low-grade building materials, which may emit higher concentrations of, or result in higher dust-borne concentrations of, VOCs and SVOCs, including phthalates, flame retardants, antimicrobials, petroleum chemicals, chlorinated solvents, and formaldehyde (Bi et al., 2018; Colton et al., 2014; Dodson et al., 2017; Wan et al., 2020). Older housing is also more likely to contain legacy pollutants, such as polychlorinated biphenyls.

Where households have aging carpets, the carpets can act as a reservoir for allergens, endotoxins, lead dust (Becher et al., 2018), and other chemicals. Observing asthma triggers among 112 low-income urban housing units, Krieger et al. (2000) found that 76.8 percent of children's bedrooms had carpeting. Sun et al. (2022) recently observed that the presence of carpeting, along with other household characteristics (e.g., age of home; wall, roof, flooring, and insulation materials; surface paints and coatings; household energy systems for cooking, heating, and lighting; air handling systems; and appliances), were correlated with the presence of multiple biocontaminants. Vinyl flooring in low-income homes was associated with increased levels of benzyl butyl phthalate and di-(2-ethylhexyl) phthalate relative to homes without vinyl flooring (Bi et al., 2018).

Energy Efficiency Factors

Energy efficiency measures can improve occupant comfort while reducing space heating demands. In both residential and nonresidential settings, building energy performance assessments can identify opportunities for energy efficiency improvements, which may be achieved through building and building envelope upgrades and retrofits. Yet, improper implementation of energy efficiency upgrades can result in over-tightening of the building envelope (Manuel, 2011) and inadequate air exchange rates, and may contribute to indoor air quality issues, including higher concentrations of chemicals emitted indoors, higher relative humidity, and microbial growth (Collins and Dempsey, 2019; Du et al., 2019). Local- to national-scale programs focused on improving building energy efficiency have historically been associated with mixed impacts on indoor air quality. A study examining the impact of energy renovation in multifamily residences found lower air exchange rates and higher concentration of carbon dioxide (CO_2), formaldehyde, and VOCs (Földváry et al., 2017). Leivo et al. (2018) also found lower air exchange rates and higher CO_2 levels (in units they observed without mechanical exhaust ventilation) after energy retrofitting. Less and Walker (2014) found that homes with dedicated outdoor ventilation systems had air change rates that were not significantly different than homes relying on natural ventilation. Problems with outdoor air ventilation systems included incorrect installations, clogged vents, and ventilation systems turned off by the occupants. However, insight into the potential but avoidable indoor air quality hazards that can be introduced through poor implementation of energy efficiency upgrades in aging workplaces, homes, and other nonresidential settings has led to improvements in the practice of weatherization and building energy efficiency performance upgrades (Fisk et al., 2020; Shrestha et al., 2019a). Several voluntary standards and rating systems are available to guide weatherization practices and energy efficiency retrofits, from industry (ASHRAE, Building Performance Institute) to government (Energy Star).

Indoor Climate Control

Indoor air chemistry is modulated by the age, efficiency, and condition of heating and cooling systems. Low-income, rural, rental, and multifamily homes tend to be lower on the energy ladder, in which lower economic status drives higher use of biomass and other solid fuels for heat and cooking (van der Kroon et al., 2013) and greater energy insecurity due to the cost burden of heating, cooling, and ventilation. In 2015, the most recent year for which data were collected, 11 percent of households surveyed reported keeping their homes at an unhealthy or unsafe temperature (EIA, 2015a). Low-income U.S. households are more likely to rely on electricity, wood, fuel oil, or propane systems as opposed to natural gas. These energy systems each have unique emissions characteristics and produce different impacts on indoor air chemistry. Whether they are more or less harmful than natural gas is influenced by many variables. Yet from a disparities lens, households without natural gas are less likely to have a central furnace or heating, ventilation, and air-conditioning (HVAC) system and associated air exchange and filtration—48.7 percent of single-family attached homes in the United States have a central furnace, as compared to 4.6 percent of multifamily units (2–4 units) and 8.7 percent of multifamily units (5 or more units) (EIA, 2015a). Disparities in indoor air quality can also arise when space-heating systems are aging, leaky, and poorly maintained by the occupants or the property manager. Poorly vented or unvented combustion indoors can result in indoor exposure to byproducts of incomplete combustion including carbon monoxide (Vicente et al., 2020) and polycyclic aromatic hydrocarbons (Tiwari et al., 2013).

Low-income households have a high energy burden: 31 percent of U.S. households report having difficulty adequately heating or cooling their homes (EIA, 2015b). Households unable to afford adequate heat are more likely to experience low indoor temperatures, which may also be associated with increased indoor humidity in some settings, as well as condensation of moisture on indoor surfaces, and microbial growth (Zhang and Yoshino, 2010). Higher or uncontrolled (i.e., more variable) temperatures in homes are also a factor in levels of relative humidity and influence chemical emission rates (Haghighat and De Bellis, 1998). Low-income homes are also less able to cool their homes than non-low-income households (U.S. Department of Health and Human Services, 2017), contributing to unsafe conditions (Clinch and Healy, 2000).

When window-opening is used for home cooling, it can contribute to higher rates of outdoor air pollution infiltration. Many low-income communities are situated within urban heat islands (UHI), where urbanization has contributed to a phenomenon in which the impervious surfaces and absence of tree canopy contribute to higher temperatures (Yang et al., 2020). UHI may be exacerbated by thermal events that are modeled to increase due to climate change (Perkins et al., 2012).

Occupancy Rates

High occupancy loads are a hallmark of low-income households (WHO, 2018), but overcrowding also occurs in areas with limited housing stock, such as rural areas and cold-climate regions. Overcrowding is a primary contributor to higher indoor temperatures and higher humidity, which in turn are positively associated with increases in formaldehyde and VOC emissions from building materials (Huangfu et al., 2019). Higher occupancy also introduces a higher concentration of human bioeffluents, including dermal (skin oils, skin flakes) and exhaled bioeffluents such as carbon dioxide and certain VOCs (Liu et al., 2016; Tsushima et al., 2018). As different latitudes and geographic regions of the globe become less tolerable due to extreme weather, human migration will place pressure on, and potentially increase occupancy rates in, existing housing stock. Coates and Norton (2021) describe how climate changes drive migration, overcrowding, and poverty, and increase opportunities for infectious disease transmission. Overcrowding is also associated with heat risks, due in part to poorly ventilated dwellings (Pelling et al., 2021).

Occupant Practices, Consumer Product Use, and Behaviors

Indoor air chemistry is influenced by occupant practices, consumer product use, and behaviors, which vary by region, culture, ethnicity, race, and family structure. Consumer and personal care products, often unique to personal and societal cultures, emit VOCs and may contain endocrine-disrupting chemicals (EDCs). For example, households using incense as a regular practice could have higher concentrations of PM (Yang et al., 2012) and inorganic gases. Certain beauty products contain parabens, preservatives, and EDCs. The majority of samples in a study of commonly used hair care products targeted for Black women showed androgen antagonist properties (James-Todd et al., 2021). Cleaning products commonly used in the United States contain phthalates and phenols, with empirical evidence indicating that the cost of "safer" or "green" alternatives may limit access to higher-income households (Finisterra do Paço and Raposo, 2010). In addition, cleaning products that might be particularly popular or common among specific groups of people or communities can be a dominant source of exposure to one or several chemicals emitted from such products. For example, Hispanic households in Boston showed higher exposures to a restricted pesticide, cyfluthrin, because products were available at local bodegas (Adamkiewicz et al., 2011). Sales of phased-out spray pesticides have been shown to persist in low-income, minority neighborhoods (Carlton et al., 2004).

Operation and Maintenance Patterns

Public policies exist to minimize disinvestment and disrepair in low-income housing (Travis, 2019), yet many states have had to enact legislation to protect tenants (Sabbeth, 2019). Deferred maintenance and neglect of both residential and nonresidential buildings influence indoor air chemistry through increases in indoor dampness, microbial contamination, contaminants from older and poorly maintained or unvented combustion appliances, higher temperatures and humidity, and lower ventilation rates. Deferred maintenance, particularly in public housing, is also linked to pest infestations (Julien et al., 2008), increasing the potential for higher indoor concentrations of insecticides and pesticides, as well as pest allergens that can be triggers for asthma exacerbation. Finally, concentrations of lead

BOX 6-1
Emergent Issues Related to Indoor Chemistry and Climate Change

- Increased time spent indoors as people seek shelter during extreme weather events, potentially increasing the concentrations of volatile organic compounds (VOCs) emitted by humans, which represent 40–57 percent of total indoor VOC concentrations in high-occupancy spaces (Tang et al., 2016);
- Higher ambient temperatures, in certain geographic regions, which will generate higher ambient concentrations of ozone (moderated in part by the reductions achieved in ozone precursor pollutants) and, therefore, higher indoor ozone concentrations due to infiltration (Moghani and Archer, 2020; Salvador et al., 2019);
- Elevated concentrations of particulate matter and combustion gases from wildfires (Shrestha et al., 2019b);
- Increased microbial contamination from extreme surface dampness due to flooding and pathogens from urban floodwaters and sewer backup; and
- Increased use of gas-powered generators during power outages, which may pose a risk to indoor air quality due to infiltration of the carbon monoxide emitted during their use (Adefeso et al., 2012; Hampson and Dunn, 2015).

in household dust are higher in low-income and multifamily housing (Benfer, 2017), where racial bias is linked to low compliance with lead-safe work practices by property managers. Rather than perform lead inspections before a child is lead poisoned, for example as a routine part of housing maintenance, U.S. lead poisoning policies and practices tend to follow a "wait and see" approach (Benfer, 2017).

Climate Change, Disparities, and Emerging Indoor Chemistry Issues

Climate change and its attendant extreme weather events are potentially widening disparities in indoor exposures among low-income households and communities of color compared to moderate-to high-income and White communities. In the 2011 Institute of Medicine report *Climate Change, the Indoor Environment, and Health*, the effect of poverty on indoor air quality is discussed in significantly greater detail than will be addressed in this report, and readers are referred to that report for in-depth consideration of these interrelated issues (IOM, 2011). Beyond poverty, EPA's 2021 report on climate change and social vulnerability models climate events and notes that race and ethnicity are associated with disproportionate impacts.

Box 6-1 summarizes several emergent issues related to indoor chemistry and climate change. All of these factors are anticipated to evolve as climate change progresses and will influence indoor pollutant concentrations (Nazaroff, 2013). In addition, factors that already influence personal exposure to chemicals indoors, such as seasonality and underlying drivers of seasonality (e.g., temperature, humidity), may drive exposure variability even further as greater extremes are reached and previously extreme values are sustained over longer time periods. For example, personal exposure measurements of flame retardants and plasticizers in adults and children in the United States revealed significant seasonal variability in exposure that was difficult to explain but also replicated in another study (Hoffman et al., 2017; Phillips et al., 2018).

THE INTERSECTION OF INDOOR CHEMISTRY AND EXPOSURE MODELING

Exposure models sit at the intersection of chemistry, human activities in microenvironments, and health impacts. Exposure models serve a wide range of uses, including filling in data gaps, quantifying exposures and the relative importance of various exposure pathways, setting indoor air quality standards for occupational and public health protection, prioritizing chemicals for risk evaluation, and evaluating approaches to exposure mitigation. Combined, these outputs of exposure models improve our understanding of how indoor air chemistry impacts human health outcomes. Earlier parts of this report describe key data and information gaps that may necessitate the use of models (e.g., to improve understanding of SVOC partitioning). The breadth of models dictates that a given model be used with care and be applied appropriately (i.e., "fit for purpose")—fully cognizant of its limitations. In the indoor air chemistry arena, near-field exposure models are most relevant. Far-field exposure models, which describe environmental behaviors of pollutants in the outdoor environment, will not be discussed here, although they are often used in the exposure and health literature to approximate personal exposures.

The Modeling Landscape

Diverse exposure models have been developed to address research and practical needs, including the ones listed above. To fully describe all exposure models in terms of their applicability, strengths, and limitations is outside the scope of this report. A set of exposure models commonly used for predicting near-field exposures is provided for reference in Appendix D.

A complete model incorporates all components necessary to accurately quantify exposure. To generalize to "any contaminant and situation," this means including all possible factors and processes, including source composition and emission rates, chemical partitioning, chemical transformations, building factors (like air exchange rates), and human factors and behaviors. In reality, a specific exposure model is designed for a particular purpose by emphasizing certain components or aspects of these components. For example, some exposure models (e.g., USEtox) incorporate detailed transport processes with just a few chemical transformations, while others (e.g., CONTAM) are capable of processing complex gas-phase chemistry but can only characterize inhalation exposure. Note that CONTAM is an exception in that, in general, only a limited number of exposure models integrate detailed chemistry and sophisticated air flows.

Exposure models can be mapped across several dimensions—most obviously, spatial and temporal. Other dimensions include exposure pathway, number of chemistries considered, and the degree to which human behavior is incorporated into the model (Isaacs and Wambaugh, 2021). Regardless of dimension, a model can be characterized by its level of complexity as it relates to model application/utility. Figure 6-3 provides a graphical representation of the model complexity continuum.

Typically, more complex models more accurately replicate physical or biological processes. But increased complexity typically comes with increased challenges in developing, using, and evaluating the model. For example, much higher spatial and temporal resolution in exposure models can be achieved but often at the expense of computational demands. While more complex models may describe processes in greater detail, users have to decide whether the cost of using a particular model is commensurate with the level of effort associated with its complexity. Model developers, practitioners, and those who use the results have to assess whether a model is the preferred option for a given purpose and weigh the tradeoffs between model granularity or resolution versus complexity, ease of use, and "fit for purpose."

One example is that air is commonly assumed to be well mixed in simpler models (Huang et al., 2017a). This allows such models to be applied to high throughput screening and other less computationally intensive applications. However, more complex models include computational fluid dynamics, which provides the detail needed to understand chemical concentrations near a stationary individual (Rim et al., 2018). To date, most indoor computational fluid dynamic models do not account for the movement of building occupants within spaces.

For indoor chemistry, several dimensions of complexity are particularly important. One dimension is the level of human activity and behavior detail included in the model. Personal activities are associated with both physicochemical processes underlying complex indoor chemistry and

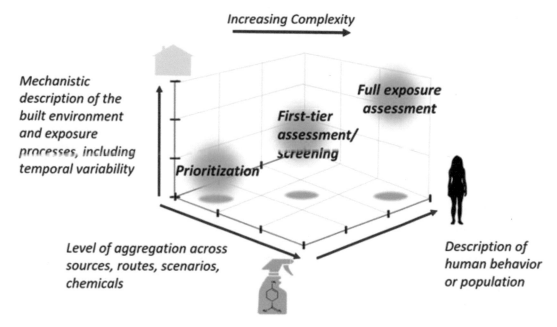

FIGURE 6-3 Models can be mapped according to various dimensions of complexity, which often ties to utility/application. SOURCE: Isaacs and Wambaugh (2021).

personal exposures. Models may integrate various levels of detail with regard to human time-activity and mobility patterns that determine exposure duration and intensity. To assess the complexity, the user might ask whether the model averages activity information over a 24-hour period or if it allows consideration of single event exposures to provide insights at shorter timescales. Does the model allow the user to specify key demographic factors or sub-population characteristics, perhaps taking into account personal characteristics (e.g., age, body mass index, and socioeconomic factors)? Increasingly, more details are being incorporated into models based on consumer product use patterns, or more sophisticated habits and practices data.

A second complexity axis is the extent to which physicochemical processes are incorporated into a given model. Although many exposure models incorporate chemistry, often this is done in a simplistic way. A notable differentiator among models is the ability of a model to handle different types of chemicals (e.g., VOCs versus SVOCs). Chapter 3 discusses recent improvements in models to help explain SVOC partitioning, and Chapter 4 discusses how parameterization of partitioning processes is embedded in some models but not in others. Chapter 4 also highlights the benefits of integrating models to capture the complexity of transformations in the indoor environment. Current exposure models have yet to incorporate some of the recent advances in chemistry models.

A third axis of complexity is that of the biological/physiological detail of the exposure mechanism in dosimetry and toxicokinetic/toxicodynamic models that accounts for metabolism, metabolites, disposition among tissues, and excretion. For instance, Lakey et al. (2017) highlight the benefit of more complicated dynamic models over steady-state models in their ability to capture transdermal uptake of SVOCs. Exposure models have also been coupled with dosimetry models and toxicokinetic/toxicodynamic models developed to predict internal dose.

Although users may decide that a high degree of complexity is not needed for a particular application, generally much could be gained from incorporating increased levels of complexity across multiple dimensions into exposure models for indoor chemistry in order to more fully simulate real-world conditions.

Several reviews of contemporary exposure models are available in the literature. Huang et al. (2017a) considered near-field exposure models suitable for life cycle assessment, chemical alternatives assessment, and high throughput screening risk assessments. This review outlines individual exposure scenarios as examples (e.g., transfer of chemicals in sprays to near-person air) and assesses strengths and weaknesses of algorithms and models for these scenarios. Of note, Huang et al. (2017a) observed that even in highly mature indoor fate transport models, the role of human occupants on chemical fate and transport is not adequately considered.

Cowan-Ellsberry et al. (2020) compare five international consumer product exposure models used for regulatory decision making. The authors outline several factors and key differences among models of which users need to be aware so that selected models are applied in a manner that is consistent with the intended use of the results (or "fit for purpose"). Efforts to simulate similar exposures using these different models were hampered by difficulties ensuring that the models used the same conditions. These challenges stemmed from inconsistencies and a lack of transparency in factors including product type definition, specifications of the exposure route for a given product, details of the exposure scenario, and output exposure metrics. Ultimately, the authors determined that additional consistency across models would be beneficial, and they highlight possible adoption of the standardized product taxonomy by the Organisation for Economic Co-operation and Development (OECD, 2017). They also stated that built-in input databases create challenges, particularly in terms of identifying the source of uncertainties as due to model algorithms versus input datasets.

The models presented above—designed for different purposes—have various levels of intrinsic complexity. They can be grouped according to the NRC's *Exposure Science in the 21st Century* exposure model groupings (NRC, 2012). The first set of models are process-based models, while others are classified as activity-based models. In the indoor chemistry context, process models include both fugacity- or equilibrium-based chemistry or mass balance-based models to characterize chemical transport and fate, exposure routes and pathways, and primary and secondary sources (e.g., emissions). Activity-based models predict exposures based on the behavior of individuals and populations. In the indoor realm, activity-based models may evaluate exposures based on consumer product use patterns, building occupancy rates, habits and practices, or occupational behaviors. Many near-field exposure models include both process with activity elements.

Exposure Model Uncertainties

When a model cannot fully characterize an exposure scenario, uncertainty arises. In the context of indoor chemistry, the primary uncertainty originates from insufficient characterization of homogeneous and heterogeneous reactions, leading to inadequate prediction of exposures to transformation products in aerosols and on surfaces (Zhou et al., 2019). Another primary uncertainty arises from inadequately describing chemical partitioning processes across complex surfaces (including clothing) in the indoor environment for SVOCs, leading to mischaracterized exposure routes. In addition, some parameters are presumed to be well known but might not be characterized adequately for a specific exposure scenario. For instance, ventilation or air exchange have been widely used and often assumed constant in exposure assessment, when, in fact, more specificity may be warranted to support modeling. For example, quantifying elevated exposures due to proximity to a source or where chemical reactions happen requires more detailed consideration of ventilation near a person, and the bulk air exchange rate may be less important. Further amplifying uncertainty are microenvironmental extremes in temperature and moisture that have an outsized impact on local chemistry (e.g., very hot or very cold surfaces in HVAC systems).

Even if a model includes all key physicochemical processes and can adequately characterize a certain exposure scenario, uncertainties remain owing to unknowns in parameterization of all parameters incorporated in the model. Major uncertainties in model parameterization include emission factors, partitioning coefficients, personal activities and product use and co-use patterns, chemical emission profiles from products, physicochemical properties for chemicals of emerging concern (e.g., per- and polyfluoroalkyl substances), and modification of physicochemical properties by environmental conditions.

Uncertainties could be addressed by using several models simultaneously to make predictions. For example, the Systematic Empirical Evaluation of Models framework was applied to develop a consensus model of chemical exposures (Ring et al., 2019). This effort employed 13 models to create a consensus-based metamodel. The article contains concise descriptions of the individual models used that provide a useful overview of several relevant fate and transport and near field exposure models, including SHEDS-HT, FINE, RAIDAR-ICE, and the product intake fraction framework. The application of multiple appropriate models to predict exposures allowed assessment of 479,926 chemicals with increased confidence in predicted outcomes.

A quantitative or qualitative uncertainty analysis can provide transparency about the uncertainties associated with a given exposure model, thus providing the user of the model or the interpreter of modeling results with the requisite understanding of the limits of the model. At the very least, a model's applicability domain needs to be documented, including the type of chemicals, the exposure scenarios, and the spatial and temporal scales over which the model can be applied.

Model integration provides the opportunity to connect models that may be more advanced on a given complexity axis to obtain more detailed understanding along the biological response pathway from sources, to emissions, to concentrations and exposure, as illustrated by Figure 6-2. Yet, existing exposure modeling approaches are fragmented. Modular structures analogous to the Modelling Consortium for Chemistry of Indoor Environments are beneficial for exposure applications but can only be undertaken if integration is considered in model design. As an example, a modular mechanistic approach has been proposed to improve predictions of SVOC exposures indoors (Eichler et al., 2021).

Furthermore, opportunities have been identified for model development in the future. Inhalation exposure models are fairly mature, but models that quantify chemical partitioning from articles to skin need improvement (Huang et al., 2017a). Cowan-Ellsberry et al. (2020) identify several opportunities to improve models. The need for standardization is a prevalent theme. Standardization of product taxonomies and chemical formulations, for example, would facilitate model integration while making it easier to gauge uncertainties and compare results. Both Ring et al. (2019) and Cowan-Ellsberry et al. (2020) call for more extensive biomonitoring (more populations, more chemicals) to provide input data, validate model results, and improve statistics for probabilistic models.

As exposure modeling matures, models may support work to better understand aggregate exposures from multiple sources of a given chemical as well as cumulative exposures from multiple chemicals. As researchers work to measure exposure to chemical mixtures and understand their impacts, modeling advances could facilitate analysis of the complex interactions of multiple chemicals on human receptors to ultimately understand health risks at the individual and population level.

A notable practical limitation of the exposure models described here is lack of integration with building thermal analysis models that are used to predict the expected energy use of buildings. Some couplings exist between tools like CONTAM and energy simulation programs, which are rarely used except by experts. Programs such as Energy Plus simulate the variation of ventilation rates, HVAC air flows, and indoor temperature and humidity, which vary as occupancy and thermal loads vary. Consequently, it is not a straightforward task to estimate annual exposures and the impact of control measures on annual energy use and cost that are highly relevant to the assessment of existing buildings and design of new buildings. Until modeling of exposures is integrated with such models of building performance in a relatively easy-to-use way, the routine quantitative analysis of indoor air quality will be impeded.

MEASUREMENT SCIENCE FOR EXPOSURE

The primary goal of exposure assessment is to obtain accurate, precise, and biologically relevant personal or population exposure estimates in the most efficient and cost-effective way. Exposure data collection frequently involves tradeoffs among accuracy and precision, time resolution, number of contaminants monitored, technological limitations, and burden and resource constraints. As noted above, exposure can be classified, measured, or modeled with direct or indirect methods, and tradeoffs exist for all methods of exposure assessment. This section addresses measurement science specific to exposure applications. Chapter 2 provides more information on measurement approaches for quantifying the chemical composition of different indoor phases.

Exposure can be classified using dichotomous values such as "exposed" or "not exposed" to a particular substance. Classifications can also incorporate multiple categories of exposure, such as "no," "low," "medium," and "high," or occupational exposure categories based on job exposure matrix questionnaires (Choi, 2020). Such classifications can also be obtained by expert assessment. As a direct method, measurements are often considered a more objective means of assessing exposure than questionnaires. The most common measurement utilized is the concentration of a given agent in a representative area. Area monitors are used to estimate exposure to individuals living within a certain proximity (e.g., ambient reference monitors) or indoor locations where individuals spend a lot of time (e.g., living rooms), and personal samplers can be used to measure individual exposure.

Indoor environmental samplers can operate passively or actively with the aid of a pumping mechanism that purposefully draws air over sampling media. Passive gas sampling relies on diffusion of gas molecules onto a sorbent medium and can measure many classes of compounds quantitatively and correlate well with measurements of internal dose or metabolites of exposure. Diffusive air sampling uses minimal testing equipment and requires minimal expertise to implement, making it an often more affordable choice for conducting ongoing air monitoring. However, diffusive sampling is not amenable to sampling aerosols and PM, since particles do not follow the same principles of diffusion as gases and vapors. With diffusive sampling, the uptake rate is fixed; in contrast, with active sampling, it is possible to vary the sampling flow rate and thus collect the required sample over a range of preferred time periods (although optimal performance of size-selective aerosol samplers requires a fixed flow rate). Diffusive samplers tend to be lighter and less obtrusive than active samplers and, therefore, potentially preferable for personal monitoring. In addition to diffusive sampling, passive samplers may instead rely on particle settling. In this case, the passive sample being collected is a particle-based sample rather than a gas sample.

Over the past several years, there has been a proliferation of consumer-grade monitors that quantify various chemicals, many of which have even been expressly developed for indoor applications (Ometov et al., 2021; Zhang and Srinivasan, 2020). Sensors that are widely used typically report concentrations of CO_2, $PM_{2.5}$, PM_{10}, and/or total VOC with time resolutions of seconds to minutes (Chojer et al., 2020). The monitors have been purchased predominantly by the average consumer interested in indoor and outdoor air quality or by researchers aiming to achieve somewhat larger-scale measurements of some common pollutants or indicators of indoor and outdoor air quality for research purposes. The lower cost of these consumer-grade sensors allows greater monitoring of personal and room scale indoor concentrations in exposure and health studies. In addition, some of these monitors incorporate web- and app-based data logging and public display that allow for real time monitoring of exposures, especially in quickly changing situations like impacts of wildfire events on indoor spaces.

However, consumer-grade monitors often use different, miniaturized, or less costly technologies than federal reference monitors to quantify the constituent of concern, resulting in potential accuracy and precision performance issues. For example, consumer-grade $PM_{2.5}$ monitors typically rely upon light scattering nephelometer sensors to count particles. These are limited by the wavelength they use and are unable to detect particles smaller than ~0.3 μm. To calibrate these consumer-grade $PM_{2.5}$ monitors, the amount of scattered light is correlated to the $PM_{2.5}$ mass concentration measured in a laboratory setting via gravimetric analysis (on a filter using a microbalance), beta attenuation, and light scattering at other wavelengths. However, the calibration equation is dependent upon the size distribution, density, chemical composition, and optical properties of the collected sample, which is not always representative of the indoor $PM_{2.5}$ mixture being measured.

While much work has been done to deliberate performance targets (Duvall et al., 2021; Williams et al., 2019) and evaluate the outdoor performance of consumer-grade sensors (Bi et al., 2020; Holder et al., 2020; Wallace et al., 2021), efforts to evaluate performance of these sensors in indoor environments are more limited. Indoor consumer-grade $PM_{2.5}$ sensors have been shown to vary from laboratory instruments by up to a factor of 2, while PM_{10} value variation can be even higher (Demanega et al., 2021; Wang et al., 2020). As a result of the variation in indoor measurements, Wang et al. (2020) conclude that indoor $PM_{2.5}$ measurements are semi-quantitative; they can be used to identify episodic events and relative changes in the same indoor setting or room but may not report accurate absolute values without additional calibration measurements co-located in the experimental setting.

Certification processes for consumer-grade sensors depend on adequate test methods that address the challenges of the indoor environment. Currently, consensus test methods for indoor $PM_{2.5}$ and CO_2 sensors are being developed (ASTM WK62732, ASTM WK74360). Application of these methods to the marketplace will hopefully produce more accurate consumer-grade instruments that can be applied to exposure measurement campaigns. Box 6-2 describes novel measurement approaches that have recently been adopted to measure children's exposure. See Chapter 2 for more discussion of approaches for gas-phase, particle-phase, dust-phase, and surface sampling.

Human behavior influences not only the chemistry present in the indoor environment but also the differences in our relative exposure to the chemicals discussed in this chapter. Development of personal wearable sensors and samplers (e.g., VOC monitors, optical particle sensors, silicone wristbands, FreshAir Band) may help us understand individual exposure profiles in a better light as opposed to relying upon measurement of chemicals in bulk samples of indoor air and dust. For example, recent studies found that levels of parabens and organophosphate ester flame retardants measured on silicone wristbands were more strongly correlated with urinary biomarkers than levels measured in house dust (Levasseur et al., 2021; Phillips et al., 2018).

BOX 6-2
Emerging Science: New Approaches to Measuring Exposure in Children

Children are recognized as a population that is vulnerable to exposure and health effects of hazardous environmental contaminants. Chemical exposures in children are different from those in adults and require different measurement approaches. These differences are one reason why children were explicitly included in the National Health and Nutrition Examination Survey starting in 1988 (National Center for Health Statistics, 2015). Through behaviors such as crawling and common hand-to-mouth activity, as well as less social inhibition, children are in more intimate contact with other humans, pets, and their environment. Infant and toddler exposures occur almost exclusively at home and daycare. Approaches to assess exposure that are typically used on adults need to be adapted for measuring or estimating children's exposures.

Some novel approaches for measuring exposure have been applied to children and include child-sized remote-controlled robots equipped with air samplers that mimic toddler movement, activities, and breathing height (Sagona et al., 2017). Infant exposure robots sample airborne particles, chemicals, and microbes while crawling on floors (Wu et al., 2018). Personal silicone wristbands or passive personal air monitors that are enclosed in lightweight wristbands can be worn by school-aged children and have been utilized to sorb semivolatile organic compounds and other chemicals from the air (Kile et al., 2016; Lin et al., 2020). Video monitoring can ascertain how frequently students touch their face or mucus membranes (Zhang et al., 2020). Most of the above approaches as well as computer simulation of vertical particle concentrations in buildings (Khare and Marr, 2015) have demonstrated that exposure to particulate matter via ingestion is higher for children than adults. The U.S. Environmental Protection Agency suggests that daily intake (mg/day) for indoor settled dust in children is 1.5 to 2.5 times greater than that of adults (EPA, 2011).

CONCLUSIONS

To date, the foremost goal of exposure science has been to identify and characterize the inhalation, ingestion, and dermal uptake by people of harmful chemicals that can cause acute or chronic health effects. The application of exposure science to the study of indoor environments and exposures that occur therein is relatively nascent but rapidly evolving. Cost-effective policies and guidance suitable for diverse indoor environments and indoor-dwelling populations demand a thorough understanding of indoor exposure profiles. Understanding large differences in indoor exposures will also require deeper insight on the societal and systemic context in which exposures occur in residential and nonresidential environments. Environmental health disparities that are persistently observed in the United States and around the world too often remain unstudied.

The evidence base and toolkits for developing a robust and comprehensive understanding of indoor exposure profiles is growing rapidly. This evidence base has grown through multiple research channels, including field-based, laboratory-based, and modeling studies. Among field-based studies, emergent tools are addressing long-standing challenges of assessing spatial and temporal resolution on concentrations of airborne hazards, as well as diversity of chemical species in indoor air. Consumer-grade measurement tools and research-grade, high-resolution instrumentation alike are achieving wider use in indoor environments.

Researchers are also working to understand exposure to chemical mixtures. These efforts complement the strategic priorities of federal agencies, like the National Institutes of Health. For example, the National Institute of Environmental Health Sciences has identified strengthening understanding of combined exposures as a strategic priority: "Study of combined exposures, or mixtures, most closely replicate the human experience, and thus may provide unique insights to environmental health sciences" (NIEHS, n.d.). Measurement science advances applied to indoor environments and personal sampling are helping to better understand discrepancies—for example, between personal exposures and stationary monitors or indoor and outdoor area concentrations. Yet, inconsistency in chemical identifiers remains a challenge. Exposure data are collected across diverse sampling platforms, ranging from very short duration and transient (e.g., 1-hour) to chronic and longitudinal (multiple years), among populations that vary greatly in size and subject composition (Tan et al., 2018). This leads to diverse data that are not standardized and therefore not readily available to support modeling efforts.

One of the most important and fundamental needs for improving the utility of models in the exposure context is to bridge the gap between physical process models and exposure models. There is a need to connect physical process models to exposure and uptake models. Integrating frameworks can be the basis for better understanding the relationship of indoor air chemistry to exposure and even to internal dosing.

RESEARCH NEEDS

On the basis of the information discussed in this chapter, the committee arrived at the following list of specific actions recommended to advance indoor exposure science and research:

- **Review current science of indoor chemistry to define gaps in current exposure assessment methods or data collection.** Examples include identification of novel chemicals or chemical reaction products to include in field exposure studies (e.g., ozonolysis intermediates), evaluation of influential behaviors (e.g., window-opening), and collection of market data for products of interest (e.g., oxidizing air cleaners or fragranced products).
- **Develop more harmonized measures to characterize indoor exposure disparities.** A sparse number of studies reproducibly demonstrate that demographic and socioeconomic factors can enhance susceptibility to chemical exposures, but the evidence base for this conclusion is incomplete and data-poor. As patterns and predictors of indoor chemical exposures and exposure variability become better understood, it will be important to

standardize and make widely available datasets that fully capture these differences. Future work that comprehensively characterizes indoor exposures across a more diverse array of settings would have significant value for a range of real-world applications, including individual-, community-, and policy-level decision making. Expanded exposure datasets could be used in concert with the NHANES biomonitoring dataset.

- **Develop methodological and technological tools to make direct measurement of exposures easier, more convenient, and lower cost, especially to chemical mixtures,** at scales that meaningfully improve the performance of exposure modeling and close gaps in understanding relationships between indoor environmental co-exposures to many chemicals and health outcomes, including persistent environmental health disparities. Tools to enhance exposure monitoring based on microenvironmental measurements should extend beyond measuring species concentrations to also track occupancy patterns in indoor environments, as these patterns can influence emissions, ventilation, and pollutant removal.

- **Grow the network of data sources on human behaviors in indoor environments to become more representative of the U.S. population** and establish criteria for standardization and harmonization across diverse sources, ranging from nationally distributed surveys by federal agencies, to market-based data, to individual- and community-based reporting. Frameworks have the potential to provide the structure and tools needed to harmonize data on exposure determinants (e.g., human behaviors, consumption patterns, time-activity, intake rates) so that they can be better integrated for modeling efforts, as well as to increase data accessibility.

- **Deepen understanding of human behavior and time-activity patterns as they relate to indoor chemistry.** Addressing this critical knowledge gap would likely contribute to greater understanding of exposure variability. For example, factors such as clothes-laundering, hand-washing, window- and door-opening, spending time indoors or outdoors, cooking, cleaning, and engaging in leisure activities can drive significant differences in chemical exposures. Detailed, representative behavioral data will be increasingly valuable for models of physical processes and exposure. In recent years, the scientific community has learned how the presence of a human body mediates indoor chemistry, including gas-phase composition, generation of VOCs, and surface reactivity. It is likely that the collection of behavioral data will accelerate in the coming years. Efforts have to be undertaken to ensure the representativeness of such data if they are to be used for model training, while protecting data privacy and sustaining the highest-caliber research ethics.

- **Improve understanding of first principles that mediate and govern exposure while continuing to build datasets that can provide empirical exposure model inputs.** At this time, it is not possible for exposure models to be fully developed based on process knowledge. Yet numerous data gaps limit the ability to substitute empirical data for process first principles. In order for exposure models to advance, the understanding of exposure factors will need to improve while continuing to build datasets that can provide empirical information to support exposure modeling efforts.

- **Improve models through better integration of an understanding of human behavior.** Human time-activity patterns (i.e., where people spend their time), habits and practices, and behavioral data associated with indoor chemistry warrant significantly more study that keeps demographic differences in mind. Opportunities also exist to support and nurture modeling consortia or modeling hubs that are cross-disciplinary and include close collaboration with experimentalists.

- **Connect physical process models to exposure and uptake models.** The utility of models in the exposure context would be greatly improved by research that facilitates the integration of physical process models and exposure models. Integrating frameworks are being used to provide insight into the relationship of indoor air chemistry to exposure and even to internal dosing, but more work is needed.

REFERENCES

Adamkiewicz, G., A. R. Zota, M. P. Fabian, T. Chahine, R. Julien, J. D. Spengler, and J. I. Levy. 2011. Moving environmental justice indoors: Understanding structural influences on residential exposure patterns in low-income communities. *American Journal of Public Health* 101(S1):S238–S245. https://doi.org/10.2105/AJPH.2011.300119.

Adefeso, I., J. Sonibare, F. Akeredolu, and A. Rabiu. 2012. Environmental impact of portable power generator on indoor air quality. *2012 International Conference on Environment, Energy and Biotechnology (IPCBEE)* 33. http://www.ipcbee.com/vol33/012-ICEEB2012-B031.pdf.

Almeida, D. L., A. Pavanello, L. P. Saavedra, T. S. Pereira, M. A. A. de Castro-Prado, and P. C. de Freitas Mathias. 2019. Environmental monitoring and the developmental origins of health and disease. *Journal of Developmental Origins of Health and Disease* 10(6):608–615. https://doi.org/10.1017/S2040174419000151.

Baxter, L. K., J. E. Clougherty, F. Laden, and J. I. Levy. 2007. Predictors of concentrations of nitrogen dioxide, fine particulate matter, and particle constituents inside of lower socioeconomic status urban homes. *Journal of Exposure Science and Environmental Epidemiology* 17(5):433–444. https://doi.org/10.1038/sj.jes.7500532.

Becher, R., J. Øvrevik, P. E. Schwarze, S. Nilsen, J. K. Hongslo, and J. V. Bakke. 2018. Do carpets impair indoor air quality and cause adverse health outcomes: A review. *International Journal of Environmental Research and Public Health* 15(2):184. https://doi.org/10.3390/ijerph15020184.

Beckie, T. M. 2012. A systematic review of allostatic load, health, and health disparities. *Biological Research for Nursing* 14(4):311–346. https://doi.org/10.1177/1099800412455688.

Bell, M. L., A. Zanobetti, and F. Dominici. 2013. Evidence on vulnerability and susceptibility to health risks associated with short-term exposure to particulate matter: A systematic review and meta-analysis. *American Journal of Epidemiology* 178(6):865–876. https://doi.org/10.1093/aje/kwt090.

Benfer, E. A. 2017. Contaminated childhood: How the United States failed to prevent the chronic lead poisoning of low-income children and communities of color. *Harvard Environmental Law Review* 491:493–561.

Bi, C., J. P. Maestre, H. Li, G. Zhang, R. Givehchi, A. Mahdavi, K. A. Kinney, J. Siegel, S. D. Horner, and Y. Xu. 2018. Phthalates and organophosphates in settled dust and HVAC filter dust of U.S. low-income homes: Association with season, building characteristics, and childhood asthma. *Environment International* 121:916–930. https://doi.org/10.1016/j.envint.2018.09.013.

Bi, J., A. Wildani, H. H. Chang, and Y. Liu. 2020. Incorporating low-cost sensor measurements into high-resolution PM$_{2.5}$ modeling at a large spatial scale. *Environmental Science & Technology* 54(4):2152–2162. https://doi.org/10.1021/acs.est.9b06046.

Block, M.L., and L. Calderón-Garcidueñas. 2009. Air pollution: Mechanisms of neuroinflammation and CNS disease. *Trends in Neurosciences* 32(9):506–516.

Bradman, A., R. Castorina, F. Gaspar, M. Nishioka, M. Colón, W. Weathers, P. P. Egeghy, R. Maddalena, J. Williams, P. L. Jenkins, and T. E. McKone. 2014. Flame retardant exposures in California early childhood education environments. *Chemosphere* 116:61–66. https://doi.org/10.1016/j.chemosphere.2014.02.072.

Bradman, A., F. Gaspar, R. Castorina, J. Williams, T. Hoang, P. L. Jenkins, T. E. McKone, and R. Maddalena. 2017. Formaldehyde and acetaldehyde exposure and risk characterization in California early childhood education environments. *Indoor Air* 27(1):104–113. https://doi.org/10.1111/ina.12283.

Branco, P. T., M. C. Alvim-Ferraz, F. G. Martins, and S. I. Sousa. 2014. The microenvironmental modelling approach to assess children's exposure to air pollution - A review. *Environmental Research* 135:317–332. https://doi.org/10.1016/j.envres.2014.10.002.

Carlson, E. D., and R. M. Chamberlain. 2005. Allostatic load and health disparities: A theoretical orientation. *Research in Nursing & Health* 28(4):306–315. https://doi.org/10.1002/nur.20084.

Carlton, E. J., H. L. Moats, M. Feinberg, P. Shepard, R. Garfinkel, R. Whyatt, and D. Evans. 2004. Pesticide sales in low-income, minority neighborhoods. *Journal of Community Health* 29(3):231–244. https://doi.org/10.1023/B:JOHE.0000022029.88626.f4.

Champion, W. M. 2017. Navajo home heating practices, their impacts on air quality and human health, and a framework to identify sustainable solutions. https://scholar.colorado.edu/concern/graduate_thesis_or_dissertations/08612n75n.

Chamseddine, A. Z. 2019. Determinants of indoor air quality in hospitals: Impact of ventilation systems with indoor-outdoor correlations and health implications. *AUB Students' Theses, Dissertations, and Projects*. http://hdl.handle.net/10938/21766

Chan, W. R., W. W. Nazaroff, P. N. Price, M. D. Sohn, and A. J. Gadgil. 2005. Analyzing a database of residential air leakage in the United States. *Atmospheric Environment* 39(19):3445–3455.

Choi, B. 2020. Developing a job exposure matrix of work organization hazards in the United States: A review on methodological issues and research protocol. *Safety and Health at Work* 11(4):397-404. https://doi.org/10.1016/j.shaw.2020.05.007.

Chojer, H., P. T. B. S. Branco, F. G. Martins, M. C. M. Alvim-Ferraz, and S. I. V. Sousa. 2020. Development of low-cost indoor air quality monitoring devices: Recent advancements. *Science of the Total Environment* 727:138385.

Clinch, J. P., and J. D. Healy. 2000. Housing standards and excess winter mortality. *Journal of Epidemiology and Community Health* 54(9):719–720. https://doi.org/10.1136/jech.54.9.719.

Coates, S. J., and S. A. Norton. 2021. The effects of climate change on infectious diseases with cutaneous manifestations. *International Journal of Women's Dermatology* 7(1):8–16. https://doi.org/10.1016/j.ijwd.2020.07.005.

Collins, M., and S. Dempsey. 2019. Residential energy efficiency retrofits: Potential unintended consequences. *Journal of Environmental Planning and Management* 62(12):2010–2025. https://doi.org/10.1080/09640568.2018.1509788.

Colton, M. D., P. MacNaughton, J. Vallarino, J. Kane, M. Bennett-Fripp, J. D. Spengler, and G. Adamkiewicz. 2014. Indoor air quality in green vs conventional multifamily low-income housing. *Environmental Science & Technology* 48(14):7833–7841. https://doi.org/10.1021/es501489u.

Cowan-Ellsberry, C., R. T. Zaleski, H. Qian, W. Greggs, and E. Jensen. 2020. Perspectives on advancing consumer product exposure models. *Journal of Exposure Science & Environmental Epidemiology* 30(5):856–865. https://doi.org/10.1038/s41370-020-0237-z.

Demanega, I., I. Mujan, B. C. Singer, A. S. Andelković, F. Babich, and D. Licina. 2021. Performance assessment of low-cost environmental monitors and single sensors under variable indoor air quality and thermal conditions. *Building and Environment* 187:107415. https://doi.org/10.1016/j.buildenv.2020.107415.

Derbez, M., G. Wyart, E. Le Ponner, O. Ramalho, J. Ribéron, and C. Mandin. 2018. Indoor air quality in energy-efficient dwellings: Levels and sources of pollutants. *Indoor Air* 28(2):318–338. https://doi.org/10.1111/ina.12431.

Diaz Lozano Patino, E., and J. A. Siegel. 2018. Indoor environmental quality in social housing: A literature review. *Building and Environment* 131:231–241. https://doi.org/10.1016/j.buildenv.2018.01.013.

Dodson, R. E., J. O. Udesky, M. D. Colton, M. McCauley, D. E. Camann, A. Y. Yau, G. Adamkiewicz, and R. A. Rudel. 2017. Chemical exposures in recently renovated low-income housing: Influence of building materials and occupant activities. *Environment International* 109:114–127. https://doi.org/10.1016/j.envint.2017.07.007.

Du, L., V. Leivo, T. Prasauskas, M. Täubel, D. Martuzevicius, and U. Haverinen-Shaughnessy. 2019. Effects of energy retrofits on indoor air quality in multifamily buildings. *Indoor Air* 29(4):686–697. https://doi.org/10.1111/ina.12555.

Duvall, R. M., G. S. W. Hagler, A. L. Clements, K. Benedict, K. Barkjohn, V. Kilaru, T. Hanley, N. Watkins, A. Kaufman, A. Kamal, S. Reece, P. Fransioli, M. Gerboles, G. Gillerman, R. Habre, M. Hannigan, Z. Ning, V. Papapostolou, R. Pope, P. J. E. Quintana, and J. Lam Snyder. 2021. Deliberating performance targets: Follow-on workshop discussing PM_{10}, NO_2, CO, and SO_2 air sensor targets. *Atmospheric Environment* 246:118099. https://doi.org/10.1016/j.atmosenv.2020.118099.

EIA (U.S. Energy Information Administration). 2015a. *Residential Energy Consumption Survey (RECS)*. https://www.eia.gov/consumption/residential/data/2015/.

EIA. 2015b. About the Residential Energy Consumption Survey (RECS)Table HC1.1 Fuels used and end uses in U.S. homes by housing unit type. https://www.eia.gov/consumption/residential/data/2015/hc/php/hc6.1.php.

Eichler, C. M. A., E. A. C. Hubal, Y. Xu, J. Cao, C. Bi, C. J. Weschler, T. Salthammer, G. C. Morrison, A. J. Koivisto, Y. Zhang, C. Mandin, W. Wei, P. Blondeau, D. Poppendieck, X. Liu, C. J. E. Delmaar, P. Fantke, O. Jolliet, H. M. Shin, M. L. Diamond, M. Shiraiwa, A. Zuend, P. K. Hopke, N. von Goetz, M. Kulmala, and J. C. Little. 2021. Assessing human exposure to SVOCs in materials, products, and articles: A modular mechanistic framework. *Environmental Science & Technology* 55(1):25–43. https://doi.org/10.1021/acs.est.0c02329.

EPA (U.S. Environmental Protection Agency). 2011. *Exposure Factors Handbook 2011 Edition (Final Report), EPA/600/R-09/052F*. Washington, DC: U.S. Environmental Protection Agency.

EPA. 2017. Chapter 5: Soil and dust ingestion. *Exposure Factors Handbook*. https://www.epa.gov/expobox/exposure-factors-handbook-chapter-5.

EPA. 2021. *Climate Change and Social Vulnerability in the United States: A Focus on Six Impacts. EPA 430-R-21-003.* Washington, DC: U.S. Environmental Protection Agency.

Erlandson, G., S. Magzamen, E. Carter, J. L. Sharp, S. J. Reynolds, and J. W. Schaeffer. 2019. Characterization of indoor air quality on a college campus: A pilot study. *International Journal of Environmental Research and Public Health* 16(15):2721. https://doi.org/10.3390/ijerph16152721.

Feld-Cook, E., and C. P. Weisel. 2021. Exposure routes and types of exposure. In *Handbook of Indoor Air Quality,* edited by Y. Zhang, P. K. Hopke, and C. Mandin. Singapore: Springer. https://doi.org/10.1007/978-981-10-5155-5_38-1.

Ferguson, L., J. Taylor, M. Davies, C. Shrubsole, P. Symonds, and S. Dimitroulopoulou. 2020. Exposure to indoor air pollution across socio-economic groups in high-income countries: A scoping review of the literature and a modelling methodology. *Environment International* 143:105748. https://doi.org/10.1016/j.envint.2020.105748.

Finisterra do Paço, A. M., and M. L. B. Raposo. 2010. Green consumer market segmentation: Empirical findings from Portugal. *International Journal of Consumer Studies* 34(4):429–436. https://doi.org/10.1111/j.1470-6431.2010.00869.x.

Fisk, W. J., B. C. Singer, and W. R. Chan. 2020. Association of residential energy efficiency retrofits with indoor environmental quality, comfort, and health: A review of empirical data. *Building and Environment* 180:107067. https://doi.org/10.1016/j.buildenv.2020.107067.

Földváry, V., G. Bekö, S. Langer, K. Arrhenius, and D. Petráš. 2017. Effect of energy renovation on indoor air quality in multifamily residential buildings in Slovakia. *Building and Environment* 122:363–372. https://doi.org/10.1016/j.buildenv.2017.06.009.

Gaspar, F. W., R. Castorina, R. L. Maddalena, M. G. Nishioka, T. E. McKone, and A. Bradman. 2014. Phthalate exposure and risk assessment in California child care facilities. *Environmental Science & Technology* 48(13):7593–7601. https://doi.org/10.1021/es501189t.

Gaspar, F. W., R. Maddalena, J. Williams, R. Castorina, Z. M. Wang, K. Kumagai, T. E. McKone, and A. Bradman. 2018. Ultrafine, fine, and black carbon particle concentrations in California child-care facilities. *Indoor Air* 28(1):102–111. https://doi.org/10.1111/ina.12408.

Gee, G. C., and D. C. Payne-Sturges. 2004. Environmental health disparities: A framework integrating psychosocial and environmental concepts. *Environmental Health Perspectives* 112(17):1645–1653. https://doi.org/10.1289/ehp.7074.

Givehchi, R., J. P. Maestre, C. Bi, D. Wylie, Y. Xu, K. A. Kinney, and J. A. Siegel. 2019. Quantitative filter forensics with residential HVAC filters to assess indoor concentrations. *Indoor Air* 29(3):390–402. https://doi.org/10.1111/ina.12536.

Haghighat, F., and L. De Bellis. 1998. Material emission rates: Literature review, and the impact of indoor air temperature and relative humidity. *Building and Environment* 33(5):261–277. https://doi.org/10.1016/S0360-1323(97)00060-7.

Hampson, N. B., and S. L. Dunn. 2015. Carbon monoxide poisoning from portable electrical generators. *The Journal of Emergency Medicine* 49(2):125–129. https://doi.org/10.1016/j.jemermed.2014.12.091.

Harley, K. G., K. Berger, S. Rauch, K. Kogut, B. Claus Henn, A. M. Calafat, K. Huen, B. Eskenazi, and N. Holland. 2017. Association of prenatal urinary phthalate metabolite concentrations and childhood BMI and obesity. *Pediatric Research* 82(3):405–415. https://doi.org/10.1038/pr.2017.112.

Haugen, A. C., T. T. Schug, G. Collman, and J. J. Heindel. 2015. Evolution of DOHaD: The impact of environmental health sciences. *Journal of Developmental Origins of Health and Disease* 6(2):55–64. https://doi.org/10.1017/S2040174414000580.

Hauptman, M., J. K. Niles, J. Gudin, and H. W. Kaufman. 2021. Individual- and community-level factors associated with detectable and elevated blood lead levels in US children: Results from a national clinical laboratory. *JAMA Pediatrics* 175(12):1252–1260.

HEI (Health Effects Institute). 2010. *Panel on the Health Effects of Traffic-Related Air Pollution. Traffic-related Air Pollution: A Critical Review of the Literature on Emissions, Exposure, and Health Effects.* HEI Special Report 17. Boston, MA.

Heyder, J., J. Gebhart, G. Rudolf, C. F. Schiller, and W. Stahlhofen. 1986. Deposition of particles in the human respiratory tract in the size range 0.005–15 μm. *Journal of Aerosol Science* 17(5):811–825. https://doi.org/10.1016/0021-8502(86)90035-2.

Hoang, T., R. Castorina, F. Gaspar, R. Maddalena, P. L. Jenkins, Q. Zhang, T. E. McKone, E. Benfenati, A. Y. Shi, and A. Bradman. 2017. VOC exposures in California early childhood education environments. *Indoor Air* 27(3):609–621. https://doi.org/10.1111/ina.12340.

Hoffman, K., C. M. Butt, T. F. Webster, E. V. Preston, S. C. Hammel, C. Makey, A. M. Lorenzo, E. M. Cooper, C. Carignan, J. D. Meeker, R. Hauser, A. Soubry, S. K. Murphy, T. M. Price, C. Hoyo, E. Mendelsohn, J. Congleton, J. L. Daniels, and H. M. Stapleton. 2017. Temporal trends in exposure to organophosphate flame retardants in the United States. *Environmental Science & Technology Letters* 4(3):112–118. https://doi.org/10.1021/acs.estlett.6b00475.

Hofmann, W. 2011. Modelling inhaled particle deposition in the human lung—A review. *Journal of Aerosol Science* 42(10):693-724.

Holder, A. L., A. K. Mebust, L. A. Maghran, M. R. McGown, K. E. Stewart, D. M. Vallano, R. A. Elleman, and K. R. Baker. 2020. Field evaluation of low-cost particulate matter sensors for measuring wildfire smoke. *Sensors* 20(17):4796. https://doi.org/10.3390/s20174796.

Huang, L., A. Ernstoff, P. Fantke, S. A. Csiszar, and O. Jolliet. 2017a. A review of models for near-field exposure pathways of chemicals in consumer products. *Science of the Total Environment* 574:1182–1208. https://doi.org/10.1016/j.scitotenv.2016.06.118.

Huang, S., W. Wei, L. B. Weschler, T. Salthammer, H. Kan, Z. Bu, and Y. Zhang. 2017b. Indoor formaldehyde concentrations in urban China: Preliminary study of some important influencing factors. *Science of the Total Environment* 590–591:394–405. https://doi.org/10.1016/j.scitotenv.2017.02.187.

Huangfu, Y., N. M. Lima, P. T. O'Keeffe, W. M. Kirk, B. K. Lamb, S. N. Pressley, B. Lin, D. J. Cook, V. P. Walden, and B. T. Jobson. 2019. Diel variation of formaldehyde levels and other VOCs in homes driven by temperature dependent infiltration and emission rates. *Building and Environment* 159:106153. https://doi.org/10.1016/j.buildenv.2019.05.031.

IOM (Institute of Medicine). 2011. *Climate Change, the Indoor Environment, and Health.* Washington, DC: The National Academies Press.

Isaacs, K., and J. Wambaugh. 2021. Modeling exposure to chemicals in indoor air. Presented at the National Academies of Sciences Workshop on Emerging Science on Indoor Chemistry and Implications. Washington, DC, April 5, 2021. https://doi.org/10.23645/epacomptox.17741114.

Jacobs, D. E. 2011. Environmental health disparities in housing. *American Journal of Public Health* 101(S1):s115–S122. https://ajph.aphapublications.org/doi/full/10.2105/AJPH.2010.300058.

James-Todd, T., L. Connolly, E. V. Preston, M. R. Quinn, M. Plotan, Y. Xie, B. Gandi, and S. Mahalingaiah. 2021. Hormonal activity in commonly used Black hair care products: Evaluating hormone disruption as a plausible contribution to health disparities. *Journal of Exposure Science & Environmental Epidemiology* 31(3):476–486. https://doi.org/10.1038/s41370-021-00335-3.

Julien, R., G. Adamkiewicz, J. I. Levy, D. Bennett, M. Nishioka, and J. D. Spengler. 2008. Pesticide loadings of select organophosphate and pyrethroid pesticides in urban public housing. *Journal of Exposure Science & Environmental Epidemiology* 18(2):167–174. https://doi.org/10.1038/sj.jes.7500576.

Khare, P., and L. C. Marr. 2015. Simulation of vertical concentration gradient of influenza viruses in dust resuspended by walking. *Indoor Air* 25(4):428–440. https://doi.org/10.1111/ina.12156.

Kile, M. L., R. P. Scott, S. G. O'Connell, S. Lipscomb, M. MacDonald, M. McClelland, and K. A. Anderson. 2016. Using silicone wristbands to evaluate preschool children's exposure to flame retardants. *Environmental Research* 147:365–372. https://doi.org/10.1016/j.envres.2016.02.034.

Klepeis, N. E., W. C. Nelson, W. R. Ott, J. P. Robinson, A. M. Tsang, P. Switzer, J. V. Behar, S. C. Hern, and W. H. Engelmann. 2001. The National Human Activity Pattern Survey (NHAPS): A resource for assessing exposure to environmental pollutants. *Journal of Exposure Analysis and Environmental Epidemiology* 11(3):231 252. https://doi.org/10.1038/sj.jea.7500165.

Kramer, A. L., L. Campbell, J. Donatuto, M. Heidt, M. Kile, and S. L. Massey Simonich. 2020. Impact of local and regional sources of PAHs on tribal reservation air quality in the U.S. Pacific Northwest. *Science of the Total Environment* 710:136412. https://doi.org/10.1016/j.scitotenv.2019.136412.

Krieger, J. W., L. Song, T. K. Takaro, and J. Stout. 2000. Asthma and the home environment of low-income urban children: Preliminary findings from the Seattle-King County healthy homes project. *Journal of Urban Health* 77(1):50–67.

Lakey, P. S. J., A. Wisthaler, T. Berkemeier, T. Mikoviny, U. Pöschl, and M. Shiraiwa. 2017. Chemical kinetics of multi-phase reactions between ozone and human skin lipids: Implications for indoor air quality and health effects. *Indoor Air* 27(4):816–828. https://doi.org/10.1111/ina.12360.

Langer, S., G. Bekö, E. Bloom, A. Widheden, and L. Ekberg. 2015. Indoor air quality in passive and conventional new houses in Sweden. *Building and Environment* 93:92–100. https://doi.org/10.1016/j.buildenv.2015.02.004.

Langer, S., O. Ramalho, M. Derbez, J. Ribéron, S. Kirchner, and C. Mandin. 2016. Indoor environmental quality in French dwellings and building characteristics. *Atmospheric Environment* 128:82–91. https://doi.org/10.1016/j.atmosenv.2015.12.060.

Leivo, V., T. Prasauskas, L. Du, M. Turunen, M. Kiviste, A. Aaltonen, D. Martuzevicius, and U. Haverinen-Shaughnessy. 2018. Indoor thermal environment, air exchange rates, and carbon dioxide concentrations before and after energy retro fits in Finnish and Lithuanian multi-family buildings. *Science of the Total Environment* 621:398–406. https://doi.org/10.1016/j.scitotenv.2017.11.227.

Less, B., and I. Walker. 2014. Indoor air quality and ventilation in residential deep energy retrofits. U.S. Department of Energy Office of Scientific and Technical Information. https://doi.org/10.2172/1167382.

Levasseur, J. L., S. C. Hammel, K. Hoffman, A. L. Phillips, S. Zhang, X. Ye, A. M. Calafat, T. F. Webster, and H. M. Stapleton. 2021. Young children's exposure to phenols in the home: Associations between house dust, hand wipes, silicone wristbands, and urinary biomarkers. *Environment International* 147:106317. https://doi.org/10.1016/j.envint.2020.106317.

Lin, E. Z., S. Esenther, M. Mascelloni, F. Irfan, and K. J. Godri Pollitt. 2020. The fresh air wristband: A wearable air pollutant sampler *Environmental Science & Technology Letters* 7(5):308-314. https://doi.org/10.1021/acs.estlett.9b00800.

Liu, S., R. Li, R. J. Wild, C. Warneke, J. A. de Gouw, S. S. Brown, S. L. Mille, J. C. Luongo, J. L. Jimenez, and P. J. Ziemann. 2016. Contribution of human-related sources to indoor volatile organic compounds in a university classroom. *Indoor Air* 26(6):925–938.

Logue, J. M., T. E. McKone, M. H. Sherman, and B. C. Singer. 2011. Hazard assessment of chemical air contaminants measured in residences. *Indoor Air* 21(2):92–109. https://doi.org/10.1111/j.1600-0668.2010.00683.x.

LoMauro, A., and A. Aliverti. 2015. Respiratory physiology of pregnancy. *Breathe* 11(4):297. https://doi.org/10.1183/20734735.008615.

Mandin, C., M. Trantallidi, A. Cattaneo, N. Canha, V. G. Mihucz, T. Szigeti, R. Mabilia, E. Perreca, A. Spinazzè, S. Fossati, Y. De Kluizenaar, E. Cornelissen, I. Sakellaris, D. Saraga, O. Hänninen, E. De Oliveira Fernandes, G. Ventura, P. Wolkoff, P. Carrer, and J. Bartzis. 2017. Assessment of indoor air quality in office buildings across Europe–The OFFICAIR study. *Science of the Total Environment* 579:169–178. https://doi.org/10.1016/j.scitotenv.2016.10.238.

Manuel, J. 2011. Avoiding health pitfalls of home energy-efficiency retrofits. *Environmental Health Perspectives* 119(2):A76–A79. https://doi.org/10.1289/ehp.119-a76.

Miranda, M. L., S. E. Edwards, M. H. Keating, and C. J. Paul. 2011. Making the environmental justice grade: The relative burden of air pollution exposure in the United States. *International Journal of Environmental Research and Public Health* 8(6):1755–1771. https://doi/org/10.3390/ijerph8061755.

Moghani, M., and C. L. Archer. 2020. The impact of emissions and climate change on future ozone concentrations in the USA. *Air Quality, Atmosphere & Health* 13(12):1465-1476. https://doi/org/10.1007/s11869-020-00900-z.

Morello-Frosch, R., M. Jerrett, B. Shamasunder, and A. D. Kyle. 2011. Understanding the cumulative impacts of in-equalities in environmental health: Implications for policy. *Health Affairs* 30(5):879–887. https://doi.org/10.1377/hlthaff.2011.0153.

Morello-Frosch, R., and R. Lopez. 2006. The riskscape and the color line: Examining the role of segregation in environmental health disparities. *Environmental Research* 102(2):181–196. https://doi/org/10.1016/j.envres.2006.05.007.

Morello-Frosch, R., and E. D. Shenassa. 2006. The environmental "riskscape" and social inequality: Implications for explaining maternal and child health disparities. *Environmental Health Perspectives* 114(8):1150–1153. https://doi.org/10.1289/ehp.8930.

Morrison, G., J. Cagle, and G. Date. 2022. A national survey of window-opening behavior in United States homes. *Indoor Air* 32(1):e12932. https://doi.org/10.1111/ina.12932.

Morrison, G. C., C. J. Weschler, and G. Bekö. 2016. Dermal uptake directly from air under transient conditions: Advances in modeling and comparisons with experimental results for human subjects. *Indoor Air* 26:913–924. https://doi.org/10.1111/ina.12277.

Nakane, H. 2012. Translocation of particles deposited in the respiratory system: A systematic review and statistical analysis. *Environmental Health and Preventive Medicine* 17(4):263–274. https://doi.org/10.1007/s12199-011-0252-8.

National Center for Health Statistics. 2015. National Health and Nutrition Examination Survey History. https://www.cdc.gov/nchs/nhanes/history.htm.

Nazaroff, W. W. 2013. Exploring the consequences of climate change for indoor air quality. *Environmental Research Letters* 8(1):015022. https://doi.org/10.1088/1748-9326/8/1/015022.

Ng, L. C., A. Musser, A. K. Persily, and S. J. Emmerich. 2012. Indoor air quality analyses of commercial reference buildings. *Building and Environment* 58:179–187. https://doi.org/10.1016/j.buildenv.2012.07.008.

NIEHS (National Institute of Environmental Health Sciences). n.d. Theme one: Advancing environmental health sciences. *Strategic Plan 2018–2021*. NIH Publication No. 18-ES-7935

Nirlo, E. L., N. Crain, R. L. Corsi, and J. A. Siegel. 2014. Volatile organic compounds in fourteen U.S. retail stores. *Indoor Air* 24(5):484–494. https://doi.org/10.1111/ina.12101.

Noonan, C. W., T. J. Ward, and E. O. Semmens. 2015. Estimating the number of vulnerable people in the United States exposed to residential wood smoke. *Environmental Health Perspectives* 123(2):A30. https://doi.org/10.1289/ehp.1409136.

NRC (National Research Council). 1991. *Human Exposure Assessment for Airborne Pollutants: Advances and Opportunities*. Washington, DC: The National Academies Press. https://doi.org/10.17226/1544.

NRC. 2012. *Exposure Science in the 21st Century: A Vision and a Strategy*. Washington, DC: The National Academies Press.

OECD (Organisation for Economic Co-operation and Development). 2017. *Internationally Harmonized Functional, Product and Article Use Categories. ENV/JM/MONO(2017)14*.

Ometov, A., V. Shubina, L. Klus, J. Skibińska, S. Saafi, P. Pascacio, L. Flueratoru, D. Q. Gaibor, N. Chukhno, O. Chukhno, A. Ali, A. Channa, E. Svertoka, W. B. Qaim, R. Casanova-Marqués, S. Holcer, J. Torres-Sospedra, S. Casteleyn, G. Ruggeri, G. Araniti, R. Burget, J. Hosek, and E. S. Lohan. 2021. A survey on wearable technology: History, state-of-the-art and current challenges. *Computer Networks* 193:108074. https://doi.org/10.1016/j.comnet.2021.108074.

Ott, W. R., A. C. Steinemann, and L. A. Wallace. 2006. *Exposure Analysis*. Boca Raton: CRC Press. https://doi.org/10.1201/9781420012637.

Pelling, M., W. T. Chow, E. Chu, R. Dawson, D. Dodman, A. Fraser, B. Hayward, L. Khirfan, T. McPhearson, and A. Prakash. 2021. A climate resilience research renewal agenda: Learning lessons from the COVID-19 pandemic for urban climate resilience. *Climate and Development*. https://doi.org/10.1080/17565529.2021.1956411.

Perez, L., C. Declercq, C. Iñiguez, I. Aguilera, C. Badaloni, F. Ballester, C. Bouland, O. Chanel, F. B. Cirarda, F. Forastiere, B. Forsberg, D. Haluza, B. Hedlund, K. Cambra, M. Lacasaña, H. Moshammer, P. Otorepec, M. Rodríguez-Barranco, S. Medina, and N. Künzli. 2013. Chronic burden of near-roadway traffic pollution in 10 European cities (APHEKOM network). *European Respiratory Journal* 42(3):594–605. https://doi.org/10.1183/09031936.00031112.

Perkins, S. E., L. V. Alexander, and J. R. Nairn. 2012. Increasing frequency, intensity and duration of observed global heat-waves and warm spells. *Geophysical Research Letters* 39(20). https://doi.org/10.1029/2012GL053361.

Phillips, A. L., S. C. Hammel, K. Hoffman, A. M. Lorenzo, A. Chen, T. F. Webster, and H. M. Stapleton. 2018. Children's residential exposure to organophosphate ester flame retardants and plasticizers: Investigating exposure pathways in the TESIE study. *Environment International* 116:176–185. https://doi.org/10.1016/j.envint.2018.04.013.

Rim, D., E. T. Gall, S. Ananth, and Y. Won. 2018. Ozone reaction with human surfaces: Influences of surface reaction probability and indoor air flow condition. *Building and Environment* 130:40–48. https://doi.org/10.1016/j.buildenv.2017.12.012.

Ring, C. L., J. A. Arnot, D. H. Bennett, P. P. Egeghy, P. Fantke, L. Huang, K. K. Isaacs, O. Jolliet, K. A. Phillips, P. S. Price, H.-M. Shin, J. N. Westgate, R. W. Setzer, and J. F. Wambaugh. 2019. Consensus modeling of median chemical intake for the US population based on predictions of exposure pathways. *Environmental Science & Technology* 53(2):719–732. https://doi.org/10.1021/acs.est.8b04056.

Rogalsky, D. K., P. Mendola, T. A. Metts, and W. J. Martin, 2nd. 2014. Estimating the number of low-income Americans exposed to household air pollution from burning solid fuels. *Environmental Health Perspectives* 122(8):806–810. https://doi.org/10.1289/ehp.1306709.

Sabbeth, K. A. 2019. (Under)enforcement of poor tenants' rights. *Faculty Publications* 466. https://scholarship.law.unc.edu/faculty_publications/466/.

Sagona, J. A., S. L. Shalat, Z. Wang, M. Ramagopal, K. Black, M. Hernandez, and G. Mainelis. 2017. Comparison of particulate matter exposure estimates in young children from personal sampling equipment and a robotic sampler. *Journal of Exposure Science & Environmental Epidemiology* 27(3):299–305. https://doi.org/10.1038/jes.2016.24.

Salvador, C. M., G. Bekö, C. J. Weschler, G. Morrison, M. Le Breton, M. Hallquist, L. Ekberg, and S. Langer. 2019. Indoor ozone/human chemistry and ventilation strategies. *Indoor Air* 29(6):913–925. https://doi.org/10.1111/ina.12594.

Seltenrich, N. 2012. Healthier tribal housing: Combining the best of old and new. *Environmental Health Perspectives* 120(12). https://ehp.niehs.nih.gov/doi/full/10.1289/ehp.120-a460.

Semmens, E. O., C. W. Noonan, R. W. Allen, E. C. Weiler, and T. J. Ward. 2015. Indoor particulate matter in rural, wood stove heated homes. *Environmental Research* 138:93–100. https://doi.org/10.1016/j.envres.2015.02.005.

Shrestha, P. M., J. L. Humphrey, K. E. Barton, E. J. Carlton, J. L. Adgate, E. D. Root, and S. L. Miller. 2019a. Impact of low-income home energy-efficiency retrofits on building air tightness and healthy home indicators. *Sustainability* 11(9):2667. https://doi.org/10.3390/su11092667.

Shrestha, P. M., J. L. Humphrey, E. J. Carlton, J. L. Adgate, K. E. Barton, E. D. Root, and S. L. Miller. 2019b. Impact of outdoor air pollution on indoor air quality in low-income homes during wildfire seasons. *International Journal of Environmental Research and Public Health* 16(19). https://doi.org/10.3390/ijerph16193535.

Śmiełowska, M., M. Marć, and B. Zabiegała. 2017. Indoor air quality in public utility environments—A review. *Environmental Science and Pollution Research* 24(12):11166–11176. https://doi.org/10.1007/s11356-017-8567-7.

Sun, L., J. D. Miller, K. Van Ryswyk, A. J. Wheeler, M.-E. Héroux, M. S. Goldberg, and G. Mallach. 2022. Household determinants of biocontaminant exposures in Canadian homes. *Indoor Air* 32(1):e12933. https://doi.org/10.1111/ina.12933.

Szanton, S. L., J. M. Gill, and J. K. Allen. 2005. Allostatic load: A mechanism of socioeconomic health disparities? *Biological Research for Nursing* 7(1):7–15. https://doi.org/10.1177/1099800405278216.

Tan, Y.-M., J. A. Leonard, S. Edwards, J. Teeguarden, and P. Egeghy. 2018. Refining the aggregate exposure pathway. *Environmental Science: Processes & Impacts* 20(3):428–436. https://doi.org/10.1039/C8EM00018B.

Tang, X., P. K. Misztal, W. W. Nazaroff, and A. H. Goldstein. 2016. Volatile organic compound emissions from humans indoors. *Environmental Science & Technology* 50(23):12686–12694. https://doi.org/10.1021/acs.est.6b04415.

Tessum, C. W., D. A. Paolella, S. E. Chambliss, J. S. Apte, J. D. Hill, and J. D. Marshall. 2021. $PM_{2.5}$ polluters disproportionately and systemically affect people of color in the United States. *Science Advances* 7(18):eabf4491. https://doi.org/10.1126/sciadv.abf4491.

Tiwari, M., S. K. Sahu, R. C. Bhangare, P. Y. Ajmal, and G. G. Pandit. 2013. Estimation of polycyclic aromatic hydrocarbons associated with size segregated combustion aerosols generated from household fuels. *Microchemical Journal* 106:79–86.

Travis, A. 2019. The organization of neglect: Limited liability companies and housing disinvestment. *American Sociological Review* 84(1):142–170. https://doi.org/10.1177/0003122418821339.

Tsushima, S., P. Wargocki, and S. Tanabe. 2018. Sensory evaluation and chemical analysis of exhaled and dermally emitted bioeffluents. *Indoor Air* 28(1):146–163.

Underhill, L. J., M. P. Fabian, K. Vermeer, M. Sandel, G. Adamkiewicz, J. H. Leibler, and J. I. Levy. 2018. Modeling the resiliency of energy-efficient retrofits in low-income multifamily housing. *Indoor Air* 28(3):459–468. https://doi.org/10.1111/ina.12446.

U.S. Department of Health and Human Services. 2017. *Low Income Home Energy Data*. https://liheappm.acf.hhs.gov/sites/default/files/private/notebooks/2017/RPT_LIHEAP_HENPart1LIHEData_No_FY2017.pdf.

U.S. Global Change Research Program. 2016. *The Impacts of Climate Change on Human Health in the United States: A Scientific Assessment*. Washington, DC: U.S. Global Change Research Program.

U.S. Office of Disease Prevention and Health Promotion. The Secretary's Advisory Committee on National Health Promotion and Disease Prevention Objectives for 2020: Phase I Report, Recommendations for the Framework and Format of Healthy People 2020. https://www.healthypeople.gov/sites/default/files/PhaseI_0.pdf.

van der Kroon, B., R. Brouwer, and P. J. H. van Beukering. 2013. The energy ladder: Theoretical myth or empirical truth? Results from a meta-analysis. *Renewable and Sustainable Energy Reviews* 20:504–513. https://doi.org/10.1016/j.rser.2012.11.045.

Vicente, E. D., A. M. Vicente, M. Evtyugina, F. I. Oduber, F. Amato, X. Querol, and C. Alves. 2020. Impact of wood combustion on indoor air quality. *Science of the Total Environment* 705:135769.

Wallace, L., J. Bi, W. R. Ott, J. Sarnat, and Y. Liu. 2021. Calibration of low-cost PurpleAir outdoor monitors using an improved method of calculating $PM_{2.5}$. *Atmospheric Environment* 256:118432. https://doi.org/10.1016/j.atmosenv.2021.118432.

Wan, Y., M. L. Diamond, and J. A. Siegel. 2020. Elevated concentrations of semivolatile organic compounds in social housing multiunit residential building apartments. *Environmental Science & Technology Letters* 7(3):191–197. https://doi.org/10.1021/acs.estlett.0c00068.

Wang, Z., W. W. Delp, and B. C. Singer. 2020. Performance of low-cost indoor air quality monitors for $PM_{2.5}$ and PM_{10} from residential sources. *Building and Environment* 171:106654. https://doi.org/10.1016/j.buildenv.2020.106654.

Wanner, A., M. Salathé, and T. G. O'Riordan. 1996. Mucociliary clearance in the airways. *American Journal of Respiratory and Critical Care Medicine* 154(6):1868–1902. https://doi.org/10.1164/ajrccm.154.6.8970383.

Weschler, C. J., and W. W. Nazaroff. 2012. SVOC exposure indoors: Fresh look at dermal pathways. *Indoor Air* 22(5):356–377. https://doi.org/10.1111/j.1600-0668.2012.00772.x.

Weschler, C. J., and W. W. Nazaroff. 2014. Dermal uptake of organic vapors commonly found in indoor air. *Environmental Science & Technology* 48(2):1230–1237. https://doi.org/10.1021/es405490a.

WHO (World Health Organization). 2018. Chapter 3: Household crowding. *WHO Housing and Health Guidelines*. Geneva: World Health Organization.

Wigle, D. T., T. E. Arbuckle, M. C. Turner, A. Bérubé, Q. Yang, S. Liu, and D. Krewski. 2008. Epidemiologic evidence of relationships between reproductive and child health outcomes and environmental chemical contaminants. *Journal of Toxicology and Environmental Health, Part B* 11(5-6):373–517. https://doi.org/10.1080/10937400801921320.

Williams, R., R. Duvall, V. Kilaru, G. Hagler, L. Hassinger, K. Benedict, J. Rice, A. Kaufman, R. Judge, G. Pierce, G. Allen, M. Bergin, R. C. Cohen, P. Fransioli, M. Gerboles, R. Habre, M. Hannigan, D. Jack, P. Louie, N. A. Martin, M. Penza, A. Polidori, R. Subramanian, K. Ray, J. Schauer, E. Seto, G. Thurston, J. Turner, A. S. Wexler, and Z. Ning. 2019. Deliberating performance targets workshop: Potential paths for emerging $PM_{2.5}$ and O_3 air sensor progress. *Atmospheric Environment: X* 2:100031. https://doi.org/10.1016/j.aeaoa.2019.100031.

Wu, T., M. Täubel, R. Holopainen, A.-K. Viitanen, S. Vainiotalo, T. Tuomi, J. Keskinen, A. Hyvärinen, K. Hämeri, S. E. Saari, and B. E. Boor. 2018. Infant and adult inhalation exposure to resuspended biological particulate matter *Environmental Science & Technology* 52(1):237–247. https://doi.org/10.1021/acs.est.7b04183.

Yang, J., Y. Wang, C. Xiu, X. Xiao, J. Xia, and C. Jin. 2020. Optimizing local climate zones to mitigate urban heat island effect in human settlements. *Journal of Cleaner Production* 275:123767. https://doi.org/10.1016/j.jclepro.2020.123767.

Yang, T. T., T. S. Lin, J. J. Wu, and F. J. Jhuang. 2012. Characteristics of polycyclic aromatic hydrocarbon emissions of particles of various sizes from smoldering incense. *Bulletin of Environmental Contamination and Toxicology* 88(2):271–276. https://doi.org/10.1007/s00128-011-0446-1.

Zaatari, M., E. Nirlo, D. Jareemit, N. Crain, J. Srebric, and J. Siegel. 2014. Ventilation and indoor air quality in retail stores: A critical review (RP-1596). *HVAC&R Research* 20(2):276–294. https://doi.org/10.1080/10789669.2013.869126.

Zhang, H., and R. Srinivasan. 2020. A systematic review of air quality sensors, guidelines, and measurement studies for indoor air quality management. *Sustainability* 12(21):9045.

Zhang, H., and H. Yoshino. 2010. Analysis of indoor humidity environment in Chinese residential buildings. *Building and Environment* 45:2132–2140. https://doi.org/10.1016/j.buildenv.2010.03.011.

Zhang, N., W. Jia, P. Wang, M.-F. King, P.-T. Chan, and Y. Li. 2020. Most self-touches are with the nondominant hand. *Scientific Reports* 10(1):10457. https://doi.org/10.1038/s41598-020-67521-5.

Zhou, S., C. H. Hwang Brian, S. J. Lakey Pascale, A. Zuend, P. D. Abbatt Jonathan, and M. Shiraiwa. 2019. Multiphase reactivity of polycyclic aromatic hydrocarbons is driven by phase separation and diffusion limitations. *Proceedings of the National Academy of Sciences* 116(24):11658–11663. https://doi.org/10.1073/pnas.1902517116.

Zota, A., G. Adamkiewicz, J. I. Levy, and J. D. Spengler. 2005. Ventilation in public housing: Implications for indoor nitrogen dioxide concentrations. *Indoor Air* 15(6):393–401. https://doi.org/10.1111/j.1600-0668.2005.00375.x.

7

A Path Forward for Indoor Chemistry

This report has focused on different aspects of indoor chemistry, including new findings related to underreported chemical species, chemical reactions, and sources of chemicals and their distribution in indoor spaces. An understanding of how indoor chemistry fits into the context of what is known about the links among chemical exposure, air quality, and human health continues to evolve. Each chapter of the report highlights key research needs related to Primary Sources and Reservoirs of Chemicals Indoors (Chapter 2), Partitioning of Chemicals in Indoor Environments (Chapter 3), Chemical Transformations (Chapter 4), Management of Chemicals in Indoor Environments (Chapter 5), and Indoor Chemistry and Exposure (Chapter 6). These needs are not reiterated here; instead, this final chapter focuses on four emerging, crosscutting issues that span the topics discussed in earlier chapters. The committee provides its recommendations for critical needs to advance research, enhance coordination and collaboration, and overcome barriers for implementation of new research findings into practice in indoor environments.

This chapter also provides the committee's vision for the future of indoor chemistry research. A critical cornerstone of this vision is increased awareness within the scientific community of the challenges and opportunities for innovation in indoor chemistry research as well as the need to fund research in indoor chemistry. It is also critical to translate the emerging knowledge on indoor chemistry into practice that benefits public health and the environment.

CHEMICAL COMPLEXITY IN THE INDOOR ENVIRONMENT

Complex Chemical Mixtures and Processes

An emerging theme in indoor chemistry is the high degree of chemical complexity in indoor environments where people spend, on average, more than 80 percent of their time. People are often in close proximity to sources and processes that emit chemicals. Recent studies have demonstrated the importance of indoor exposure to, for example, polychlorinated biphenyls (Meyer et al., 2013), tris(1,3-dichloro-2-propyl) phosphate (Meeker et al., 2013), Firemaster 550 (Hoffman et al., 2014), and di(n-butyl) phthalate (Lorber et al., 2017). Additional data come from the National Health and

Nutrition Examination Survey, an effort undertaken by the Centers for Disease Control and Prevention to collect survey and biomonitoring data that many researchers use to characterize exposures to important pollutants. Indoor exposure is discussed in Chapter 6 of this report.

Despite the importance of indoor exposure, very little is known about how humans get exposed to multiple chemicals across phases and pathways, how these joint exposures interact across timescales, and the cumulative and long-term impacts of the indoor chemical environment on human health. Humans are rarely exposed to single chemicals and instead are usually in contact with mixtures of chemicals that may have additive, synergistic, or antagonistic modes of action and effects on health. Many of these mixtures are not chemically characterized or quantified. Studies of exposure to mixtures in the indoor environment and their health effects are lacking, in part owing to the complexity and dynamics of indoor chemistry.

Early indoor chemistry studies typically focused on a small number of chemical contaminants, such as the mass of fine particulate matter ($PM_{2.5}$) and ozone, small aldehydes, lead, and polycyclic aromatic hydrocarbon concentrations. Yet recent research has demonstrated that a much higher diversity of chemical species is present than previously recognized, including many highly functionalized organic compounds. As described in Chapter 2, these contaminants arise from different primary sources, such as influx of outdoor air and emissions from building materials and consumer products. The study of indoor chemistry is further complicated by the role human occupants themselves play and how their behaviors and time-activity patterns influence or modify their exposures. For example, human activities such as cooking and cleaning lead to significant and varied primary emissions. Most studies examining human influences on indoor environments have focused on chemical exposures in developed countries and with communities of higher socioeconomic status, or in underdeveloped countries relying on solid or fossil fuels for indoor heating and cooking (termed "household air pollution" by the World Health Organization). The full range of indoor settings is understudied with respect to indoor chemistry and indoor exposures.

> **Recommendation 1:** Researchers should further investigate the chemical composition of complex mixtures present indoors in a wide range of residential and nonresidential settings and how these mixtures impact chemical exposure and health.

Chemical Reactivity

A second major emerging theme is the considerable degree to which many indoor contaminants are chemically reactive, largely via oxidative processes but also via photochemistry, hydrolysis, and other reaction mechanisms (see Chapter 4). These reactions can occur via interactions of gas-phase species with indoor surfaces, or homogeneously within surface reservoirs. As is now known for gas-surface ozonolysis reactions, the transformations of one precursor molecule lead to numerous reaction products, increasing the chemical complexity of the indoor environment multifold beyond that generated by the primary sources alone. Yet our current understanding of the relative magnitude and duration of exposure to these reaction byproducts and their toxicity relative to parent chemicals in the indoor environment is very limited. This is especially important because reaction products often have higher redox activity or oxidative potential and may elicit larger effects on health on their own, enhance or exacerbate the toxicity of the overall mixture, or even explain adverse health effects attributed to precursors or other chemicals indoors.

The dependence of multiphase reaction kinetics on oxidant concentrations, condensed-phase water abundance, light levels, and substrate chemical composition is poorly understood. Moreover, as reaction products form within different surface reservoirs, they may react with each other, forming even more complex products with less well-known exposure and health consequences. In contrast, due to factors such as lower concentrations of oxidants indoors, the persistence of some chemicals may be higher indoors.

Recommendation 2: Researchers should focus on understanding chemical transformations that occur indoors, using advanced analytical techniques to decipher the underlying fundamental reaction kinetics and mechanisms both in the laboratory and in indoor environments.

Distribution of Indoor Chemicals

Another challenge is to accurately describe the phase distribution of chemical contaminants (see Chapter 3). Whereas the field of indoor chemistry has largely focused on the measurement of gas-phase and aerosol particle abundance, it is now clear that most molecules are largely present in a variety of surface reservoirs. Our incomplete quantitative understanding of the partitioning of semivolatile molecules limits the ability of models to accurately describe the removal rate and exposure levels, especially for near-field exposures. Additional measurements are needed of the spatially and temporally dependent abundance of contaminants in all surface reservoirs in a range of indoor environments, and of the rates at which such gas-surface partitioning processes occur. This information then needs to be integrated into partitioning models to more accurately predict the phase distributions of a wide variety of contaminants. Paired with knowledge of human behavior and time-activity patterns indoors, this can greatly enhance our understanding of exposure across phases and pathways. In the case of a single chemical of interest or concern, this would advance our understanding of aggregate exposure and complete health effects across exposure pathways. In the case of multiple chemicals or mixtures of concern, these advances would bring us closer to a cumulative risk assessment paradigm to better understand the combined and overall toxicity of multiphase, multipathway contributions of the indoor chemical environment on health.

Recommendation 3: Researchers should prioritize understanding the phase distribution of indoor chemicals between all indoor reservoirs and incorporate these findings into exposure models.

Overall, integration of our knowledge of partitioning processes, transformation chemistry, environmental conditions, human influences, and building and heating, ventilation, and air-conditioning (HVAC) parameters is essential to more accurately represent these complex processes and enable more accurate chemical exposure and health risk assessments. This field is ripe for multidisciplinary collaboration to significantly advance knowledge of the indoor chemical environment and its importance for human exposures, health, and well-being. It requires expertise from a variety of disciplines, including chemistry, engineering, building science, toxicology, exposure science, epidemiology, social sciences, urban planning, environmental regulation, and risk assessment. The complexity of the indoor environment and how it interacts with the outdoor environment and with humans is too diverse and broad for siloed approaches to make a significant contribution to advancing the field.

Recommendation 4: All stakeholders should proactively engage across disciplines to further the development of knowledge on the fundamental aspects of complex indoor chemistry and its impact on indoor environmental quality, exposure assessment, and human health.

INDOOR CHEMISTRY IN A CHANGING WORLD

There are unprecedented changes occurring to the outdoor environment due to climate change, wildfires, and urbanization, standing in contrast to improvements derived from environmental regulations and advancements in technology. This section examines the impacts of those changes on indoor environments.

Outdoor-Indoor Chemistry Interactions

Outdoor air pollution is associated with a wide range of adverse human health impacts. Epidemiological evidence on the relationship between outdoor air pollution and adverse health outcomes at the population level has driven regulatory policies setting outdoor air quality standards. Yet most exposure to outdoor air pollution actually occurs indoors. Large population-based epidemiological studies aiming to understand the health effects of outdoor air pollution generally rely on outdoor air quality measurements. Outdoor concentrations are treated as surrogates of personal exposure to air pollution of outdoor origin, yet this treatment introduces exposure misclassification and is sometimes not clearly described. Indoor chemistry transforms some outdoor air contaminants that enter indoor spaces into new chemical hazards, altering what people are exposed to. For example, ozone concentrations are lower indoors than outdoors, but many oxidation products of ozone are substantially higher indoors. The size distribution of outdoor particulate matter (PM) changes as the ultrafine and coarse particles are preferentially removed by infiltration and deposition, leaving more of the fine particles that most efficiently penetrate deeply into the lungs. Outdoor particles release some volatile components into the gas phase and take up indoor airborne chemicals such as phthalates, plasticizers, flame retardants, and cooking emissions. This chemical complexity contributes to PM's variable toxicity and impact on health and implies that, for the same $PM_{2.5}$ mass concentration, health risk could vary based on the origin and chemical composition of $PM_{2.5}$. It stands to reason that indoor chemistry, both transformations and partitioning, plays a significant—and currently overlooked—role in modifying the outdoor pollution that primarily concerns recent epidemiological studies.

> **Recommendation 5:** Researchers who study toxicology and epidemiology and their funders should prioritize resources toward understanding indoor exposures to contaminants, including those of outdoor origin that undergo subsequent transformations indoors.

Similarly, indoor chemistry influences outdoor air pollution. Indoor emissions from building materials, cooking, and consumer products markedly impact outdoor air pollution. Volatile chemical products (VCPs) made from petrochemical feedstocks are important contributors to ambient photochemical air pollution, comprising half of all petrochemical pollution in industrialized cities, yet many of those products are used and emitted indoors (McDonald et al., 2018). As cities and their resident populations continue to grow, and outdoor pollution regulations continue to reduce other sources, indoor emissions are poised to make up increasingly higher proportions of primary chemicals found outdoors. Buildings may alter the local outdoor air in subtle ways that may be important as outdoor air pollution continues to improve. Beyond emissions from VCPs, other consumer products, building materials, and cooking, other processes may meaningfully alter outdoor air pollution, such as chemical transformations of gases and particles (of outdoor origin) associated with partitioning, hot surfaces, combustion, and use of air and surface cleaners or other activities. Our understanding of the specific impact of indoor chemistry on outdoor air pollution is limited, but it has never been more important to characterize the sources, chemistries, and eventual exposures described in this report.

> **Recommendation 6:** Researchers and their funders should devote resources to creating emissions inventories specific to building types and to identifying indoor transformations that impact outdoor air quality.

A Resilient Built Environment

The way that buildings are designed and operated has to respond to the changing climate and trends in energy efficiency and energy sources. Energy efficiency has often been considered to be in conflict with the desire to improve indoor air quality through increased ventilation. This is especially true in home weatherization programs and in newer construction in the United States with tighter building envelopes. It is possible, however, to reject the energy-ventilation tradeoff paradigm. Tight building envelopes, purposeful ventilation, and reduced energy consumption through heat recovery ventilation are possible routes to improving energy efficiency without impacting ventilation rates. Ventilation with clean outdoor air is an important modifier of human exposure to all the chemical sources present indoors. Epidemiological evidence has consistently demonstrated improved health outcomes in building occupants under increasing outdoor air ventilation rates (Sundell et al., 2011), and increased ventilation has been observed to reduce symptoms associated with sick building syndrome (Fisk et al., 2009). Future building design will need to account for indoor chemistry and contend with environments that are continuously changing due to trends in building regulations and energy choices, including regulated decreases in the installation of natural gas appliances in new construction.

The availability of healthy outdoor air for ventilation cannot be taken for granted, especially with the increasing impacts of climate change. Recent events have shown that exchanging indoor air with outdoor air can be problematic when the concentrations of chemical species and PM in outdoor air pose a health risk. This conundrum traditionally occurs when ventilation is used to reduce indoor contaminant exposures in urban areas with elevated outdoor levels of pollution. A recent and more acute scenario required residents of the western United States to weigh the known benefits of outdoor air ventilation on reducing the risk of airborne transmission of COVID-19 with the known dangers of exposure to elevated $PM_{2.5}$ concentrations from wildfires. This also extends to the potential for poor COVID-19 outcomes in patients with increased exposures to air polluted by wildfires. Finding solutions will require a continued commitment to improving outdoor air quality. In parallel, it may also include improved capacity for building envelopes to remove contaminants from outdoor air and continued reductions in indoor chemical sources. In the absence of wildfire impacts, as outdoor air quality continues to improve due to more stringent regulations and adoption of electric vehicles and renewable energy sources, its impact on the indoor environment will be less important. In this scenario, the relative importance of indoor sources on human health will become proportionally larger.

Another central impact of climate change in some areas is increased presence of dampness in buildings due to extreme precipitation events, rising sea levels, and frequent flooding. Moisture in building materials and the building envelope may initiate or accelerate chemical emissions. Increased moisture in buildings also results in mold and pests ranging from dust mites to cockroaches. In addition to the mycotoxins and volatile metabolites produced by mold and pests, occupants respond to their presence by using more disinfectants and pesticides with implications for indoor chemistry and air quality.

Recommendation 7: Researchers and engineers should integrate indoor chemistry considerations into their building system design and mitigation approaches. This can be accomplished in different ways, including by consulting with indoor air scientists.

FUTURE INVESTMENTS IN RESEARCH

Looking towards the future of indoor chemistry research, it is important to acknowledge the need for strategic investments and coordination among funding agencies (Box 7-1). This section highlights several areas where such a coordinated effort could have a major impact on advancing the science.

BOX 7-1
Investing in Coordinated, Interdisciplinary Research

This report calls for interdisciplinary research that will require involvement from scientists across many fields. Executing this research agenda effectively will also require major investments by a wide range of federal agencies and other organizations that fund research in the areas of basic science, public health, and implementation science. The federal sponsors of this report (the U.S. Environmental Protection Agency, National Institute of Environmental Health Sciences, and Centers for Disease Control and Prevention), along with the National Science Foundation, the National Aeronautics and Space Administration, the Department of Defense, and other parts of the National Institutes of Health, may all have a role to play.

A comprehensive national initiative targeting indoor chemistry will require interagency coordination to share infrastructure and resources; avoid unnecessary duplication; and establish shared goals, priorities, and strategies that complement agency-specific missions and activities. Several examples of interagency coordination that may serve as useful models include the large-scale U.S. National Nanotechnology Initiative as well as more-focused outdoor atmospheric chemistry field campaigns (Fire Influence on Regional and Global Environments Experiment [FIREX] and Atmospheric Emissions and Reactions Observed from Megacities to Marine Areas [AEROMMA]).

Investing in Top-Down and Bottom-Up Approaches

Intersecting and integrating top-down and bottom-up approaches is essential for addressing complex chemical processes that cover a wide range of temporal and spatial scales in indoor environments. Top-down approaches may include indoor field observations that use a suite of analytical tools to characterize gas-phase compounds, aerosol particles, and indoor surfaces. Such observations and experimental analysis can provide quantitative information on indoor species and may lead to a discovery of unknown phenomena or shed light on poorly understood processes. However, top-down approaches by themselves would not fully elucidate fundamental and molecular-level understanding of specific processes. Bottom-up approaches are necessary to achieve full process-level understanding. Efforts to integrate laboratory experiments, indoor field observations, and modeling need to be a high priority and have strong potential for impact when coupled with exposure and health studies. This "three-legged stool" approach has been successfully applied in environmental science and needs to continue to be implemented for indoor environments. Applications of indoor models are a critical component of understanding indoor chemistry to develop an in-depth understanding of complex processes. Models can guide measurements through the identification of key chemical species and predictions of expected concentrations as well as to assist in the interpretation of laboratory experiments and field observations. Controlled laboratory measurements are needed to determine kinetics for emerging reactions and to elucidate chemical mechanisms of heterogeneous processes. Quantum chemical calculations and molecular dynamics simulations provide a full molecular picture of complex surface interactions and estimates for physicochemical parameters that may be compared with measurements and used in models. Chemical kinetic and thermodynamic models that resolve gas- and multiphase chemistry can be applied to gain mechanistic and quantitative interpretations of indoor observations by testing current knowledge and different hypotheses. Computational fluid dynamics coupled with mechanistic chemistry models can help resolve the spatial and temporal evolution of species in indoor environments. Application of these indoor models can help extrapolate field observations and experimental results to other indoor conditions and provide reasonable estimates of concentrations of different chemical species that may be inaccessible by measurements.

Recommendation 8: Given the challenges, complexity, knowledge gaps, and importance of indoor chemistry, federal agencies and others that fund research should make the study of indoor chemistry and its impact on indoor air quality and public health a national priority.

Analytical Tools

Characterizing the chemically complex indoor environment presents a wide array of both challenges and opportunities. The recent development of advanced analytical techniques has permitted the identification of a much greater number of indoor contaminants. For example, non-targeted high-resolution mass spectrometry approaches are revealing the myriad chemicals present in indoor samples, such as settled dust. A current limitation to these approaches is that the data are primarily qualitative and focus on accurately identifying chemical features present in the samples, although these data can be used in a semi-quantitative basis in some cases. Accurate quantification can be problematic when authentic standards do not exist (which is the case for many chemical transformation products) or instruments are not calibrated for various types of chemical classes. This imposes a potential limitation on our ability to quantitatively assess chemical exposures when instrument response cannot be "translated" into a concentration. Additionally, integrating on-line methods to quantify indoor versus outdoor sources and how they contribute to indoor air chemistry is important. This can be done by measuring gas-phase and aerosol composition indoors and outdoors at the same time. Investments in novel methods and chemoinformatic resources that increase our ability to identify and quantify the abundances of wide classes of indoor chemicals are needed in order to understand the impact of these exposures on human health. Future research efforts that combine non-targeted analyses with in vitro and in vivo toxicity assessments will provide more insight into the potential health impacts of these complex chemical mixtures.

Recommendation 9: Researchers and their funders should invest in developing novel methods and chemoinformatic resources that increase our ability to identify and quantify the abundances of wide classes of indoor chemicals, both primary emissions and secondary chemical reaction products.

Our emerging picture of indoor environments indicates chemical complexity in gas, particle, and surface phases. Although new analytical tools have been instrumental in improving our understanding of indoor chemistry, several key challenges remain. For example, indoor surface measurements of real-world samples have been limited to off-line measurements, typically involving sample collection and subsequent analysis. Development of real-time surface measurements is an emerging direction. Comprehensive chemical characterization with real-time instrumentation can provide insight into indoor systems, but the technical expertise, logistical constraints, and instrument costs restrict these measurements to short-term measurements in a limited number of settings. Because of these constraints, researchers commonly conduct field studies in manipulated test houses or in convenient and accessible locations, such as university classrooms or academic employee residences. Future studies will need to include an array of buildings that are more representative of the U.S. population's experience. Two important directions include the following: (1) expanding comprehensive indoor chemical studies to different buildings and environments across building use, age, and location to increase the diversity of stakeholders and occupants; and (2) developing and applying lower-cost chemical sensors for more widespread research applications. While measurement tools for atmospheric chemistry have developed rapidly over the past decade, the complexity of the indoor environment warrants both targeted and non-targeted analyses, with particular attention paid to potential instrument interferences and calibrations to the novel compounds present in indoor settings.

Recommendation 10: Researchers measuring indoor environments should apply and develop new analytical tools that can probe the chemical complexity of gases, aerosols, and surfaces.

Human Activity Data and Indoor Chemistry

Indoor chemistry is not only heavily influenced by chemicals used in indoor spaces and building characteristics but also influenced by the activities of daily living. Human activities in indoor environments are incredibly diverse, and many of these activities contribute to indoor chemistry. Therefore, it is important to gather human activity data, in combination with other metadata regarding buildings, to capture contributions of human activity to indoor chemistry in a wide-ranging manner (Li et al., 2019; Nguyen et al., 2014). Nationally representative surveys (e.g., the American Census Survey, American Housing Survey, and Residential Energy Consumption Survey) already exist to collect probability-based samples of the U.S. population and include some questions that provide insight on indoor activities of daily living that are associated with environmental exposures. These surveys are deployed at varying frequencies and scales, and none of them expressly include questions designed to characterize activities associated with indoor chemistry (e.g., sources, sinks, and patterns of occupancy). The National Human Activity Patterns Survey (NHAPS) was purposefully designed to document the time, location, and characteristics of activities relevant to estimating contaminant exposures, but this survey was only implemented from 1992 to 1994 as a probability-based national telephone interview survey of approximately 10,000 people. Detailed surveys of occupancy patterns, activities, and materials are essential to link activities with chemical implications, so efforts such as the NHAPS warrant more sustained implementation moving forward. In addition to national-level surveys, there are numerous other ways to gather information on people's activities in indoor environments as they relate to indoor chemistry and indoor chemical exposures, including in-home sensing of environmental and air quality parameters, energy and water use monitoring in homes and buildings, and data on consumer choices and expenditures. One challenge with these types of data is that they vary with respect to their availability and the extent to which privacy and confidentiality issues arise.

> **Recommendation 11:** Federal agencies should design and regularly implement an updated National Human Activity Patterns Survey. Federal and state agencies should add survey questions in existing surveys that capture people's activities in indoor environments as they relate to indoor chemistry and indoor chemical exposures.

COMMUNICATING SCIENCE AND RISKS: INDOOR CHEMISTRY AND ENVIRONMENTAL QUALITY

In this section, the committee seeks to convey the importance of communicating emerging information about indoor chemistry to stakeholders. To many stakeholders, science concerns itself mainly with discovery ("what?" and "how?") and leaves questions of relevance ("why does it matter?") and application ("how can it be used?") to others. The process of creating scientific knowledge and transferring it into the spheres of practice and policy can be inefficient and slow. There is also a need for science to address research questions that are derived from practical concerns. The monumental effort during the COVID-19 pandemic to apply scientific tools to mitigate its effects exemplifies what can happen when this connection is made. Making the same connection between science and application is essential in indoor chemistry.

Actively Engaging Stakeholders

Stakeholders with an interest in emerging issues in indoor chemistry include building occupants; researchers in ambient and indoor air quality, indoor chemistry, toxicology, epidemiology, and environmental and public health; practitioners in health care, healthy housing, air quality, HVAC, weatherization, and construction; government agencies that address public health; communities impacted by indoor chemistry-related exposures; and policy analysts and legislators. Some stakeholders are involved in creating scientific knowledge, while others focus on applying this knowledge in practice and to regulation. This section focuses on how those in the former group can most effectively dialog with the latter.

The stakeholder who most affects, and who is most affected by, indoor chemistry is the general public. Effectively engaging this audience requires scientific and technical professionals to communicate two overarching messages: first, how indoor air quality and indoor chemistry contribute to exposure and their personal health outcomes; and second, how their own actions and behaviors could mitigate or exacerbate exposure. For example, the increased use of consumer-grade indoor air quality monitors may help improve indoor air quality and present opportunities for citizen science but not unless users are also equipped with enough knowledge to interpret the information they provide. Empowering the public with this knowledge is essential, but community engagement is challenging and requires resources (see Chapter 8 in NASEM, 2016). Conveying these often technical messages in a way that brings about changes in behavior will require risk communication from front-line professionals who work directly with the public. Health care providers can increase air quality awareness among patients through consultation; yet, the health care industry has not widely adopted the practice of consulting patients about health risks from ambient or indoor air quality, suggesting that the scientific community could do more to engage with these important intermediaries. Communication with stakeholders in accessible terms that emphasize the personal relevance of indoor chemistry is central to translation of new research findings into practice.

Recommendation 12: Researchers should proactively engage in links that connect research to application throughout the indoor chemistry research process—for example, at the dissemination stage, by engaging with technical and standard-writing committees, presenting at conferences attended by practitioners, and disseminating the significance of research findings in social and mass media.

Environmental Justice and Indoor Chemistry

Indoor chemistry is a complex and emergent field of research. Moving forward, careful consideration is warranted for the potentially unique indoor environments documented in low-income, rural, and cold-climate areas as well as in communities of color. Foremost, researchers and practitioners can directly engage environmental justice (EJ) communities in formulating future research priorities and forge collaborations with social scientists who work with EJ communities. In addition, the development of a conceptual framework that holistically captures the scope and scale of variables that are unique to substandard housing may be useful. Such a framework would contextualize and ground the broader field of indoor chemistry within an equity lens. Factors and variables for consideration in the framework include

- unique sources and distribution of chemical contaminants in substandard housing and the influence of different types and variable quality of heating, cooling, ventilation, and filtration systems;
- health effects of indoor exposures in substandard versus standard housing due to building materials and maintenance practices;
- source, proximity, and scale of outdoor contaminants to which EJ communities may have greater indoor exposure;
- unique behaviors, chemical emissions, and chemical interactions that may occur due to a variety of reasons including locus of controls, differences in activity patterns, occupant density, and type of housing, such as multifamily versus single family; and
- how chemical interactions in a changing world may affect EJ communities in unique ways.

Recommendation 13: Researchers and practitioners should include environmental justice communities in the wide range of indoor environments they study and engage these communities in formulating research priorities and recommendations for future indoor air quality standards.

Consumer Products, Sensors, and Services

A wide variety of products and services are marketed to consumers to improve indoor air quality. This report has highlighted several types of products that are known or suspected to influence indoor chemistry. For example, there are oxidant-emitting air cleaners and ultraviolet light-based air disinfection systems. There are also services that remediate buildings with smoke damage or mold that use strong oxidants in the process. Low-cost carbon dioxide or particle sensors inform consumers of conditions that may prompt action (e.g., opening a window). The impacts of these kinds of products, services, and sensor-prompted behaviors on indoor chemistry are poorly understood.

Recommendation 14: Funding agencies should support interdisciplinary research to investigate the impact of products and services on indoor chemistry, especially under realistic conditions. There is also a need to determine how occupant access to air quality data leads to behavior that influences indoor chemistry.

It is important to note that manufacturers continue to market novel air-cleaning products and remediation services that are of uncertain value and that may adversely impact indoor air quality and occupant health. This has been especially the case during the COVID-19 pandemic. These products may range from useful to useless and from beneficial to harmful. Some claims about these products misrepresent benefits or worse. Indoor chemistry is complex, and consumers and facility managers currently do not have useful tools for evaluating these products, services, or related manufacturer claims. Even the experts on this committee do not yet have sufficient information about most of these products and services to make informed decisions about their utility or potential for harm.

Conclusion 1: Standardized consensus test methods could enable potential certification programs for air-cleaning products and services. Such test methods could help regulators determine whether action on these products and services is warranted.

Standards for Indoor Environmental Quality

Unlike regulation of outdoor chemistry, the management of indoor chemistry is at a nascent stage. Regulation of outdoor air has followed scientific discoveries for years. Many outdoor air hazards have been greatly reduced based on effective, science-based approaches to regulation and public risk communication. In the United States, the National Ambient Air Quality Standards regulate the concentration of carbon monoxide, lead, nitrogen dioxide, sulfur dioxide, ozone, and $PM_{2.5}$ in outdoor air, leading to demonstrable reductions in morbidity and mortality.

An effective regulatory framework for indoor air has not been established in the United States. The committee recognizes the inherent challenges in regulating non-occupational indoor air quality, such as privacy, personal liberty, and property rights. Indoor spaces contain thousands of chemical species at a wide range of concentrations where little to no data exist on acute or chronic health impacts, especially for children and other vulnerable populations. There is also a broad diversity of building stock and uses, occupant behavior, and other factors that further contribute to the complexity of indoor chemistry.

Globally, regulators are currently working to determine how best to use indoor chemistry findings to create guidance for a range of stakeholders. For example, in 2021, within its resolution on the implementation of the Ambient Air Quality Directives, the European Parliament called on the European Commission to regulate indoor air quality. Canada has enacted indoor air quality guidelines for numerous contaminants, as well as Indoor Air Reference Levels. The World Health Organization has published guidelines for nine contaminants, and ASHRAE recently published threshold values for 14 compounds and $PM_{2.5}$.

While these recent steps indicate a growing desire to reduce contaminants found indoors, it is important to note that, just as the challenges in implementing regulations differ from those for outdoor air quality, threshold levels may also differ. For example, those deciding where to set a maximum indoor standard for ozone may want to consider not only the health effects of ozone itself but also the products formed by ozone-initiated indoor chemistry (Xiang et al., 2019).

Establishing standards for the indoor chemical environment that are protective of public health across multiple settings requires crosscutting, multipronged collaborations and solutions. For example, while the U.S. Environmental Protection Agency does not have the authority to regulate indoor concentrations of chemicals in air or dust as it does for outdoor air, it has the authority to regulate emission factors of new and recycled products introduced indoors. Building codes, standards, and guidelines all have a role to play in eliminating hazards resulting from the indoor chemical environment.

> **Recommendation 15:** Researchers and their funders should prioritize understanding the health impacts from exposure to specific classes and mixtures of chemicals in a wide range of indoor settings. Such understanding is needed to inform any future standards, guidelines, or regulatory efforts.

CLOSING COMMENTS

There is a growing awareness that exposure to environmental contaminants contributes to the burden of human disease. For decades, much of the attention of the scientific and regulatory community has focused on ambient (outdoor) air quality and drinking water. These efforts have contributed to improvements in outdoor air and water quality that have measurably protected human health and the environment. An important contributor to human health, namely the indoor environment, has received far less attention, although its potential consequences have been documented.

The paucity of advanced chemical studies on the indoor environment directly hinders our ability to predict and mitigate both health and environmental effects. A better understanding of sources, sinks, and transformations of chemicals found indoors is needed. It is clear that the chemical complexity of the indoor environment—present in its sources, chemical transformations, and loss mechanisms—currently precludes accurate and complete assessments of the exposure rates for many of the chemicals present indoors. Effective integration of laboratory experiments, indoor measurements, and modeling is necessary to determine the impacts of this chemistry on indoor environmental quality and chemical exposures. As more attention is deservedly focused on indoor chemistry, more chemicals and their reactions will be identified, adding to an already complex problem. Many of these chemicals may have little to no information regarding their toxicity, either as individual agents or in combination with other chemicals present in the environment. Mitigating chemical hazards will depend on many factors and needs to be done in a manner that considers the impacts of any mitigation strategy itself on the indoor environment. This will require efforts in changing building design and operation, altering the use of products and materials, and minimizing the impact of human activity on indoor chemistry.

The immensity of this daunting task need not lead to inaction. Rather, investment at this time in a holistic approach that considers chemistry, biology, and social contributions to health will pay dividends in the future.

REFERENCES

Fisk, W. J., A. G. Mirer, and M. J. Mendell. 2009. Quantitative relationship of sick building syndrome symptoms with ventilation rates. *Indoor Air* 19(2):159–165. https://doi.org/10.1111/j.1600-0668.2008.00575.x.

Hoffman, K., M. Fang, B. Horman, H. B. Patisaul, S. Garantziotis, L. S. Birnbaum, and H. M. Stapleton. 2014. Urinary tetrabromobenzoic acid (TBBA) as a biomarker of exposure to the flame retardant mixture Firemaster® 550. *Environmental Health Perspectives* 122(9):963–969. https://doi.org/10.1289/ehp.1308028.

Li, T., M. Alavy, and J. A. Siegel. 2019. Measurement of residential HVAC system runtime. Building and Environment 150:99–107. https://doi.org/10.1016/j.buildenv.2019.01.004.

Lorber, M., C. J. Weschler, G. Morrison, G. Bekö, M. Gong, H. M. Koch, T. Salthammer, T. Schripp, J. Toftum, and G. Clausen. 2017. Linking a dermal permeation and an inhalation model to a simple pharmacokinetic model to study airborne exposure to di(n-butyl) phthalate. *Journal of Exposure Science and Environmental Epidemiology* 27(6): 601–609. https://doi.org/10.1038/jes.2016.48.

McDonald, B. C., J. A. d. Gouw, J. B. Gilman, S. H. Jathar, A. Akherati, C. D. Cappa, J. L. Jimenez, J. Lee-Taylor, P. L. Hayes, S. A. McKeen, Y. Y. Cui, S.-W. Kim, D. R. Gentner, G. Isaacman-VanWertz, A. H. Goldstein, R. A. Harley, G. J. Frost, J. M. Roberts, T. B. Ryerson, and M. Trainer. 2018. Volatile chemical products emerging as largest petrochemical source of urban organic emissions. *Science* 359(6377):760–764. https://doi.org/10.1126/science.aaq0524.

Meeker, J. D., E. M. Cooper, H. M. Stapleton, and R. Hauser. 2013. Urinary metabolites of organophosphate flame retardants: Temporal variability and correlations with house dust concentrations. *Environmental Health Perspectives* 121(5):580–585. https://doi.org/10.1289/ehp.1205907.

Meyer, H. W., M. Frederiksen, T. Göen, N. E. Ebbehøj, L. Gunnarsen, C. Brauer, B. Kolarik, J. Müller, and P. Jacobsen. 2013. Plasma polychlorinated biphenyls in residents of 91 PCB-contaminated and 108 non-contaminated dwellings— An exposure study. *International Journal of Hygiene and Environmental Health* 216(6):755–762. https://doi.org/10.1016/j.ijheh.2013.02.008.

NASEM (National Academies of Sciences, Engineering, and Medicine). 2016. *Health Risks of Indoor Exposure to Particulate Matter: Workshop Summary*. Washington, DC: The National Academies Press. https://doi.org/10.17226/23531.

Nguyen, J. L., J. Schwartz, and D. W. Dockery. 2014. The relationship between indoor and outdoor temperature, apparent temperature, relative humidity, and absolute humidity. *Indoor Air* 24(1):103–112. https://doi.org/10.1111/ina.12052.

Sundell, J., H. Levin, W. W. Nazaroff, W. S. Cain, W. J. Fisk, D. T. Grimsrud, F. Gyntelberg, Y. Li, A. K. Persily, A. C. Pickering, J. M. Samet, J. D. Spengler, S. T. Taylor, and C. J. Weschler. 2011. Ventilation rates and health: Multidisciplinary review of the scientific literature. *Indoor Air* 21:191–204. https://doi.org/10.1111/j.1600-0668.2010.00703.x.

Xiang, J., C. J. Weschler, J. Zhang, L. Zhang, Z. Sun, X. Duan, and Y. Zhang. 2019. Ozone in urban China: Impact on mortalities and approaches for establishing indoor guideline concentrations. *Indoor Air* 29(4):604–615. https://doi.org/10.1111/ina.12565.

Appendix A

Glossary

Aerosol: A stable suspension of solid or liquid particles in air.

Air change rate: The volumetric flow rate of air entering a room divided by the net volume of air in that room (1/h).

Biological particle: Solid-phase material derived from biological organisms (e.g., bacteria and viruses, pollen, fungi, dust mites, endotoxins, and skin dander).

Bottom-up approach: A method of inventorying the chemical composition of the indoor environment by determining the chemical composition and/or emissions of every individual source object or material.

Consumer-grade sensor: An air quality sensor available to the public that measures, typically indirectly and with low sensitivity and/or precision, the concentration of a pollutant in the surrounding air. Sometimes called a "low-cost" sensor.

Dust: A material that builds up on surfaces, made up of settled particles, fibers, and biological matter that can be mechanically removed.

Exposome: The record of all exposures, both internal and external, that a person receives throughout a lifetime.

Indoor chemistry: Reactions involving indoor pollutants, occurring either in the gas phase or on surfaces.

Infiltration: Uncontrolled inward air leakage to conditioned spaces through unintentional openings in ceilings, floors, and walls from unconditioned spaces or the outdoors, caused by differences in pressure.

Intake rate: The amount of carrier medium crossing into the body per unit time for a given route of exposure (e.g., volume of inhaled air per unit time).

Interface: A surface forming a common boundary of two bodies, spaces, or phases.

Interstitial space: A hidden space within a building that is infrequently accessed by occupants (e.g., spaces where mechanical systems are located).

Mechanical ventilation: Ventilation provided by mechanically powered equipment, such as motor-driven fans and blowers, but not by devices such as wind-driven turbine ventilators and mechanically operated windows.

Microbial volatile organic compound (mVOC): Microbial organic compounds in the vapor state present in an indoor atmosphere.

Natural ventilation: Movement of air into and out of a space primarily through intentionally provided openings (e.g., windows and doors) or by infiltration.

Particle: A tiny droplet or fragment of condensed-phase matter suspended in air or settled on a surface; sometimes referred to in bulk as particulate matter (PM).

Partitioning: The transfer of molecules from one phase to another (e.g., from air to a surface).

Primary source: An item or material that directly emits or releases chemicals into the indoor environment.

Reservoir: Any surface or volume to which molecules can partition. Can act as both sources and sinks of chemicals.

Secondary source: A process or mechanism that creates new chemicals based on reactions between primary precursors.

Semivolatile organic compound (SVOC): An organic compound of intermediate volatility. SVOC definitions vary between organizations and fields of research. Detailed definitions define an SVOC based on chemical characteristics such as boiling point, vapor pressure, and/or molecular weight.

Sink: A reservoir that accumulates and stores chemicals for a long period of time.

Sorption: The process of a substance becoming captured by a condensed phase, or adsorption and absorption considered as a single process.

Stack effect: The movement of air into and out of buildings, chimneys, flue gas stacks, or other containers, driven by density differences between indoor and outdoor air (caused by differences in temperature). The stack effect is also referred to as the chimney effect, and it helps drive natural ventilation and infiltration.

Surface: The top-most (or outermost) region of any solid or liquid material accessible to chemicals present in air.

Top-down approach: A method of inventorying the chemical composition of the indoor environment by measuring the total chemical composition of the gas, particle, dust, and surface phases.

Under-reported chemicals: Species that have historically received less attention from researchers and are not extensively documented in the scientific literature (for the purposes of this report, not to be confused with another common definition, which is the deliberate act of reporting a lower level than what is present).

Volatile chemical product (VCP): Consumer products made from petrochemicals, including pesticides, coatings, printing inks, adhesives, cleaning agents, and personal care products, that contain volatile organic compounds.

Volatile organic compound (VOC): An organic compound that easily evaporates at ambient temperatures. VOC definitions vary between organizations and fields of research. Detailed definitions define a VOC based on chemical characteristics such as boiling point, vapor pressure, and/or molecular weight.

Appendix B

Committee Biosketches

David C. Dorman (*Chair*) is a professor of toxicology in the Department of Molecular Biomedical Sciences at North Carolina State University. Dr. Dorman's research interests include neurotoxicology, nasal toxicology, pharmacokinetics, and cognition and olfaction in animals. Dr. Dorman is an elected fellow of the Academy of Toxicological Sciences and a fellow of the American Association for the Advancement of Sciences. Dr. Dorman is a diplomate of the American Board of Veterinary Toxicology and the American Board of Toxicology. He has chaired or served on several National Research Council committees and is a National Associate of the National Academies of Sciences, Engineering, and Medicine. He completed a combined PhD and veterinary toxicology residency program at the University of Illinois at Urbana-Champaign and holds a Doctor of Veterinary Medicine from Colorado State University.

Jonathan Abbatt is a chemistry professor at the University of Toronto. Dr. Abbatt's research activities lie broadly in the areas of atmospheric aerosol and multiphase chemistry, using both laboratory and field measurement techniques. His current research interests include the chemistry of polluted atmospheres, the Arctic, wildfires, and indoor chemistry. Dr. Abbatt was a member of the National Aeronautics and Space Administration/Jet Propulsion Laboratory Data Evaluation Panel for Atmosphere Modeling and was co-chair of the Gordon Conference on Atmospheric Chemistry. He has served on the scientific steering committee of the International Global Atmospheric Chemistry project. He was given the Canadian Institute for Chemistry Environmental Research Award, is a fellow of the American Geophysical Union, was elected to the Royal Society of Canada, and was awarded a Killam Research Fellowship. He received his PhD from Harvard University in atmospheric chemistry.

William P. Bahnfleth is a professor of architectural engineering at Pennsylvania State University. His primary expertise is in building heating, ventilating, and air-conditioning systems from the perspective of both indoor air quality control and energy efficiency. His research interests include the control of chemical, particulate, and biological indoor air contaminants with ventilation, filtration, and air cleaners. He is a past president of the American Society of Heating, Refrigerating and Air-Conditioning Engineers (ASHRAE), and a fellow of ASHRAE, the American Society of Mechanical

Engineers, and the International Society for Indoor Air Quality and Climate. He has served previously on the National Research Council's Committee on Protecting Occupants of Department of Defense Buildings from Chemical and Biological Release and Committee on Safe Buildings Program. He received his PhD in mechanical engineering from the University of Illinois at Urbana-Champaign.

Ellison Carter is an assistant professor in the Department of Civil and Environmental Engineering at Colorado State University. Dr. Carter has expertise in indoor air quality, exposure science, and the residential built environment. She leads field-based assessments of personal, indoor, and outdoor air quality; human behaviors in the home; and health that directly relate to the development and implementation of healthy housing and indoor environmental interventions in diverse domestic and international settings. She earned her PhD in civil engineering, focusing on indoor environmental science and engineering, from the University of Texas at Austin.

Delphine Farmer is an associate professor in the Department of Chemistry at Colorado State University. Her research focuses on outdoor atmospheric and indoor chemistry, with an emphasis on understanding the sources and sinks of reactive trace gases and particles and their impacts on climate, ecosystems, and human health. Her recent work has focused on air chemistry in residential environments; Dr. Farmer was a co-lead of the House Observations of Microbial and Environmental Chemistry study, and is the co-lead of the Chemical Assessment of Surfaces and Air indoor study. Dr. Farmer received the Arnold and Mabel Beckman Young Investigator Award. She received her PhD in chemistry from the University of California, Berkeley.

Gillian Gawne-Mittelstaedt is director of the Partnership for Air Matters/Tribal Heathy Homes Network, a consortium that works to prevent exposure to indoor air hazards through training, community-based research, and culturally tailored interventions. Dr. Gawne-Mittelstaedt also leads the Partnership for Air Matters, a nonprofit that provides low-cost indoor air toolkits to engage and empower at-risk families. Her research focuses on developing a more cohesive national risk communication strategy around fine particle air pollution in the indoor environment, driving risk-informed decisions that reduce exposure, inequities, and costs to society. Dr. Gawne-Mittelstaedt currently serves as co-chair of the U.S. Environmental Protection Agency's (EPA's) Clean Air Act 50th Anniversary workgroup, as an indoor air appointee on EPA's Clean Air Act Advisory Committee, as a co-chair of the National Coalition on Safe and Healthy Housing, and formerly chaired the Washington State Asthma Initiative and the Washington Leadership Council for the American Lung Association. Dr. Gawne-Mittlestaedt holds an MPA from the Maxwell School at Syracuse University and a PhD in public health leadership from the University of Illinois at Chicago.

Allen H. Goldstein is the MacArthur Foundation Chair and distinguished professor in the Department of Environmental Science, Policy, and Management and in the Department of Civil and Environmental Engineering at the University of California, Berkeley. Additionally, Dr. Goldstein is associate dean for academic affairs in the Rausser College of Natural Resources. Dr. Goldstein's research program focuses on anthropogenic air pollution, biosphere-atmosphere exchange of radiatively and chemically active trace gases, and the development and application of novel instrumentation to investigate the organic chemistry of Earth's atmosphere. Additionally, he engages in field measurement campaigns; controlled laboratory experiments; and modeling activities covering indoor, urban, rural, regional, intercontinental, and global scale studies of aerosols and their gas-phase precursors. Dr. Goldstein previously served as co-chair of the International Global Atmospheric Chemistry program. He previously served on the National Academies of Sciences, Engineering, and Medicine's Committee on the Future of Atmospheric Chemistry Research. Dr. Goldstein received his PhD in chemistry from Harvard University.

Vicki H. Grassian is chair of the Department of Chemistry and Biochemistry at the University of California, San Diego. Additionally, she is a distinguished professor with appointments in the Departments of Chemistry and Biochemistry, Nanoengineering, and Scripps Institution of Oceanography, and is the Distinguished Chair of Physical Chemistry within the Department of Chemistry and Biochemistry. Dr. Grassian's group focuses on environmental interfaces as they relate to atmospheric aerosols, engineered and geochemical nanoparticles, and indoor surfaces. Dr. Grassian currently leads the SURFace Consortium for the Chemistry of the Indoor Environment Program. She is a recipient of the American Chemical Society National Award in Surface Chemistry and received the William H. Nichols Medal Award for her contributions to the chemistry of environmental interfaces. She is an elected member of the American Academy of Arts and Sciences and a fellow of the Royal Society of Chemistry. She received her PhD in chemistry from the University of California, Berkeley. Research funding is provided by P&G, and Dr. Grassian is an unpaid member of the P&G Global Hygiene Scientific Advisory Board.

Rima Habre is an associate professor of environmental health and spatial sciences at the University of Southern California (USC). She leads the Exposure Sciences Research Program in the USC National Institute of Environmental Health P30 Center. Her expertise lies in environmental health, air pollution, and exposure sciences. Her research aims to understand the effects of complex air pollution mixtures in the indoor and outdoor environment on the health of vulnerable populations across the life course. Dr. Habre's expertise spans measurement, spatiotemporal and geographic information system–based modeling, and mobile health approaches to assessing personal exposures and health risk. She co-chairs the Geospatial Working Group in the nationwide National Institute of Health's Environmental Influences on Child Health Outcomes program. Dr. Habre received her ScD in environmental health from the Harvard T.H. Chan School of Public Health.

Glenn Morrison is a professor of environmental sciences and engineering at the University of North Carolina, Chapel Hill. His research is related to the chemistry and physics of indoor air pollution and its influence on human exposure to contaminants. He has a particular interest in interfacial chemistry, ozone-surface chemistry, acid-base chemistry, methamphetamine contamination in buildings, and the interactions of chemicals with occupant surfaces, including skin. His group has studied how chemicals can be transported by fine particles indoors and has measured reactive oxygen species in homes. In recent years, he has focused on how clothing influences indoor chemistry and occupant exposure to chemicals. He is a fellow of the International Society of Indoor Air Quality and Climate and has served on the board and as the president from 2014 to 2016. He received his PhD from the University of California, Berkeley.

Jordan Peccia is the Thomas E. Golden Jr. Professor of Environmental Engineering at Yale University. His research integrates genetics with engineering to study human exposure to microbes in buildings. Dr. Peccia is a member of the Connecticut Academy of Science and Engineering and is an associate editor for the journal *Indoor Air*. Previously, Dr. Peccia served on the National Academies of Sciences, Engineering, and Medicine's study on Microbiomes of the Built Environment. He earned his PhD in environmental engineering from the University of Colorado.

Dustin Poppendieck is an environmental engineer at the National Institute of Standards and Technology. Dr. Poppendieck has a unique perspective on how building materials, building envelopes, dynamic infiltration, varied scheduled mechanical ventilation, low energy building designs, and heating, ventilation, and air-conditioning (HVAC) system operation can interact and impact indoor chemistry. Dr. Poppendieck's research involves characterizing primary emission sources and heterogeneous reactions at material surfaces. He has investigated emissions from kerosene can lamps, spray polyurethane foam, and non-smoldering cigarette butts. Additionally, Dr. Poppendieck has studied the disinfection of biologically contaminated building materials (i.e., anthrax) using high concentrations of ozone, chlorine dioxide, hydrogen peroxide, and methyl bromide. He received his PhD in civil and environmental engineering from the University of Texas at Austin.

Kimberly A. Prather holds a joint appointment as a professor in chemistry and biochemistry at Scripps Institution of Oceanography at the University of California, San Diego. Dr. Prather is involved in aerosol source apportionment studies, and her group is working to better understand the impact of specific aerosol sources on health and climate. Dr. Prather was formerly a member of the Fine Particle Monitoring Subcommittee of the U.S. Environmental Protection Agency's Clean Air Scientific Advisory Committee. She is on the editorial boards of several journals, including *Aerosol Science and Technology*. Dr. Prather is also a member of many professional societies including the American Association for Aerosol Research, the American Chemical Society, and the American Geophysical Union. She is an elected member of the National Academy of Engineering and the National Academy of Sciences. Dr. Prather received her PhD in chemistry from the University of California, Davis.

Manabu Shiraiwa is an associate professor of chemistry at the University of California, Irvine. He serves as the principal investigator of the Modelling Consortium for Chemistry of Indoor Environments (MOCCIE). MOCCIE connects models over a range of spatial and temporal scales with an ultimate aim to develop integrated physical-chemical models that include a realistic representation of gas-phase, aerosol-phase, and surface chemistry and how occupants, indoor activities, and buildings influence indoor processes. Dr. Shiraiwa has received numerous awards, including the Kenneth T. Whitby Award from the American Association for Aerosol Research, Paul J. Crutzen Award from the International Commission on Atmospheric Chemistry and Global Pollution, Walter A. Rosenblith Award from the Health Effects Institute, National Science Foundation CAREER Award, Sheldon K. Friedlander Award from the American Association for Aerosol Research, Paul Crutzen Prize from the German Chemical Society, and Otto Hahn Medal from the Max Planck Society. He received his PhD from the Max Planck Institute for Chemistry.

Heather M. Stapleton is the Ronie-Richele Garcia-Johnson Distinguished Professor in the Pratt School of Engineering at Duke University. Dr. Stapleton's research focuses on understanding the fate and transformation of organic contaminants in aquatic systems and in indoor environments. Her main focus has been on the bioaccumulation and biotransformation of brominated flame retardants, and specifically polybrominated diphenyl ethers. Her current research projects explore the routes of human exposure to flame retardant chemicals and examine the way these compounds are photodegraded and metabolized using mass spectrometry to identify breakdown products/metabolites. She uses both in vivo techniques with fish and in vitro techniques with cell cultures to examine metabolism of this varied class of chemicals. Dr. Stapleton earned her PhD from the University of Maryland, College Park.

Meredith Williams was appointed director of the California Department of Toxic Substances Control (DTSC) by Governor Gavin Newsom in 2019. She joined DTSC in 2013 as deputy director of the department's Safer Consumer Products Program to lead the implementation of California's groundbreaking effort to reduce toxic chemicals in consumer products. She received her PhD in physics from North Carolina State University in 1994 and a BA from Yale University in 1984. She was a speaker and panelist for the National Academies of Sciences, Engineering, and Medicine's Standing Committee for Emerging Science for Environmental Health Decisions' 2017 Workshop on Understanding Pathways to a Paradigm Shift in Toxicity Testing and Decision Making. And in 2019, she participated in the Environmental Health Matters Initiative's workshop on Understanding, Controlling, and Preventing Exposure to PFAS.

Appendix C

Open Session Agendas

COMMITTEE ON EMERGING SCIENCE ON INDOOR CHEMISTRY
1st Meeting—Virtual

1:00 Purpose of Open Session and Introduction of Committee Members

David Dorman
Chair, Committee on Emerging Science on Indoor Chemistry
Professor, North Carolina State University

1:10 Sponsors' Perspectives on the Committee Task

Yulia Carroll
Senior Medical Officer, Centers for Disease Control and Prevention

Christopher P. Weis
Toxicology Liaison, National Institute of Environmental Health Sciences and National Toxicology Program

Laura Kolb
Director, Center for Scientific Analysis, Indoor Environments Division, Office of Radiation and Indoor Air, Environmental Protection Agency

Evan S. Michelson
Program Director, Alfred P. Sloan Foundation

2:00 Discussion—Sponsors and Committee

2:30 Open Microphone—Opportunity for Public Comment

3:00 *End of Open Session*

EMERGING SCIENCE ON INDOOR CHEMISTRY AND IMPLICATIONS:
AN INFORMATION-GATHERING WORKSHOP
Monday, April 5, 2021
Virtual

I. Introduction and Overview

9:30   Opening Remarks and Goals of the Workshop

David Dorman
Committee Chair

II. Emerging Science on Indoor Air Chemistry
Session Chair: Allen Goldstein

9:40   Overview of Emerging Research and Discoveries

Charles Weschler
Rutgers Environmental and Occupational Health Sciences Institute

10:05   Indoor Chemistry of Building Materials

Tunga Salthammer
Fraunhofer WKI

10:30   Consumer Products in the Home Environment: Considerations for Indoor Air

Kathie Dionisio
U.S. Environmental Protection Agency

10:55 *Break*

11:05   Impact of Air Cleaners and Indoor Chemistry

Richard Corsi
Portland State University

11:30   Research Needs in Indoor Surface Chemistry

Hugo Destaillats
Lawrence Berkeley National Laboratory

11:55 *Lunch Break*

III. Monitoring and Exposure
Session Chair: Rima Habre

12:45   Emerging Sensor Technologies to Enhance Understanding of Indoor Air Pollutants

Andrea Polidori
SCAQMD

1:10   Future Chemicals of Concern: Semivolatile Compounds in the Indoor Environment

Deborah Bennett
UC Davis

1:35   **Modeling Exposure to Chemicals in Indoor Air**

John Wambaugh and Kristin Isaacs
U.S. Environmental Protection Agency

2:00   *Break*

IV. Crosscutting Issues
Session Chair: Jonathan Abbatt

2:10   **Indoor Chemistry Has Exposure Consequences That Are (Potentially) Health-Relevant**

Bill Nazaroff
UC Berkeley

2:35   **Residential Indoor Air Exposure Disparities**

Gary Adamkiewicz
Harvard University

3:00   **Relationships between the Building and Indoor Chemistry**

Jeff Siegel
University of Toronto

3:25   **Combined Modeling of Organic Chemical Fate and Human Exposure**

Frank Wania
University of Toronto

3:50   *Break*

V. Data Gaps and Research Needs
Session Chairs: Delphine Farmer and Bill Bahnfleth

4:00   **A Panel Discussion to Explore Research Needs and Data Gaps**

Panel Members:

- Bill Nazaroff, *UC Berkeley*
- Charles Weschler, *Rutgers Environmental and Occupational Health Sciences Institute*
- Deborah Bennett, *UC Davis*
- Paul Wennberg, *Caltech*

Potential Discussion Questions:

- What are the opportunities for incorporating existing research findings into practice, and what are the barriers?
- What research topics are most important to advancing our understanding of the composition of indoor air?
- What are the major gaps in our current understanding of the building factors (e.g., ventilation rate, temperature, relative humidity, and others) that impact indoor chemistry?
- What research is most needed on indoor exposures and their effects?

- What research is needed to define indoor air quality (i.e., which components matter in terms of their effects on metrics ranging from safety and comfort to health and productivity)?
- What methodological or technological barriers are most strongly hindering our understanding of indoor chemistry?
- What types of coordination across fields or collaboration are most important for advancing our understanding of indoor chemistry and interactions between chemical, particulate, and microbial components of the indoor environment?
- There is a lot already known about linkages between chemical exposure, air quality, and human health. What do we need to consider in order to place indoor chemistry into that context?

5:00   *Adjourn*

Appendix D

Summary Table of Available Exposure Models

Model Name	Model Description	Emission and Transport	Transformation	Exposure Routes and Pathways
The Stochastic Human Exposure and Dose Simulation Model for Multimedia, Multipathway Chemicals: Residential Module (SHEDS-Residential) (EPA, 2021)	Probabilistic model to simulate population exposures to single or multiple chemicals in air, water, and surfaces through multiple exposure route. Model results can be used as input in physiological-based pharmacokinetic models. Variability and uncertainty can be characterized at the same time through Monte Carlo simulations.	User-defined and scenario-based chemical emission. A single decay rate for each scenario, based on fugacity model.	No built-in chemical transformation mechanisms.	Multimedia exposure through inhalation, ingestion, and dermal absorption.
Stochastic Human Exposure and Dose Simulation–High Throughput (SHEDS-HT) (Isaacs et al., 2014)	A "screening-tier" model in the SHEDS family for prioritization purpose. The model structure is similar to SHEDS-Residential (or Multimedia), but the model is more computationally efficient. SHEDS-HT is a probabilistic model to simulate population exposures to multiple chemicals in microenvironments through multiple exposure routes. Technically, SHEDS-HT estimates variability but does not characterize uncertainty directly.	Two scenarios for chemical release—instantaneous release and constant release. One type of indoor surface (floor) is included. Decay rates are based on fugacity models.	No built-in chemical transformation mechanisms.	Multimedia exposure through inhalation, ingestion, and dermal absorption.
USEtox (Rosenbaum et al., 2011)	Deterministic model with multiple scales (far field and near field) to characterize the impact of chemicals throughout their life cycles on human and freshwater systems. Variability across a population and the uncertainties are not simultaneously characterized by the model.	Emission rate is user defined based on exposure scenarios. Chemical transfers among compartments are characterized by partitioning coefficients.	Built-in equations to account for gas-phase degradation through reactions involving ozone, hydroxyl radicals, and nitrate radicals.	Multimedia and multipathway exposures through inhalation, ingestion, and dermal absorption.

continued

Model		Chemical transformation	Exposure
Consumer Exposure Model (CEM) (EPA, 2019a)	Deterministic model for regulatory purpose to characterize human exposures in indoor environments to chemicals in consumer products and articles. Variability across a population and the uncertainties are not simultaneously characterized by the model.	Emission rate is user defined based on exposure scenarios. Chemical transfers among compartments are characterized by partitioning coefficients.	No built-in chemical transformation mechanisms.	Multimedia user and non-user exposure through inhalation, ingestion, and dermal absorption.
Indoor Environmental Concentrations in Buildings with Conditioned and Unconditioned Zones (IECCU) (EPA, 2019b)	Deterministic model for regulatory purpose to characterize the concentration of airborne chemicals in a multizone indoor environment, due to emissions from articles, paints, etc. Variability across a population and the uncertainties are not simultaneously characterized by the model.	Constant and variable source emissions are defined by user. Chemical transfers among compartments are characterized by partitioning coefficients.	Built-in equations to account for limited number of gas-phase reactions.	Inhalation exposure.
CONTAM (NIST, 2018)	A multizone indoor air quality model to predict ventilation, indoor air pollutant concentration, and inhalation exposure.	Source emission profiles can be specified in the model. Ventilations, including indoor-outdoor air exchange and zone-to-zone airflows, due to pressure difference, temperature difference, and mechanical forces, can be modeled. Sorption and desorption are also accounted for in the model.	Chemical transformations are modeled with kinetic reaction simulation. Radioactive species are modeled based on exponential decay constants.	Inhalation exposure.
Consumer Exposure and Uptake Models (ConsExpo) (National Institute for Public Health and the Environment, 2017)	A suite of models (the latest version is a web application), ranging from screening-level models to probabilistic models, to estimate exposure and resulting absorbed dose of chemicals due to consumer products' use under various conditions, including emissions from vapor, spray, and solid materials; direct contact to products; and chemical migration from food packaging materials.	Chemical release is modeled based on product use conditions and chemical properties. There is limited consideration of chemical partition among compartments. Bulk air exchange rate and mass transfer coefficients are employed to account for chemical migration.	Chemical transformation is not modeled.	Estimate exposure and absorbed dose through inhalation, ingestion, and dermal absorption.

Model Name	Model Description	Emission and Transport	Transformation	Exposure Routes and Pathways
Risk Assessment IDentification And Ranking Indoor and Consumer Exposure (RAIDAR-ICE) (Li et al., 2018)	A steady-state screening model, formulated in fugacity notation, predicting human exposure in the indoor environment to neutral organics, through estimating human external exposure and internal dose and comparing these metrics with health benchmark.	Chemical release is characterized by either a steady-state emission rate or a chemical application rate. Chemical transport among seven indoor compartments, including indoor air, polyurethane foam, carpet, vinyl floor, and organic films on vertical, up-facing, and down-facing surfaces.	No built-in chemical transformation mechanisms.	Estimate exposure and internal dose through inhalation, ingestion, and dermal absorption.
Production-To-Exposure Model (PROTEX) (Health Environmental Assessment Team, n.d.)	A multipathway and multiscale model coupling near-field and far-field exposure with the capability to predict exposure and toxicokinetics.	User-defined emission rate; indoor environment comprises four compartments, including indoor air, carpet, vinyl floor, and organic film on indoor surfaces; mass transfer across indoor, urban, and rural environments.	No built-in chemical transformation mechanisms, but transformation can be quantified with a half-life time.	Estimate exposure and internal dose through inhalation, ingestion, and dermal absorption.
Fugacity-based INdoor Exposure (FINE) (Bennett and Furtaw, 2004)	A flexible framework to build multicompartment dynamic mass balance model in fugacity notation to predict near-field exposures to organics released in the indoor environment.	User-defined emission scenarios; diffusive and advective mass transfer across compartments, including but not limited to air, carpet, vinyl floor, ceiling, and wall.	No built-in chemical transformation mechanisms.	Estimate exposure through inhalation, hand-to-mouth ingestion, and dermal absorption.
Activity-Based Indoor Chemical Assessment Model (ABICAM) (Kvasnicka et al., 2020)	Multimedia residential indoor model, based on mass balance and in fugacity notation, to predict transient exposures and internal dose to semivolatile organic compounds.	User-defined personal activity–related emissions. Chemical mass transfer among nine compartments, including "skin-surface lipids in hands and the rest of skin, internal body, indoor air, polyurethane-foam (PUF) mattress pad, hard floor, carpet, upward-oriented surfaces coated by a thin layer of organic film and other film-coated surfaces of vertical and downward orientations."	No built-in chemical transformation mechanisms.	Estimate exposure and internal dose through inhalation, hand-to-mouth ingestion, and dermal absorption.

REFERENCES

Bennett, D. H., and E. J. Furtaw. 2004. Fugacity-based indoor residential pesticide fate model. *Environmental Science & Technology* 38(7):2142–2152. https://doi.org/10.1021/es034287m.

EPA (U.S. Environmental Protection Agency). 2019a. *Consumer Exposure Model (CEM) User Guide.* https://www.epa.gov/sites/default/files/2019-06/documents/cem_2.1_user_guide.pdf.

EPA. 2019b. Indoor environmental concentrations in buildings with conditioned and unconditioned zones (IECCU): Simulation program for estimating chemical emissions from sources and related changes to indoor environmental concentrations in buildings with conditioned and unconditioned zones. In *IECCU User's Guide.* https://www.epa.gov/sites/default/files/2019-06/documents/ieccu_1.1_users_guide.pdf.

EPA. 2021. Stochastic Human Exposure and Dose Simulation (SHEDS) to estimate human exposure to chemicals. https://www.epa.gov/chemical-research/stochastic-human-exposure-and-dose-simulation-sheds-estimate-human-exposure.

Health Environmental Assessment Team. n.d. Comprehensive Exposure Model, in PROTEX (PROduction-To-Exposure). Accessed February 2, 2022. https://lilienv.weebly.com/protex.html.

Isaacs, K. K., W. G. Glen, P. Egeghy, M. R. Goldsmith, L. Smith, D. Vallero, and H. Özkaynak. 2014. SHEDS-HT: An integrated probabilistic exposure model for prioritizing exposures to chemicals with near-field and dietary sources. *Environmental Science & Technology* 48(21):12750–12759. https://doi.org/10.1021/es502513w.

Kvasnicka, J., E. Cohen Hubal, J. Ladan, X. Zhang, and M. L. Diamond. 2020. Transient multimedia model for investigating the influence of indoor human activities on exposure to SVOCs. *Environmental Science & Technology* 54(17):10772–10782. https://doi.org/10.1021/acs.est.0c03268.

Li, L., J. N. Westgate, L. Hughes, X. Zhang, B. Givehchi, L. Toose, and J. A. Arnot. 2018. A model for risk-based screening and prioritization of human exposure to chemicals from near-field sources. *Environmental Science & Technology* 52(24):14235–14244. https://doi.org/10.1021/acs.est.8b04059.

National Institute for Public Health and the Environment. 2017. *ConsExpo Consumer Exposure Models Model Documentation.* RIVM Report 2017–0197. https://www.rivm.nl/bibliotheek/rapporten/2017-0197.pdf.

NIST (National Institute of Standards and Technology). 2018. IAQ analysis and contaminant transport. https://www.nist.gov/el/energy-and-environment-division-73200/nist-multizone-modeling/applications-contam/iaq-analysis.

Rosenbaum, R. K., M. A. Huijbregts, A. D. Henderson, M. Margni, T. E. McKone, D. Van De Meent, and O. Jolliet. 2011. USEtox human exposure and toxicity factors for comparative assessment of toxic emissions in life cycle analysis: Sensitivity to key chemical properties. *The International Journal of Life Cycle Assessment* 16(8):710–727. https://doi.org/10.1007/s11367-011-0316-4.